Pragmatic Legal and Policy Implications of Environmental Lawmaking

Hussein Movahedian
Islamic Azad University, UAE & Department of Private Law, Islamic Studies and Law Faculty, Imam Sadiq University, Tehran, Iran

Nima Norouzi
Islamic Azad University, UAE & Law and Political Science Department, University of Tehran, Tehran, Iran

A volume in the Practice, Progress, and Proficiency in Sustainability (PPPS) Book Series

Published in the United States of America by
    IGI Global
    Information Science Reference (an imprint of IGI Global)
    701 E. Chocolate Avenue
    Hershey PA, USA 17033
    Tel: 717-533-8845
    Fax: 717-533-8661
    E-mail: cust@igi-global.com
    Web site: http://www.igi-global.com

Library of Congress Cataloging-in-Publication Data

Names: Norouzi, Nima, 1994- editor. | Movahedian, Hussein, 1991- author.
Title: Pragmatic legal and policy implications of environmental lawmaking /
  Nima Norouzi, and Hussein Movahedian.
Description: Hershey, PA : Information Science Reference, an imprint of IGI
  Global, 2022. | Includes bibliographical references and index. |
  Summary: "The monograph covers a wide range of topical issues in
  environmental and energy law, from technology innovation and transfer,
  climate and energy regulation, to pollution control and environmental
  governance and enforcement while addressing more general topics within
  environmental and energy law by outlining key sectoral or environmental
  'media-specific (air, water, land) legal regimes"-- Provided by
  publisher.
Identifiers: LCCN 2022015773 (print) | LCCN 2022015774 (ebook) | ISBN
  9781668441589 (hardcover) | ISBN 9781668441596 (paperback) | ISBN
  9781668441602 (ebook)
Subjects: LCSH: Environmental law, International.
Classification: LCC K3585 .N67 2022 (print) | LCC K3585 (ebook) | DDC
  344.04/6--dc23/eng/20220630
LC record available at https://lccn.loc.gov/2022015773
LC ebook record available at https://lccn.loc.gov/2022015774

This book is published in the IGI Global book series Practice, Progress, and Proficiency in Sustainability (PPPS) (ISSN: 2330-3271; eISSN: 2330-328X)

British Cataloguing in Publication Data
A Cataloguing in Publication record for this book is available from the British Library.

All work contributed to this book is new, previously-unpublished material. The views expressed in this book are those of the authors, but not necessarily of the publisher.

For electronic access to this publication, please contact: eresources@igi-global.com.

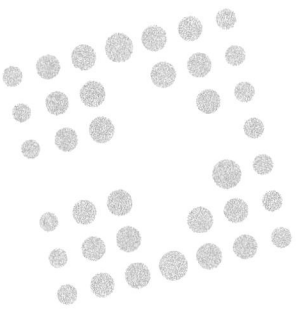

Practice, Progress, and Proficiency in Sustainability (PPPS) Book Series

Ayman Batisha
International Sustainability Institute, Egypt

ISSN:2330-3271
EISSN:2330-328X

MISSION

In a world where traditional business practices are reconsidered and economic activity is performed in a global context, new areas of economic developments are recognized as the key enablers of wealth and income production. This knowledge of information technologies provides infrastructures, systems, and services towards sustainable development.

The **Practices, Progress, and Proficiency in Sustainability (PPPS) Book Series** focuses on the local and global challenges, business opportunities, and societal needs surrounding international collaboration and sustainable development of technology. This series brings together academics, researchers, entrepreneurs, policy makers and government officers aiming to contribute to the progress and proficiency in sustainability.

COVERAGE

- Strategic Management of IT
- Environmental informatics
- Global Content and Knowledge Repositories
- Technological learning
- Eco-Innovation
- E-Development
- Innovation Networks
- Intellectual Capital
- Knowledge clusters
- Global Business

IGI Global is currently accepting manuscripts for publication within this series. To submit a proposal for a volume in this series, please contact our Acquisition Editors at acquisitions@igi-global.com or visit: https://www.igi-global.com/publish/.

Titles in this Series

For a list of additional titles in this series, please visit: www.igi-global.com/book-series

Innovations in Environmental Legislation and Justice Environmental and Water-Energy-Food Nexus Laws
Nima Norouzi (Islamic Azad University, UAE & Law and Political Science Department, University of Tehran, Tehran, Iran) and Hussein Movahedian (Islamic Azad University, UAE & Department of Private Law, Islamic Studies and Law Faculty, Imam Sadiq University, Tehran, Iran)
Information Science Reference • © 2023 • 268pp • H/C (ISBN: 9781668471883) • US $215.00

Sustainable Approaches and Strategies for E-Waste Management and Utilization
A. M. Rawani (National Institute of Technology, Raipur, India) Mithilesh Kumar Sahu (O.P. Jindal University, India) Siddharth S. Chakarabarti (O.P. Jindal University, India) and Ajit Kumar Singh (GLA University, India)
Engineering Science Reference • © 2023 • 300pp • H/C (ISBN: 9781668475737) • US $240.00

Handbook of Research on Solving Societal Challenges Through Sustainability-Oriented Innovation
Luísa Cagica Carvalho (Instituto Politécnico de Setúbal, Portugal & CEFAGE, University of Évora, Portugal) Paulo Bogas (Instituto Politécnico de Setúbal, Portugal) Jordana Kneipp (Federal University of Santa Maria, Brazil) Lucas Avila (Federal University of Santa Maria, Brazil) and Elis Ossmane (Universidade Aberta, Portugal)
Engineering Science Reference • © 2023 • 419pp • H/C (ISBN: 9781668461235) • US $295.00

Handbook of Research on Socio-Economic Sustainability in the Post-Pandemic Era
Jozef Oleński (High School Technology and Economics, Jaroslaw, Poland) Jeffrey Sachs (Columbia University, USA) Masayuki Susai (Nagasaki University, Japan) Yannis Tsekouras (University of Macedonia, Greece) and Arjan Gjonça (London School of Economics and Political Science, UK)
Information Science Reference • © 2023 • 400pp • H/C (ISBN: 9781799897606) • US $270.00

Handbook of Research on Promoting Sustainable Public Transportation Strategies in Urban Environments
Zafer Yilmaz (TED University, Turkey) Silvia Golem (University of Split, Croatia) and Dorinela Costescu (Polytechnic University of Bucharest, Romania)
Engineering Science Reference • © 2023 • 401pp • H/C (ISBN: 9781668459966) • US $295.00

Energy Transition in the African Economy Post 2050
Olayinka Ohunakin (Covenant University, Nigeria)
Engineering Science Reference • © 2023 • 300pp • H/C (ISBN: 9781799886389) • US $215.00

Implications of Industry 5.0 on Environmental Sustainability
Muhammad Jawad Sajid (Xuzhou University of Technology, China) Syed Abdul Rehman Khan (Xuzhou University of Technology, China) and Zhang Yu (ILMA University, Pakistan)
Business Science Reference • © 2023 • 328pp • H/C (ISBN: 9781668461136) • US $250.00

701 East Chocolate Avenue, Hershey, PA 17033, USA
Tel: 717-533-8845 x100 • Fax: 717-533-8661
E-Mail: cust@igi-global.com • www.igi-global.com

Table of Contents

Preface vii

Introduction xii

Chapter 1
Environmental Policy and Law 1

Chapter 2
Principles of Environmental Policy and Legislation 18

Chapter 3
Principles and Concepts of Environmental Law 32

Chapter 4
Systems of Environmental Protection Responsibility 45

Chapter 5
Legal Authorities of Environmental Law 64

Chapter 6
Environmental and International Trade Legislation 84

Chapter 7
Environmental Public Finance Law 103

Chapter 8
Environmental Intellectual Property Law 117

Chapter 9
Environmental and Competition Law 136

Chapter 10
Environmental and Contract Law 151

Chapter 11
Environmental and Criminal Law 167

Chapter 12
Environmental and Economic Law 179

Chapter 13
Environmental and Theological Law 197

Chapter 14
Environmental and Energy Law 215

Chapter 15
Environmental and Islamic Law and Jurisprudence 228

Conclusion 244

Compilation of References 251

About the Authors 266

Index 267

Preface

Ecology refers to the habitats of all plants, animals, and other creatures with natural environmental conditions in a particular area. A habitat with specific types of animal life is determined by location, climatic conditions, latitude, and altitude. A language comprises many ecosystems with a small population of living organisms and their living conditions (the natural parts that affect them). While the boundaries of a language are determined by climate, the demographic boundaries of an ecosystem are separated by natural features such as mountain ranges or upper river walls. A particular language ecosystem tends to have growth patterns like plants and animals with similar eating habits. The main habitats of the planet include tropical rainforests, northern coniferous forests, tundra, desert, savannah pastures, and short oak. However, the term ecosystem is sometimes used interchangeably concerning marine environments, referring to marine ecosystems such as the coastal and shallow water areas, the open water area of the open oceans, and the deep ocean floor area. Of course, these divisions are relative, and some geographers consider other ecosystems; the temperate zone's grasslands are considered.

Human impact on the environment or artificial impacts on the environment includes changes in biophysical environments and ecosystems, biodiversity and natural resources directly or indirectly generated by humans, including global warming, environmental degradation such as ocean acidification, widespread extinction And the loss of biodiversity, the environmental crisis and the collapse of the environment. Changing the environment according to the needs of society has devastating effects. Some human activities that directly or indirectly harm the global environment on a global scale include population growth, over-exploitation, pollution, and deforestation. Some problems, such as global warming and biodiversity loss, pose a vital threat to humanity. Some experts attribute the crisis to an increase in the human population.

The term anthropogenic defines an effect or object caused by human activity. The term was first used technically by the Russian geologist Alexei Pavlov. It was first used in English by the English ecologist Arthur Tansley in connection with the human impact on climax plant communities. Geologist Paul Krutzen coined the term "Anthropocene" in the mid-1970s. The term is sometimes used in pollution from human activities since the beginning of the Green Revolution but is also widely used for all significant human impacts on the environment. Many of the actions taken by humans to help warm the environment comes from burning fossil fuels from a variety of sources, such as electricity, cars, airplanes, space heaters, or building or destroying forests.

Environmental law, or the law of the environment and natural resources, is a collective term to describe a set of common laws, treaties, and regulations, which refers to the effects of human activities on the natural environment. The main regimes in environmental laws and laws often include environmental pollution. A related but distinct set of regimes heavily influenced by legal principles and environmental laws focuses on managing specific natural resources such as forests and mines. Other areas, such as environmental impact assessment, may not be regularly relevant to each category; however, they are an essential part of environmental law.

Throughout history, prototypes of legal enactments designed to protect the human environment consciously have been found. Customary rights related to environmental damage are also permitted only in cases of privacy or due to court rulings. Still, in public cases such as preventing odors from keeping pigs, dumping garbage, or explosion damage to dams, it is an urgent public responsibility. You are coming. However, law enforcement intervention is permitted to deal with serious environmental threats. In 1858, the stench of sewage on the River Times in London forced the House to evacuate. The Metropolitan Sewerage Commission Act of 1848 allowed the Sewerage Commission to block ditches around the city to clear rivers, but this also caused people to pollute the river. Nineteen days later, Parliament passed another law to build London's sewer system. London also suffers from terrible air pollution, with heavy coal smoke peaking in 1982 and the Clean Air Act being passed in 1956.

Archaeological research has shown that Mesopotamia, Egypt, India, Iran, and China were among the first cradles of human civilization. Worship of natural elements has been shared among the inhabitants of these areas. The Iranians were pious and considered Mithras or the sun to symbolize power. After the emergence of Zoroastrianism, polluting water, soil, and the air was considered a sin, and rules were stated in the Avesta. The Iranians were the first to establish clear rules for protecting natural resources and the environment. Other ancient writings and books, such as the Rig Veda, the Torah, and the Law of Hammurabi, contain hints about preserving natural resources and avoiding polluting the environment.

Suppose we want to judge human behavior towards the environment throughout history. In that case, this behavior falls into the spectrum of abuse, so we can say that humans have been abusive to the environment throughout their lives. In other words, human abuse has not only caused a lot of damage to its fellow human beings but has also led to damage to nature, which human beings themselves have suffered the most in this regard. On this basis, we can talk about international environmental law on the world stage, but not in the sense that governments are committed to all environmental do's and don'ts within the framework of the international legal order, but to custom rules. International and those treaties to which they have acceded and accepted commitments.

Today, the world has reached the discourse of human rights, which has now become a discourse, after experiencing different approaches and approaches from academic fields such as economics and political philosophy, and the defeat of great ideas with universal claims and metaphysical or completely global views such as Marxism which has dominated the world.

An essential component of human rights is attention to a healthy environment. Therefore, it is considered globally. The presence of governments, international organizations, and individuals in conferences and scientific meetings and decision-making in this regard is evaluated. The 24th World Conference on Climate Change and another example of this is the annual meeting held from December 2 to December 14, 2018, in Katowice, Poland.

The publication of environmental degradation statistics worldwide and the announcement of scientific assessments of the process of its destruction have attracted everyone's attention to the environment. For this reason, environmental issues and environmentalists have been considered by practitioners (technocrats). As the environment today has a prominent role, face, and symbol in politics, in addition to the realization of a healthy global environment, it is a concern for many Others are concerned, either out of compassion or that they are only interested in protecting the environment and achieving a higher degree of politics.

Given the importance of the environment, which undoubtedly plays a vital role for all human beings, regardless of race, religion, color, ethnicity, and nationality. That everyone will play a role in protecting a healthy environment. Thus, environmental law aims to formulate laws and regulations to protect the environment from realizing the right to a healthy environment for the present and future generations. This book is aimed to study the innovations and new concepts of environmental law.

The purpose of chapter 1 is to explain the role of law as a facilitator of the executive structure of environmental regulation inappropriate conditions in line with social interaction. It also seeks to explain the importance of regulation and regulation. Regulation is one of the most important social standards and guarantees the strong implementation of legal obligations in society. This fundamental standard has been established in public law and seems to be an important approach to protecting the environment and citizens' adherence to environmental obligations.

Chapter 2 first examines the concept of environmental governance system and its features and then the shortcomings and challenges of global environmental governance and then seeks to evaluate the impact of cooperation and coordination on the global environmental governance system by examining various sources and examine the hypothesis that whatever The higher the level and level of coordination and cooperation between international actors, especially governments, as the foremost and most influential actors, the better the protection of the environment and sustainable development will be achieved. Finally, the present article seeks to show what characteristics the global environmental governance system should have to achieve the result, namely protecting the global environment and sustainable development.

Environmental law has developed in response to emerging awareness of and concern over issues impacting the entire world. While laws have developed piecemeal and for various reasons, some effort has gone into identifying key concepts and guiding principles common to environmental law as a whole. The principles discussed below are not exhaustive and are not universally recognized or accepted. Nonetheless, they represent essential principles for understanding environmental law around the world. Chapter 3 describes the fundamental principles of environmental law on which the discipline is based. Also, explain the basic concepts of environmental law and the relationship between principles and concepts.

Environmental protection protects the natural environment by individuals, organizations, and governments. Its objectives are to conserve natural resources and the existing natural environment and, where possible, to repair damage and reverse trends. Due to the pressures of overconsumption, population growth, and technology, the biophysical environment is being degraded, sometimes permanently. This has been recognized, and governments have begun restricting activities that cause environmental degradation. Since the 1960s, environmental movements have created more awareness of the multiple environmental problems. The purpose of chapter 4 is to explain that it is essential that all actors on earth, including humans, governments, institutions, transnational corporations, international organizations, and other actors, be held accountable for any action that leads to environmental degradation. Responsible for existing cracks that result in damage to ecosystems.

Chapter 5 describes domestic and international legal authorities that are somehow related to the environment and can influence it with their decisions and actions.

In chapter 6, first, a view that shows the conflict between international trade law and international environmental law. Then another view that expresses the attention to the environment in international trade law will be presented.

Chapter 7 describes the status of green tax and its history in Iran as a case and discusses new tools in environmental policy implementation.

Chapter 8 aims to analyze the existing relationship between Industrial Property and access to environmentally friendly technologies; without proposing formulas, outline some elements to be taken into account by developing countries for their access. In the first part of the present work, a recount of the means of protection of technology that respects the environment will be made; in the second part, it will be explained how the use of the Intellectual Property system contributes to access to technology that respects the environment; in the third part, the different ways of accessing environmentally friendly technologies will be analyzed, and, finally, in the fourth part, the negotiations under debate in the framework of the United Nations Framework Convention on Climate Change will be presented.

Chapter 9 explores the interaction between competition law and sustainability in the EU in a historical context that includes 'competition law as a shield' and 'competition law as a sword' perspectives. While the European Commission may have indicated its intention to pursue more sustainable competitive practices, it has not yet explicitly expressed a position (except in its policy brief). This chapter charts trailblazing national initiatives being proposed by the EU.

In chapter 10, with a comparative study of oil contracts, including the new generation of oil contracts of the Islamic Republic of Iran (IPC), the environmental conditions contained in these contracts have been studied and analyzed. The main purpose of this study is to review the solutions provided in international agreements, analyze the current status of laws and regulations of the Islamic Republic of Iran, and provide the necessary solutions according to the current situation of oil operations areas in Iran. The results of this paper besides its ethical value, can be useful for policymakers and lawyers practicing in the field of oil and gas contracts.

An overview of the evolution of environmental protection at a global level is the purpose of chapter 11. At the beginning, we discuss the emergence of environmental concerns, followed by the second point, which addresses environmental protection at the international level. Third, the phenomenon of criminalizing harmful conduct against the environment is explored, with a special focus on the role of the European Union in this regard. We will review the most important social movements, international events with the most impact and the legislative evolution of Community Law in this area.

In chapter 12, the concept, legal nature of the green economy, and its role in achieving sustainable development have been studied. While describing the actions of the international community to protect the environment, the analysis of the green economy and its vital role in achieving sustainable development has been discussed. Today, sustainable development can no longer be seen as a choice but as a commitment that all governmental and non-governmental actors must make every effort to achieve a sustainable economy to transition to a green economy. Paying attention to natural resources is valuable for any society, and short-term, medium-term, and long-term plans in the light of a green economy will bring growth and social welfare to the society.

In Chapter 13, some aspects of environmental law and theology will be addressed. Accordingly, first we will conceptually analyze the doctrinal foundations of the right to the environment (1) and then discuss the general doctrinal principles of this right from the perspective of Theological law (2). The basic presumption of this article is based on the grounds that the right to the environment, can not achieve their goals regardless of religious and metaphysical status to ensure a healthy environment.

Given that the human being cannot live or develop without using energy, it is that in chapter 14, we seek to investigate aspects of encounter between both paradigms for a more adequate relationship between them, considering them with their sights set on the goal of sustainable development. In the first term, we will consider central aspects of the vision of Energy Law and then attend to the vision of Environmental Law in energy matters.

Although there is no independent discipline of Islamic environmental law, Islamic law is proclaimed as a source of the legal system in the constitutions of Muslim countries. Thus, we can find in Islamic law a theoretical and practical foundation for environmental law. All sources of Islamic law can be used for this purpose as they all have a potential ecological application. Chapter 15 explores the sources of Islamic law in order to find avenues to Islamic environmental law. In this regard, it analyzes its potential, sources and paradigms. Then according to results of these analysis general Islamic principles introduced with a comparative view over the globally accepted environmental law. These general principles can be used in those countries with Islamic jurisprudence as a reference to improve their environment conservation lawmaking.

Nima Norouzi
Islamic Azad University, UAE & Law and Political Science Department, University of Tehran, Tehran, Iran

Hussein Movahedian
Islamic Azad University, UAE & Department of Private Law, Islamic Studies and Law Faculty, Imam Sadiq University, Tehran, Iran

Introduction

Archaeological research has shown that Mesopotamia, Egypt, India, Iran, and China were among the first cradles of human civilization. Worship of natural elements has been common among the inhabitants of these areas. The Iranians were pious and considered Mithras or the sun as a symbol of power. After the emergence of Zoroastrianism, polluting water, soil, and the air was considered a sin, and rules were stated in the Avesta. The Iranians were one of the first peoples to establish clear rules for protecting natural resources and the environment. Other ancient writings and books, such as the Rig Veda[1], the Torah, and the Law of Hammurabi, contain hints about preserving natural resources and avoiding polluting the environment.[2]

If we want to judge human behavior towards the environment throughout history, this behavior falls into the spectrum of abuse, so we can say that humans have been abusive to the environment throughout their lives. In other words, human abuse has not only caused a lot of damage to its fellow human beings but has also led to damage to nature, which human beings themselves have suffered the most in this regard. On this basis, we can talk about international environmental law[3] on the world stage, but not in the sense that governments are committed to all environmental do's and don'ts within the framework of the international legal order, but to custom rules. International and those treaties to which they have acceded and accepted commitments.[4]

Today, the world has reached the discourse of human rights, which has now become a discourse, after experiencing different approaches and approaches from intellectual fields such as economics and political philosophy, and the defeat of great ideas with universal claims and metaphysical or completely global views such as Marxism which has dominated the world.

An important component of human rights is attention to a healthy environment. Therefore, it is considered globally. The presence of governments, international organizations, and individuals in conferences and scientific meetings and decision-making in this regard is evaluated. The 24th World Conference on Climate Change[5] and another example of this is the annual meeting held from December 2 to December 14, 2018, in Katowice[6], Poland.[7]

The publication of environmental degradation statistics worldwide and the announcement of scientific assessments of the process of its destruction have attracted everyone's attention to the environment. For this reason, environmental issues and environmentalists have been considered by practitioners (technocrats). As the environment, today has a prominent role, face, and symbol in politics, in addition to the realization of a healthy global environment, it is a concern for many Others are concerned, either out of compassion or that they are only interested in protecting the environment and achieving a higher degree of politics.

Given the importance of the environment, which undoubtedly plays a vital role for all human beings, regardless of race, religion, color, ethnicity, and nationality. That everyone will play a role in protecting a healthy environment. Thus, environmental law's goal is to formulate laws and regulations to protect the environment from realizing the right to a healthy environment for the present and future generations.[8]

ENDNOTES

[1]   Hindu Bible, Hassan Naha, Sadegh, 1998, Creation of Life, Vol. 2, Tehran: Iranian University Press, p. 211.

[2]   In books and articles related to the nature, instead of the word "environment", the word "environmental" is used due to its more eloquence in the sentence, which means the same word "environment".

[3]   Jus Cogens - peremptory norm are always binding rules that should not be violated under any circumstances. Such as the prohibition of environmental degradation and the prohibition of torture.

[4]   Hassan Naha, Sadegh, 1998, Creation of Life, Vol. 2, Tehran: Iranian University Press, p. 211.

[5]   COP24 refers to the 24rd annual Conference of the Parties to the 1992 United Nations Framework Convention on Climate Change.

[6]   United Nations Climate Change Conference (Katowice Poland – December 2018).

[7]   The 23rd World Climate Change Summit was held from November 6 to November 17, 2017 in Bonn, Germany.

[8]   In this regard, it should be noted that "Angela Merkel" became the Chancellor of Germany with the support of the Green Party.

Chapter 1
Environmental Policy and Law

ABSTRACT

Environmental regulation is one of the most important subsets of social regulation. Regulation is a framework for implementing the rules adopted in society, and legal standards guarantee this framework. The purpose of this chapter is to explain the role of law as a facilitator of the executive structure of environmental regulation inappropriate conditions in line with social interaction. It also seeks to explain the importance of regulation. Regulation is one of the most important social standards and guarantees the strong implementation of legal obligations in society. This fundamental standard has been established in public law and seems to be an important approach to protecting the environment and citizens' adherence to environmental obligations.

INTRODUCTION

One of the basic tasks of governments is to regulate public affairs through regulation. Regulation is a tool or process that can largely reflect the mechanisms of government involvement and intervention in social affairs. However, the progress and achievements of regulation in recent years have led to the extension of regulation logic from economics to other areas such as the environment, labor and employer relations, consumer protection, etc.

Regulation regulates all social spheres and mechanisms of social control, whether conscious, intentional, or unintentional. Regulation can include any attempt by the government to influence human behavior. It is often claimed that we live in an age of "regulatory government." In other words, regulation is the developed part of the modern government (Kiss & Shelton, 1997).

Social regulation forces individuals and companies, or lower levels of government, to take certain measures to improve public welfare. In a law-abiding society with a free system, social frameworks are established by considering citizenship situations. This means that the decisions of the general public can implement economic, social, cultural, health, political and environmental tools, and government interventions are less visible and, if they exist, are more to facilitate matters. In this regard, one of the most important social levels is environmental protection and public health, and having a healthy environment. In this way, to achieve the goals in the framework of law-abiding citizen relations, the original-

DOI: 10.4018/978-1-6684-4158-9.ch001

ity of this is important in the light of customary rules and norms. However, environmental regulation indicates the need for civil society to adopt a social regulation approach. Although regulation takes place at various levels upstream, including the executive and the legislature, and sometimes beyond the decision-making powers, and sometimes intermediate, including the ministry or the directorate, in this process, environmental regulators, especially at the legislative level, play a role. They play a very important role(Kiss et al., 1991).

On the one hand, it seems that environmental regulation itself is a comprehensive framework and is at the top due to its great importance. On the other hand, we may look down on the environment and see this as a subset of social regulation. In any case, government interventions are justified despite social weaknesses. Despite all these issues, environmental regulation is important, politically, economically, culturally, and socially. In this regard, the aspect of environmental regulation legally leads us to examine the environmental effects. This chapter tries to deal with one of the important tools for resolving social and health, and environmental disasters from a legal perspective under environmental regulation.

Environment and Policy

About the regulation principles, the community needs legal frameworks for the action and implementation of legislative norms by citizens. Therefore, regulation can be done in all social matters, including environmental issues, and communities have the right to ask the government to achieve some degree of social justice. The most important theory in the field of regulation is the theory of public interest. Accordingly, the government regulates places where economic activity has external and environmental effects or an asymmetry of information in which the government enters. Therefore, in this regard, regulatory institutions are considered experts who correct the existing shortcomings and actions in the public interest(Lazarus, 2008).

Social regulation examines the effects of economic activities on the social welfare of citizens. Due to the increasing complexity of public affairs in today's world, legislators must be aware of the objectives they have set for their enactment and be fully aware of the various effects that an enactment has on the economic, social, social, cultural, and environmental sectors. In addition, experience shows that creating harmony and balance between different policies and regulations can be more effective in integrating the overall regulatory system of the country(Nanda, 1995).

With the formation of the regulatory government as the governing paradigm in public sector management, increasing attention has been paid to the preparation and implementation of good regulation, which is one of the important tools. Achieving the goals of the regulatory system is considered. The description of goodness and desirability of regulations can be focused on the aspect of quantity, quality, or both. Quantity oversees the number of regulations, bylaws, and sections of applicable regulations. Numerous laws and regulations are considered a factor that impedes access to the law and its transparency and sometimes efficiency. The quantity of regulations harms its quality. Recently, this focus on the quality of regulations has been used in a broader sense, and the idea has been put forward that if one authority approves a regulation, in any case, another authority must improve its quality. Therefore, optimal regulation regarding quantity and quality should consider the social growth cycle and social standards. In this regard, the excellence of social growth depends on ensuring the participation of development and achieving it towards sustainable development. Its realization depends on protecting human rights, including the right to a healthy environment and its protection(Orts, 1994).

The Environmental Protection Agency is considered a cross-sectoral organization at the national level as a regulatory body. Municipalities also regulate as a structure. In this regard, regulation is determining the manner of implementation of regulations and approvals. The general policies of the system are regulated when they are implemented. The Sixth Development Plan and the Charter of Civil Rights, which also consider environmental aspects, are, in fact, policies that take place after the adoption of existing regulations.

Among the most important powers vested in regulatory bodies is the regulatory authority, licensing of regulatory companies, imposing fines or suspensions on infringing companies, enforcing rules, and in some cases resolving disputes. However, most of the literature on the competence of regulatory bodies is related to their regulatory competence. Therefore, regulation is the most important competence of regulatory bodies.

Although the environment is the most important component of human life, its true value and assistance have not yet been properly identified, so that one of the most important challenges facing governments in the 21st century is environmental crises. That is why governments try to adopt

Various policies and programs are based on field surveys' results to overcome environmental problems or reduce the negative effects of human functions on the environment. For example, in Iran, in recent years, in the form of regulations and five-year plans, efforts have been made to protect the environment and reduce environmental pollution(Palmer, 1992).

Environmental protection involves social and economic regulation because the source of the problems associated with it is usually economic, and their negative effects are often social. In this regard, the government has the role of paving the way for realizing the desired structure.

With this description, regulation based on social conditions should be comprehensive, and the government will make the most effective measures in this regard. This regulation in the field of the environment is undoubtedly undeniable.

One of the shortcomings in today's society is the weak rule of law in the light of its relational practices. Lack of regulation in this direction has taken the citizens out of the circle of the rule of law and lawlessness. This anomaly has caused people to not react well to the rules and norms around them, and the founders of regulation have not done their job properly. In the opinion

It seems that the spirit of cooperation and interaction in decision-making structures has been greatly degraded and has not been significant. Due to administrative bureaucracies and a lack of discipline, regulation is difficult. However, sometimes this problem arises from structural anomalies and heterogeneities. This has led to crises such as environmental crises in society. Accordingly, regulatory failures are undesirable because of the heterogeneity between decisions and do not maintain a social normative balance. In this regard, this failure also causes excessive costs due to adverse activities and causes pollution and destruction of the environment(Sands & Peel, 2012).

In the context of legal review, regulation is one aspect of legislation, which takes place at the supra-sectoral (parliamentary) and sectoral (environmental protection organization) and inter-sectoral (municipal) levels, and the other aspect is structural monitoring and evaluation and its consequences. Evaluating the effects of technical regulations to improve the quality of regulations is based on experience, which is done through systematic and continuous assessment of their potential effects, in other words, it is a policy tool that evaluates the effects in terms of costs, benefits, and risks of each proposed law. This evaluation method has been pursued following the expansion of the tendency to improve and reform the regulatory system after the "inefficiencies" of government intervention in public affairs in some countries of the world. Therefore, it seems that the legal standard will be a reliable cornerstone for environmental regulation, and this framework should be maintained. However, environmental regulation is also the subject of various legal challenges from multiple legal sources. The main issue in this regard is the observance of the ritual and legal process of regulation. However, legal considerations influence the decision to use socio-environmental regulation as a tool of governance.

Challenges of Environmental Law

The development and codification of international law are among the main and fundamental tasks of the United Nations Commission on International Law, specified in paragraph 1 of Article 13 of the UN Charter with the same title. In addition, the United Nations Environment Program has played an important role in the development and drafting of international environmental law, and since its inception, it has proposed several binding and non-binding instruments to governments for ratification. In addition, governments and other international organizations, both governmental and non-governmental, have played important roles in developing and codifying international environmental law in recent years. Environmental protection, at least in its modern sense, which began in the 1970s and continues to this day, encompassed various areas such as air, soil, forest, and water and ratified various conventions in this area. The context reflects the importance of contemporary international law in this regard. Undoubtedly, environmental protection is one of the concerns and concerns of today's human societies. Explosive population growth, unreasonable exploitation of natural resources, degradation and degradation of biodiversity, and increasing pollution have affected the air, soil, and water of the world in various ways, finally degrading the natural quality of human life. As a result, the disturbance of the balance and appropriateness of the environment has led governments, organizations, and international organizations to formulate and implement laws and regulations to prevent pollution and environmental degradation. The formulation of binding environmental principles and rules has gradually led to the development of environmental law, both nationally and internationally. In non-governmental organizations, global public opinion calls on their governments to take the necessary measures to protect the region and the environment. In practice, environmental laws can be considered mandatory criteria, criteria that It has been accepted as a valid criterion by the competent authority in the society. The obligatory nature of the law and the guarantee of implementation provide legal rules to prevent harmful behavior to the environment. Thus, efforts to regulate environmental protection were made in the 1970s, especially after environmentalists warned of its widespread destruction, especially by industrialized countries because international environmental law for the global protection of the environment has been developed by the international community through the enactment and implementation of binding and non-binding legal rules and has developed in terms of content, form, and structure in recent decades. However, despite international efforts to protect the environment, environmental challenges persist and have increased significantly in many areas. Today's world's environmental concerns and threats go beyond experts' and scientists' predictions at the first Stockholm International Human and Environment Conference in 1972. Ozone depletion, climate change, air pollution, soil and water, biodiversity loss, deforestation, and desertification are the most important problems facing human beings today. Thus, despite global efforts to enact and enforce international environmental law, failure to reduce environmental degradation and increase environmental pollution indicates the inefficiency and inefficiency of this legal field. Using a critical approach, this article seeks to address the obstacles and shortcomings of international environmental law while showing the existing capacities for development and proposing legal and executive gaps to address them. There are three important challenges to the substantive development of international environmental law(Salzman & Thompson, 2010):

RELUCTANCE OF GOVERNMENTS TO DELEGATE NATIONAL SOVEREIGNTY

Regardless of the concept of sovereignty in public international law and international environmental law, one of the main obstacles to the development of international environmental law is the reluctance of governments to delegate or limit sovereignty to the benefit of international environmental organizations. The political structure of government always tends to focus and has no interest in delegating it to other centers of power and decision-making. On the other hand, the conflict of interests between the main actors of international law (states) regarding environmental protection, which arises from the national sovereignty of states, challenges the development and expansion of international law. This conflict of interests can include political, economic, commercial interests, etc. In addition, the conflict of interest of developing countries and developed countries in the application and enforcement of the rules and regulations of international environmental law is also one of the limitations of this legal field. Despite the principle of shared but distinct responsibility, in many sources of international environmental law, conflict is seen between these two groups of developed and developing states. The International Court of Justice's advisory ruling on the legitimacy of the threat and use of nuclear weapons in 1996 is a clear example of the pressure exerted on this international body by nuclear-weapon states. However, one of the most important principles enshrined in the Stockholm Declaration of 1972 so far in international environmental declarations is the principle of sovereignty, which redefines the sovereignty of states and a new definition of the concept of sovereignty in international environmental law based on the concept of "use." Fair and reasonable "of the land, but the shadow of absolute sovereignty still weighs heavily on international environmental law. But the fundamental problem is how governments participate in exercising this kind of sovereignty. Countries' differing positions on the world's environmental landscape, including the Nordic countries, which can influence and direct international budgets and accelerate various processes, are a source of concern, such as the United States' refusal to ratify a major agreement. In recent governments, as a result of tensions with Europe and Japan and the protests of southern countries, can have a significant impact on the process of international environmental governance. Thus, the refusal of a country like the United States from its international environmental obligations has catastrophic consequences in the validity and application of environmental governance policies and policies invented by other northern countries. The legitimacy of these countries among the southern countries to the recipient of environmental aid will be lost due to a lack of cooperation and coordination among donor countries and will lead to the risk of non-cooperation between institutions and the suspension of aid. This is evident in the Convention on Biological Diversity and the Kyoto Protocol, and the United Nations. The crises and challenges of international environmental governance Governance in pluralistic management should consider specific social and environmental policies and activities by integrating experience and knowledge in various institutions and effective social factors. And apply. Increasing environmental problems in terms of climate change, biodiversity loss and natural ecosystem degradation, increased pollution, precautionary principle, and transgenic organisms, the risk of nuclear radiation, freshwater scarcity on a very large scale as threats should be effective factors in exacerbating the crisis and Economic development prospects in different countries and regions are limited, controlled and managed. The measures taken and ongoing to protect the environment against the scientific community's warnings are insufficient, and the necessary reforms to improve this protection process require time, energy, financial resources, and most importantly, diplomatic negotiations and consultations. The serious crisis of environmental degradation and the inability to deal with it unanimously have become a topic for all countries. But persistent differences and slow progress in organizing this crisis are serious challenges to governance(Tarlock, 1993).

MULTIPLICITY AND DIVERSITY OF SOURCES OF INTERNATIONAL ENVIRONMENTAL LAW

Another substantive challenge to the development of international environmental law is the multiplicity and diversity of binding and non-binding resources. Soft or non-binding rights are resources that are not binding and do not guarantee specific implementation, which include: statements, resolutions, agendas, action plans, and so on. Their main purpose is to express the principles and rules that guide governments. Although these resources are not binding in themselves, they have a very important impact on the development of international environmental law. The most important examples of these sources are the 1972 Stockholm Declaration on Man and the Environment, the World Charter of Nature adopted by the 1982 UN General Assembly, and the Rio Declarations of 1992, Johannesburg 2002, and finally, the Rio + 20 Declaration of 2012. They had a law. In addition, Agenda 21, adopted in 1992 on the sidelines of the Rio Summit, is a non-binding legal instrument containing important guidelines for protecting, protecting, and managing the environment. United Nations General Assembly resolutions and statements by the UNP Board of Governors; should also be considered a non-binding instrument in international environmental law. During this period, many international conventions were formed, including the Convention on International Trade in Endangered Species of Wild Fauna and Flora, the 1973 Washington Convention on the Law of the Sea, the 1982 Monte Gobi Convention, and many other conventions. They usually had an organization and structure for that convention. Today, most of the sources of international law are contained in the first paragraph of Article 38 of the Statute of the International Court of Justice. According to paragraph 1 of Article 38 of the Statute, the Court has to settle disputes according to international law and implement the following standards. International treaties, public (private) or private (private), which establish rules (legal) and are explicitly recognized by the parties to the dispute; International custom as the reason for the general procedure that has been accepted as a legal rule; General principles of law recognized by civilized nations; According to the provisions of Article 59 of the Articles of Association, judicial decisions and the opinions of the most competent public law experts of different nations are considered as a subsidiary (auxiliary) knowledge of legal rules; Article 38 also stipulates that the provisions of this article do not impair the discretion of the Court in issuing a judgment following the (rule of fairness and discretion), provided that the litigants agree to it. However, today Article 38 of the Statute of the Court does not provide a complete picture of the sources of international law, but rather the international rules, unilateral legal actions of countries, and binding resolutions of international organizations, each of which is effective in regulating international legal relations and can serve as a source. International law to play a role. Despite their thematic diversity and geographical scope, international environmental treaties have common features, use similar legal techniques, and are often interrelated. Features such as the absence of retaliation in commitments, related materials or references from one document to another, structural protocol agreements, the creation of new institutions or the facilitation of existing or former institutions to promote continuous cooperation, the existence of rituals and innovative practices related to adherence and Non-compliance of the contracting members is the existence of a mechanism to correct and review their common funds. However, more than 300 multilateral treaties and 700 bilateral environmental agreements have led to the sequence and multiplicity of this source of rights, which is considered one of the essential challenges in the development of international environmental law. Custom also plays an important role as another source of international environmental law, and many issues, such as the international responsibility of governments, especially in the field of transboundary pollution, have customary roots. The number of other international treaties and instruments that enshrine the same legal rules related to the environment in international treaties is increasing. The work of the UN Commission on International Law shows that the inclusion of customary rules in many international instruments can be considered a substantial development of international environmental law. General principles of international law, although not sufficient to protect the environment, but due to unsystematic development, treaty law remains in force. Principles of unity, resources, and responsibilities of the common heritage of humanity is also one of the basic rules in this branch of law. On the other hand, the issuance of judicial rulings plays an important role in developing this field of law. International environmental law, in its short life, has been able to study and study solutions in terms of the potential importance of the environment; Issues such as the international responsibility of polluters, the obligations of governments towards their ratified treaties, and finally, the criminal liability for environmental degradation are among the important issues of international environmental law. But in international environmental law, despite the large volume of binding and non-binding documents and adherence to international custom, to fulfill the international responsibility of governments is not without difficulty. As many legal opinions and procedures have shown, adherence is not customary in limited cases(Wiener, 2000).

Thus, the custom seems to have lost its traditional effectiveness in international environmental law, and for customers to be effective in international law, it is necessary to consider the behaviors and practices of governments in environmental protection. Therefore, due to the novelty of this field of law, sources in international environmental law, and these sources, there are other sources such as international rules of procedure, international obligations of states and UN resolutions, and UN Security Council resolutions and resolutions are cited. Are. This multiplicity of resources, especially in the large number of international treaties that sometimes have similar or sometimes conflicting obligations, has challenged the implementation of international environmental law.

INSUFFICIENT GUARANTEE FOR THE IMPLEMENTATION OF INTERNATIONAL ENVIRONMENTAL LAW

Another essential challenge of international environmental law is the inadequacy of the enforcement guarantee. To implement a legal system, after going through the rules, it must be thought that its implementation guarantees an important role. This issue in the world community; has different forms because governments are never willing to jeopardize their national interests against each other. Even in some cases, there are shortcomings in the domestic laws to support this issue. Due to the sovereignty of governments and their role, there is an obstacle to the effective allocation and enforceability of these documents.

On the other hand, the existing enforcement mechanisms will sometimes be forced to retreat in the face of pressure from governments. Unlike the domestic law system of countries, international law lacks coherent enforcement mechanisms such as the executive, the administrative police, or the judiciary. Existing classical safeguards, such as criminal or civil warranties, do not work, and many environmental issues, despite being cross-border and non-refundable(Nanda, 1995).

STRUCTURAL CHALLENGES OF INTERNATIONAL ENVIRONMENTAL LAW

With the rise of global environmental protection since the 1970s, this mission has been directly or indirectly reflected in many international organizations' goals. In the meantime, international governmental organizations, both global and regional, and even non-governmental organizations have tried to play their role in environmental protection and commit their members to international environmental regulations to support Take a step away from the environment. Undoubtedly, as active subjects of international law, international organizations have a serious task to achieve in this regard. The number of these organizations that today have some kind of environmental competence is important. Almost every international governmental organization has somehow gotten into the subject of the environment. The world community is aware that environmental protection is very important for the well-being of the world community and ecosystems and sustainable development. For this reason, environmental protection and sustainable economic development are at the center of the agenda of international and regional international organizations. In addition to the United Nations, other international organizations also offer programs in this direction. In this regard, one of the important reflections of the UN Conference on Man and the Environment, Stockholm, 1972, was the establishment of the United Nations Environment Program as the global and executive arm of the organization in the field of environment and its preservation in A subset of this organization was established. Accordingly, it was tasked with assessing the state of the global environment, formulating global environmental programs, and providing the necessary funding for these matters. It is one of the United Nations-affiliated bodies, in other words, the only UN mechanism that has the authority to policy, coordinate and encourage various environmental issues between governments and specialized agencies of the United Nations and other organizations. It has international governmental and non-governmental organizations. The United Nations Environment Program was established to monitor the state of the global environment and the effects of national and international environmental policies and measures. The General Assembly in 1997, while emphasizing UNEP as the United Nations Chief Agent in the field of environment, called on it to be the main custodian of the global environment and to set the world environmental agenda and to promote and coherently implement the environmental dimensions of sustainable development. The UNEP program is headquartered in Nairobi, the capital of Kenya, and from there is responsible for directing programs and managing the work of other offices that were later established in other countries. The United Nations Environment Program's headquarters in Africa will help understand the pattern of environmental issues facing developing countries. The main goal of the United Nations Environment Program is to establish and encourage international coordination and cooperation to protect the environment. UNEP is also a center for informing the governments and people of the world and a means to improve their living standards without harming the rights of future generations. In other words, UNEP is an institution that advocates the prudent use of the environment and sustainable development. UNEP also works with the United Nations, international organizations, governments, non-governmental organizations, the private sector, and civil society to achieve its goals. Accordingly, the United Nations Environment Program, as one of the most fundamental structures of global environmental protection, has a distinct position compared to other international organizations. However, according to their field of work in implementing environmental programs, other organizations act following their authority, all of which have been created with one purpose: to protect the environment. The current challenges facing the environmental system at the international level reveal the need to review the institutions' structure. Although all these efforts aim to address environmental sustainability concerns, they are still witnessing environmental degradation and the non-implementation of decisions and aspirations of related institutions. The lack of coordination in the environmental system is due to the structural weakness of international environmental law. However, the structural development of international environmental law has been challenged by international competition and conflict between governments, especially developed and developing countries. Although the United Nations Environment Program is a United Nations body that oversees and directs the environmental activities of its members, it continues to function as a program to become the "World Environment Organization," which has Universal and more complete powers that face limitations. The United Nations Environment Program, as the United Nations Main Program for Environmental Protection, despite its many successes in protecting the global environment, does not have the authority to enforce the rules and regulations of international environmental law on a global scale. Therefore, the need to revise this program to upgrade its current position or change its dependence on the United Nations is inevitable(Rosenbaum, 1993).

ENVIRONMENTAL JUSTICE AND SUSTAINABLE DEVELOPMENT

The environment is the first human right. Without a secure environment, one cannot claim to have other political, social, or economic rights. The environment is defined as all external factors affecting the life and activity of living things, plants, and humans. So the environment refers to everything around us that can affect human and non-human life. For example, X's environment has trees, water, sun, air, other people, etc. The idea of X's life and existence, assuming the support of others for the realization and flourishing of his humanity, means that he needs the government and the relationship in the social framework, he needs plants to provide food, clothing, and oxygen, he needs to travel to the sea and energy and to set the clock requires the sun. In principle, a man needs his environment and needs to be familiar with its functions(Arrow et al., 1996).

As we have said, not only human life is affected by the environment, and it is not only human beings who are in the environment and not only human beings who live in the environment, but also the lives of other living beings and plants are affected by the activities around them. For example, plant and animal life are affected when humans cut down trees. Environmental studies have always proven the interdependence of living organisms and their interaction with the environment. Therefore, the treatment of the environment needs a lot of attention. The normative questions that arise are whether human behavior with the environment is fair. Does non-human life (e.g., animals) have a right to the environment? Are the past and present lifestyles of global citizens fair to future citizens of the world? These are just some of the goal-setting sharewares that you can use. These questions become more violent when we try to understand the interaction between human development and the environment.

Development and the Environment

The history of economic development has hardly been formed without its impact on the environment. Technically, development refers to the process of change or transformation. The development indicates improvement, progress, and development. Hence, economic development means the development of economic affairs or activities. According to economic development researchers, improved tools and techniques are necessary to develop or promote economic activities. The history of human economic activity has traversed the agricultural age, the industrial age, and the information age. Prehistoric agriculture included hunting and feeding with simple tools such as a shovel and a stick. Agriculture became a scientific trade through the scientific revolution and the industrial revolution of the 17th century and emerged as an industry. The advent of industrial agriculture marked a turning point in economic development. Hence, large-scale agriculture has led to the use of artificial fertilizers and pesticides, the formation of large-scale livestock farms, antibiotics and hormones, and dependence on machinery have become current issues.

The continuation of the industrial revolution, which brought about industrial development, especially industrial agriculture, has had a detrimental effect on the environment in various ways. Economic development has also had negative and detrimental environmental consequences. The gift of industrialization has been the uncontrolled consumption of water resources. That is, groundwater and watersheds are consumed faster than they can be regenerated. Irregular energy consumption is also characterized by economics. For example, with special reference to agricultural development, it should be said that heavy agricultural machinery needs a lot of energy, and a lot of energy is spent on the production of nitrogen fertilizers and pesticides. In addition, the transportation of food over long distances requires a lot of energy. These are all air pollution topics due to the consumption of large amounts of fossil fuels, which in turn causes global warming(Gillies, 2013).

Due to the development of biotechnology, farmers have resorted to the use of chemical fertilizers to increase production. The effect is a significant reduction in soil ability to retain moisture and a strong dependence on irrigation systems. Moreover, groundwater and surface water are contaminated by the use of herbicides and insecticides. The result of these anthropogenic activities of agriculture and large-scale industrialization is a constant threat to human life and other living things. This question is appropriate here; Is it fair for human beings to treat the environment in any way they want in the name of development? And it is precisely this question that has created the need for this study to examine the issue of environmental justice in the face of sustainable development concerns with particular reference to developing economies.

Human Needs and Environmental Justice

Man needs food, clean water, education, income, and good health. Poverty is the cause of the inability to meet these basic needs. So poverty is a threat to the sustainability of human life. Human rights are derived from these basic human needs. In other words, human beings have the right to food, water, education, health, and so on. The absence of any of them can cause negative chain reactions. For example, lack of food weakens the human immune system, and the weakened immune system causes malnutrition; malnutrition opens the door to diseases and illnesses.

Human needs for food, housing, clothing exist in any socio-political and economic system. Therefore, the right to the necessities of life, which is an indicator of human rights, is universal and inalienable. Therefore, talking about the right to the environment, equality or environmental equality, and environmental justice is logical, and the phenomenon of environmental justice means fair behavior and meaningful participation of all human beings regardless of race, color, gender, nationality, or income concerning the development and implementation of laws., Environmental regulations and law. Therefore, the right to the environment is an undeniable necessity.

According to researchers, environmental justice is associated with a social change towards meeting basic human needs and improving the quality of human life, in other words, the quality of education, health care, housing, human rights, environmental protection, and democracy. By analyzing the various definitions and concepts of environmental justice, David Schlossberg introduces the four main issues in the category of environmental justice as follows:

- Equitable distribution of environmental risks and benefits;
- Fair and meaningful participation in the environmental decision-making process;
- Recognition of social lifestyle, local knowledge, and cultural differences;
- The ability of communities and individuals to do work and flourish in society.

Therefore, environmental justice in the search for fair behavior is related at two levels: fair human behavior and fair environmental behavior.

People need a rehabilitative environment to realize their right to the necessities of life. For example, the fair thing that X should do is to create an enabling environment for others to enjoy their rights, and others should compensate him. X's environment includes living, working, playing and learning, and so on. On the contrary, what is fair about the environment is that everyone, including X, protects the environment and keeps it safe and sound.

Bojan Bryant has expressed his concern about a proper understanding of environmental justice, that when people reach their full potential, they have served environmental justice, especially since environmental justice is the right relationship between humans and the planet. Environmental justice focuses on humans and other vulnerable creatures at risk of destructive and greedy human activities. Environmental justice, which is necessarily human-centered from the point of view of ethics, seeks to promote human well-being and social equality while at the same time not underestimating endangered species and environmental health.

Proponents of environmental justice have sought to compensate for the unequal distribution of environmental burden through a proposal to guarantee equal rights to the environment. However, environmental burdens that can prevent people from realizing their potential include pollution, industries, industrial facilities, and crimes [. Hence, what is considered environmental injustice has several causes.

The main cause of environmental injustice is illegal changes in land, water, energy, and air. For example, when the abusers of land, water, energy, and air are not dealt with without discrimination, wildlife and human life will suffer the consequences. In this regard, the vast deforestation in the name of human development has caused incalculable damage to the world's natural resources. Industrial advances have always threatened the environmental rights of early inhabitants around the world. Aquatic life has not been spared from dealing with water users without discrimination. Coastal communities around the world depend on water for livelihoods, transportation, and so on. In particular, fishing is a real source of income for beach communities. So, if humans do not treat water with respect, the livelihood of some people will be jeopardized.

Energy change and climate change are similar to the causes of environmental injustice. For example, burning gasoline and exploring fossil fuels have contributed significantly to the degeneration of the environment in many parts of the world. Areas such as the Persian Gulf or the Caspian region, and the Niger Delta in Nigeria are rich in crude oil. Today, however, indigenous peoples in most crude oil exploration areas appear to be facing environmental injustice. Simultaneous land, water, energy, and air has brought about environmental degradation and associated injustice.

Among other reasons, the failure of governments to formulate responsible and accountable government policies and regulations has been recognized as a cause of environmental injustice. The lack of recognition of the rights of others to a clean, safe and healthy environment has also been attributed to governments failing to address the need for thoughtful government policies and regulations on environmental pollution, especially in developing countries. The mindset of developing countries where environmental injustice is ostensibly common is that governments should wait for environmental pollution to occur and then look for a solution.

Institutionalized racism also threatens environmental justice. The tendency to consider the environment or region of some people inferior to others also causes injustice. The truth is that the poor of the world are mostly found in black communities. The intellectual tendency, then, is that local black communities are polluted and polluted by the environment. This explains why black communities around the world are an accessible place to dispose of waste. And because poor communities (or races) lack the interests and power to fight advanced industrial communities or races, they succumb to fate.

Environmental racism, which seeks to polarize the world and create new communities, a lower environmental community, and a superior environmental community, causes environmental injustice. From what has been said, it follows that not everyone enjoys equal protection against environmental and health hazards, and at another level of injustice, not everyone has equal access to the decision-making process for a healthy environment(Bolla & McDorman, 1999).

The same is evident at the level of international relations. The relationship between developed and developing countries regarding the global environment's treatment indicates environmental inequality and injustice. Advanced or industrialized economies are the worst polluters. Advanced economies do not respect the environment of developing countries because toxic industrial wastes often make their way to landfills in developing countries.

Sustainable Development and Environmental Injustice

Sustainable development and environmental justice phenomena come into play during the search for growth and development in developing economies. Sustainable development means meeting the present generation's economic, environmental, and socio-political needs without endangering future generations.

Sustainable development is a single phenomenon with three dimensions: from one socio-political dimension, the second economic dimension (economic sustainability), and the third environmental dimension (environmental sustainability). But these three dimensions of sustainable development are intertwined in the form of an intertwined coil. So, it becomes almost difficult to place these areas of sustainable development in impermeable enclosures. Thus, for example, socio-political activities (such as war) affect economic activity (e.g., wartime budgets) as well as the environment (heavy use of weapons that destroy the environment).

Sustainable development comes to mind in the sense that resource consumption should be such that it does not run out in the short term. So, for example, the question we ask about the West Asian region is what contribution has West Asia, and the Middle East made to global efforts towards the sustainable use of natural resources? And if the goal of sustainable development is to meet the needs of the people while maintaining the comprehensiveness of the environment; Do most Middle Eastern countries pass the Sustainable Development Exam? The answers to these questions include themes for sustainable development and environmental justice in the region.

Undoubtedly, the region of West Asia enjoys the blessings of natural resources: for example, vast arable lands, rivers and minerals, and so on. Africa is one of the most prosperous regions on the continent, but it is also home to extreme poverty, unimaginable misery, and unprecedented environmental degradation. Environmental justice is a huge vacuum in rich regions like Africa, and it can be said that the African environment does not support environmental justice. The environmental rights of the average African - for example, the right to food, clean water, and a healthy environment - are often threatened by a lack of environmental education based on environmental issues.

Sustainable development, which guarantees these rights, is based on a clear understanding of the place of the environment in the human yard. Humans are each born in a place (in the environment); They grow up in an environment, and their beliefs and values are influenced by their relationship with those around them (the environment). Human beings' failure to respect the environment leads to food crises, floods, and threats to humans and wildlife. For example, countless birds, animals, and plants are on the endangered species list in wildlife. All over the world, most of the wildlife and the most beautiful natural resources on the planet have been abandoned in the name of development. And many animals are dying because of the loss of their natural habitat. The extinction of wildlife contributes to the poverty of human existence. The chain of ecosystems that preserve human life is broken every time a species becomes extinct(Carson, 1962).

Today, this view of environmental justice is considered a problem and a limitation, which is the best approach to this important internationalization of environmental protection and respect. In any case, to achieve environmental justice, a fundamental change in worldview, we need and must have the positive impact of change to challenge human communities to understand a more comprehensive view of things. Environmental justice is embedded in a vision for broader justice, and efforts to promote economic justice and political power do not marginalize environmental justice.

CONCLUSION

Regulations generally include rules that specify permitted and prohibited activities concerning individuals, companies, or government administrative bodies and are accompanied by penalties or rewards, or both. Social regulation also aims to limit or prohibit behaviors that directly threaten public health, safety, and public welfare. These threats include environmental pollution, unsafe work environments, unsanitary living conditions, and social deprivations, and social regulation addresses the behavior of individuals, companies, and lower levels of government. In this regard, the new approach of regulatory organizations is to facilitate regulations and promote community acceptance and compliance with regulations.

Governance decisions have the same effect as a regulation. Although some believe that all decision-making is sovereign and citizens have no role, this is a wrong view because wherever the general public disobeys the law, it is a sign of a violation of sovereignty in decision-making. Therefore, public diplomacy is still ongoing, and this is also reflected in environmental decision-making. This process takes place in a standard context and follows environmental regulation. Now, this environmental regulation includes various dimensions. This trend shows the coherent efforts of civil society on the circuit of environmental decision-making.

In this regard, it is suggested that environmental regulation, which certainly reduces the gap between commitment and compliance with environmental law, should be prioritized by competent organizations, which is important when its effects are realized. And a clear example of this is institutional arrangements for environmental decisions at the national level. Thus, environmental regulation operates in all national and international sectors and thus reflects the transnational nature of the environment and the vastness and diversity of individuals and organizations, including countries affected by environmental decision-making in the light of public and environmental diplomacy.

In recent decades, the gradual development of international environmental law has been based more on human-environmental needs and necessities than anything else. Economic growth and increasing technological advances in the contemporary period have caused fundamental damage to the environment. In response to these needs and requirements, international law has sought to bind governments to environmental protection by enacting international laws and regulations. Although the international community has made great efforts to enforce binding and non-binding legal instruments for global environmental protection, these efforts have failed to stop the widespread environmental degradation in the world. In terms of the content of the law, creating a comprehensive system of international environmental law can reduce the legal gaps created by the multiplicity of environmental legal documents. In addition, one of the ways to reduce the existing challenges is to adopt an approach of "internationalization" of environmental protection, which is based on two legal bases.

On the one hand, the basic rules and regulations of international environmental law bind on members of the international community. The legal basis of these rules and regulations is mainly "customary" and is based on customary law, some of which are now known as "international rules." On the other hand, some of the basic principles of international environmental law, such as the principle of "prohibition of harm to other lands," the principle of "fair and rational use of land," the principle of "cooperation," and the principle of "prevention" should be considered in global environmental protection. Therefore, international law's conceptual and substantive development can be considered an effective solution to the limitations and obstacles to international environmental law.

On the other hand, "internationalization" of environmental protection on the principle of "institutionalization" of international law is based on the contemporary environment. Institutionalization of international law means creating and expanding international organizations and institutions for the effective and efficient protection of the environment. Also, participation in the domestic dimension includes public participation in environmental decision-making and implementation of environmental decisions and participation in the international dimension overseeing the participation of all governments, both rich and poor, developed and developing, industrial and non-industrial in decision-making. Therefore, internationalization can be one of the important strategies for the development of international environmental law.

Also, the criminalization of large-scale environmental degradation and the passage of "crime against humanity" through the recognition of "crime against future generations" can be a sign of progress in the sense of extra responsibility for the future, future conditions, and protection of other living species and the environment is considered. International criminal law does not yet have clear legal solutions to criminalize environmental degradation as international crimes; however, recognizing "crime against future generations" as an international crime could extend to international environmental law.

Structurally, the transformation of the United Nations Environment Program, which currently has limited capacity, into a "global environmental protection organization" with the necessary authority and power, could be an important step in the structural development of international law. The issues related to establishing this global organization will not be outside the conflict of developing and developed countries.

REFERENCES

Arrow, K., Jodha, N., Jentoft, S., McCay, B., McKean, M., Sanderson, S., & Young, O. (1996). *Rights to nature: ecological, economic, cultural, and political principles of institutions for the environment*. Island Press.

Bolla, A. J., & McDorman, T. L. (1999). *Comparative Asian environmental law anthology*. Carolina Academic Press.

Carson, R. (1962). Silent spring III. *New Yorker (New York, N.Y.)*, 23.

Gillies, D. (2013). A guide to EC []. Routledge.]. *Environmental Law (Northwestern School of Law)*, 9.

Kiss, A., & Shelton, D. (1997). *Manual of European environmental law*. Cambridge University Press.

Kiss, A. C., Shelton, D., & Shelton, D. (1991). *International environmental law* (Vol. 3). Transnational Publishers.

Lazarus, R. J. (2008). *The making of environmental law*. University of Chicago Press.

Nanda, V. (1995). *International environmental law & policy*. Brill Nijhoff.

Orts, E. W. (1994). Reflexive environmental law. *Nw. UL Rev.*, *89*, 1227.

Palmer, G. (1992). New ways to make international environmental law. *The American Journal of International Law*, *86*(2), 259–283. doi:10.2307/2203234

Rosenbaum, K. L. (1993). Sustainable Environmental Law: Integrating Natural Resource and Pollution Abatement Law from Resources to Recovery. Environmental Law Institute, 575-674.

Salzman, J., & Thompson, B. H. Jr. (2010). *Environmental Law and Policy: Concepts and Insights*. Foundation Press.

Sands, P., & Peel, J. (2012). *Principles of international environmental law*. Cambridge University Press. doi:10.1017/CBO9781139019842

Tarlock, A. D. (1993). The nonequilibrium paradigm in ecology and the partial unraveling of environmental law. *Loy. LAL Rev.*, *27*, 1121.

Wiener, J. B. (2000). Something borrowed for something blue: Legal transplants and the evolution of global environmental law. *Ecology Law Quarterly*, *27*, 1295.

ADDITIONAL READING

Kiss, A., & Shelton, D. (1997). *Manual of European environmental law*. Cambridge University Press.

Kiss, A. C., Shelton, D., & Shelton, D. (1991). *International environmental law* (Vol. 3). Transnational Publishers.

Lazarus, R. J. (2008). *The making of environmental law*. University of Chicago Press.

Nanda, V. (1995). *International environmental law & policy*. Brill Nijhoff.

Orts, E. W. (1994). Reflexive environmental law. *Nw. UL Rev.*, *89*, 1227.

Palmer, G. (1992). New ways to make international environmental law. *The American Journal of International Law*, *86*(2), 259–283. doi:10.2307/2203234

Salzman, J., & Thompson, B. H. Jr. (2010). *Environmental Law and Policy: Concepts and Insights*. Foundation Press.

Sands, P., & Peel, J. (2012). *Principles of international environmental law*. Cambridge University Press. doi:10.1017/CBO9781139019842

Tarlock, A. D. (1993). The nonequilibrium paradigm in ecology and the partial unraveling of environmental law. *Loy. LAL Rev.*, *27*, 1121.

Wiener, J. B. (2000). Something borrowed for something blue: Legal transplants and the evolution of global environmental law. *Ecology Law Quarterly*, *27*, 1295.

KEY TERMS AND DEFINITIONS

Equity: Defined by UNEP to include intergenerational equity - "the right of future generations to enjoy a fair level of the common patrimony" - and intragenerational equity - "the right of all people within the current generation to fair access to the current generation's entitlement to the Earth's natural resources" - environmental equity considers the present generation under an obligation to account for long-term impacts of activities and to act to sustain the global environment and resource base for future generations. Pollution control and resource management laws may be assessed against this principle.

Polluter pays principle: The polluter pays principle stands for the idea that "the environmental costs of economic activities, including the cost of preventing potential harm, should be internalized rather than imposed upon society at large." All issues related to responsibility for cost for environmental remediation and compliance with pollution control regulations involve this principle.

Precautionary principle: One of the most commonly encountered and controversial principles of environmental law, the Rio Declaration formulated the precautionary principle: In order to protect the environment, the precautionary approach shall be widely applied by States according to their capabilities. Where there are threats of serious or irreversible damage, lack of complete scientific certainty shall not be used as a reason for postponing cost-effective measures to prevent environmental degradation. The principle may play a role in any debate over the need for environmental regulation.

Prevention: The concept of prevention can perhaps better be considered an overarching aim that gives rise to a multitude of legal mechanisms, including prior assessment of environmental harm, licensing or authorization that set out the conditions for operation and the consequences for violation of the conditions, as well as the adoption of strategies and policies. Emission limits and other product or process standards, the use of best available techniques, and similar techniques can all be seen as applications of the concept of prevention.

Public participation and transparency: identified as necessary conditions for "accountable governments,... industrial concerns," and organizations generally, public participation and transparency are presented by UNEP as requiring "effective protection of the human right to hold and express opinions and to seek, receive and impart ideas,... a right of access to appropriate, comprehensible and timely information held by governments and industrial concerns on economic and social policies regarding the sustainable use of natural resources and the protection of the environment, without imposing undue financial burdens upon the applicants and with adequate protection of privacy and business confidentiality," and "effective judicial and administrative proceedings." These principles are present in environmental impact assessment, laws requiring publication and access to relevant environmental data, and administrative procedures.

Sustainable development: Defined by the United Nations Environment Programme as "development that meets the needs of the present without compromising the ability of future generations to meet their own needs," sustainable development may be considered together with the concepts of "integration" (development cannot be considered in isolation from sustainability) and "interdependence" (social and economic development, and environmental protection, are interdependent). Laws mandating environmental impact assessment and requiring or encouraging development to minimize environmental impacts may be assessed against this principle.

Transboundary responsibility: Defined in the international law context as an obligation to protect one's environment and prevent damage to neighboring environments, UNEP considers transboundary responsibility at the international level as a potential limitation on the sovereign state's rights. Laws that limit externalities imposed upon human health and the environment may be assessed against this principle.

Chapter 2
Principles of Environmental Policy and Legislation

ABSTRACT

The global environmental governance system has not successfully achieved its primary goal: to protect the global environment and achieve sustainable development. Among the reasons for this is the incoherence and lack of cooperation of international actors; inflation of multilateral environmental agreements, inappropriate international structures, and institutions; failure to follow the documents and non-implementation of the obligations contained in them; inefficient allocation of resources; unpopular way of making decisions; policy-making and decision-making outside the system of global environmental governance, and the absence of non-governmental actors in a state-centered system. This chapter aims to identify the causes of the inefficiency of the global environmental governance system and examine the necessary measures to address the shortcomings of this system.

INTRODUCTION

The "Global Environmental Governance System" consists of documents, organizations and institutions, decisions and policies, financing mechanisms, rules, rituals, norms, and implementation of the provisions of the documents through which the protection of the global environment and sustainable development is to be achieved (Koh, 2008). But has the system been able to achieve these goals sufficiently? If not, what is the main reason or reason? And what can be done to address these shortcomings? Policymaking, organizing, establishing institutions, financing, setting rules and rituals, and ultimately implementing principled policies through relevant rules require cooperation and coordination among international actors, including governments, civil society, and the private sector (Boyle, 2002). Therefore, it is necessary to examine the role of coordination and cooperation among international actors in the global environmental governance system to achieve environmental protection and sustainable development goals(Alexandroff & Cooper, 2010). In a division, we can say Global environmental governance consists of three pillars: first, a process that ultimately aims to regulate the outcome of various meetings and decisions in the

DOI: 10.4018/978-1-6684-4158-9.ch002

form of documents inappropriate language that the political representatives approve of governments, approved by the parliaments of the countries, and finally In the legal system of the signatory states, it takes the form of essential documents; A total of formal and informal institutions, both private and public; And third, the implementation and fulfillment of the environmental commitments of States and members of the international community, which, according to the international instruments referred to in the preceding component (Meganck & Saunier, 2012), are necessary—existing to be implemented using current structures (Najam, 2003). The effect of human actions on the ecosystem; Deep is usually borderline and sometimes irreversible. In the most optimistic case, assuming that governments have the necessary political will to protect the environment, we can still not expect the phenomenon of border crossings to be controlled by occasional and uncoordinated governments' responses without the cooperation of all international actors. In practice, the measures taken by the countries of the world to protect the environment, due to their inconsistency and lack of cooperation with each other, did not have acceptable results and could not adequately respond to environmental degradation, including climate change, air pollution, soil and oceans, hazards: nuclear or genetic manipulation, rampant reduction of natural resources, and destruction of biodiversity and landscapes(Risse, 2004). Much of the provisions of the hundreds of multilateral environmental agreements that have taken so much energy, time, and financial resources to build have never been implemented due to inconsistencies and lack of cooperation from international actors, and this has led many of these documents never to achieve their goals because they were created not to achieve (Najam, 2003). All this shows how necessary it is to create an efficient global environmental governance system in which all actors in the international community, especially governments, are organized in a coherent system. Only in the shadow of such a system will the international community's actors, especially the governments, react in a timely and efficient manner to the factors and forces that harm the environment while preventing the destruction of the environment and avoiding environmental destruction anti-environmental actions. This article first examines the concept of environmental governance system and its features and then the shortcomings and challenges of global environmental governance and then seeks to evaluate the impact of cooperation and coordination on the global environmental governance system by examining various sources and examine the hypothesis that whatever The higher the level and level of coordination and cooperation between international actors, especially governments, as the foremost and most influential actors, the better the protection of the environment and sustainable development will be achieved. Finally, the present article seeks to show what characteristics the global environmental governance system should have to achieve the result, namely protecting the global environment and sustainable development(Miller, 2007).

METHODOLOGY

Leading research has been conducted using literature review and authoritative internet resources. The sources of this research are mainly articles and books of researchers who have previously researched global environmental governance. In this article, while studying the basic concepts such as governance and, more specifically, global environmental governance, the current state of the global environmental governance system has been examined. Given the overall rate of environmental degradation, both nationally and at the regional and global levels, some of the causes of the inefficiency of the existing system of global environmental governance have been studied. In the end, several proposed solutions have been discussed(Thomas, 2001).

CONCEPT OF GLOBAL ENVIRONMENTAL GOVERNANCE

Global environmental governance consists of organizations and institutions, decisions and policies, financing mechanisms, rules, rituals, and norms that seek to protect the world's environment and guarantee sustainable development. According to another definition, governance is the interaction between formal and informal institutions and social actors that affect how environmental problems are identified and framed (Koh, 2008).

In addition to the above definitions, scholars have given various other definitions of world governance. Since global environmental governance is also a type of world governance, these definitions apply to the latter issue; Moreover, many of these definitions are of environmental background. For example, Patricia Byrne and Alan Boyle, authors of International Law and the Environment, argue that global governance is a "continuous process involving the harmonization of conflicting or diverse interests and cooperation." Global governance encompasses formal institutions and regimes created to follow informal treaties and arrangements ... and has no specific model or form, or structure, or set of structures, but involves any broad, complex, and dynamic decision-making process. In this definition, cooperation, the existence of formal institutions, and an uncertain decision-making structure are important features of global governance.

They add that the international community is a more inclusive concept than an international community made up of governments alone in world governance. Thus, many concepts derived from governance involve the cooperation and participation of other institutions, such as NGOs, industry and commerce, and civil society (Birnie & Boyle, 1995).

Global governance is also the political interaction between cross-border actors to resolve problems that affect more than one state or region without executive power. According to another theory, in its most straightforward and comprehensive sense, global governance refers to all regulations that aim to organize human societies globally (Podder, 2014). Political interaction between cross-border actors and solving environmental problems mentioned in the recent definition also requires, above all, the cooperation and coordination of these actors(Birnie, 2002).

Adil Najam, a global governance expert at Boston University and co-author of Global Environmental Governance, provides a straightforward definition of global governance: "Managing Global Processes in the Absence of Global Governance" (Riazati, 2006) Thomas G. Weiss. Ralph Buncher, director of the Institute for International Studies, said, "Global governance is a collective effort to identify, understand or address global problems beyond the control of individual governments(Weiss & Thakur, 2010)."

"Global governance" is not a normative term that implies excellent or destructive behaviors but a descriptive term that refers to arrangements that work together to solve problems. These arrangements may be formal and in the form of legal rights or founding institutions for various actors - such as officials, international organizations, NGOs, the private sector, other civil society actors, or individuals - to manage their collective affairs (Podder, 2014). These arrangements can include informal procedures or guidelines or occasional institutions, such as coalitions (Podder, 2014).

Global governance can be defined as complex formal or informal institutions, mechanisms, communications, and processes between governments, markets, citizens, and intergovernmental or non-governmental organizations. Rights and obligations arise through collective interests in the international arena, and differences are mediated (Podder, 2014). The history of global environmental governance dates back to the establishment of UNEP in 1972. Since then, the global environmental governance system has undergone many changes, and the international community has produced a large number of multilateral environmental agreements, with more than 600 multilateral environmental agreements registered with the United Nations over the past 50 years, including 61 on the atmosphere; 155 cases in the field of biodiversity; 179 cases in the field of chemicals, toxins, and wastes; 46 items about land and 196 cases was about water. Some researchers believe that environmental policy means representatives of civil society with each other; the interaction between negotiation processes at national, regional, and international levels; and the role of international institutions in the formation of legal regimes and policies (Charnovitz, 2002).

Some experts consider at least three essential components to the global environmental governance system: process, structure, and implementation. In its ideal form, this system reflects the interrelationships between the mentioned components in a sequential manner and in perfect harmony with each other, which is often not achieved. The first component, the process, includes assemblies, conferences, congresses, and meetings. Some of the most recent examples are the 1972 United Nations Conference on the Human Environment, the 1992 United Nations Conference on Environment and Development, the 2002 Summit on Sustainable Development, and the 2012 United Nations Conference on Sustainable Development, known as the Rio + 20. As the 1995 World Summit on Social Development, the 1996 World Summit on Sustainable Development, the 2001 Doha Ministerial Conference, the WTO is part of the global environmental governance process.

On the other hand, for each of these meetings or sessions, hundreds of introductory sessions are held in which the various interests of the various negotiators are discussed. These interests are sometimes divided into smaller groups to address specific issues and sometimes merged to form coalitions based on geography, economics, culture, ideology, and current affairs. Reviews, conversations, changes, and subsequent actions are recorded. A significant part of this process is to write the cases in the appropriate language to become documents that will later become internationally binding documents with the signature of the executive and the legislature of a sufficient number of countries or other institutions. Meetings convened at the call of the global environmental governance system can be large or small, formal or informal, inclusive or private; They may also represent considerations about specific sectors, geographies, or cultures. Also at the core of the formalities reached in the negotiation process are the agreements reached by States to negotiate treaties or conventions through a series of international negotiating committees, often before these formal meetings and in corridors, breathing space and rest between parties, meetings. Work that is scheduled out of schedule and lasts until late is achieved. To these should be added the meetings of NGOs and, more recently, business communities, which reflect the complexity, cost, variety of benefits, amount and accuracy of the information, and confidentiality(Karns & Mingst, 2010).

In short, the process of global environmental governance is mainly an attempt to organize and compile the results of various meetings and various decisions in the form of documents inappropriate language that the political representatives approve of governments, approved by the parliaments of the countries, and finally entered the legal system in the form of required documents (Meganck & Saunier, 2012). It was evident that, in all stages of global environmental governance, from the preparation of these documents to the ratification and entry of the provisions of these documents in the legal system of member countries, it requires close coordination and cooperation of governments other international elements and actors.

The structure of global environmental governance is the agreements, whether declarations or declarations (soft law) or conventions, treaties and protocols (complicated law), and the institutions necessary to interpret and manage governance instruments, consisting of formal and informal, private and public institutions. Which should eventually lead to the implementation of various formal and informal advisory committees

(Boards, subsidiaries, offices, etc.) and conferences or meetings of members negotiating or signing a treaty or protocol of organizations around an agreement. Secretariats, programs, and commissions are also part of this structure. There are four types of conventions in global environmental governance. Framework conventions, such as the Climate Change Convention or the Biodiversity Convention, require separate protocols to achieve their goals. The government may sign and even ratify the main convention but not sign or ratify any of the annexed protocols. Umbrella agreements, such as the Convention on Wild Migratory Species of Animals, allow other family agreements to fall under its umbrella. Open agreements to be signed by governments and non-governmental organizations are the fourth type, of which the Ramsar Convention is one example(O'brien et al., 2000).

Implementation is the third component of the global environmental governance structure. The United Nations Environment Program's Board of Governors defines implementation as follows in its "Guidelines for National Implementation, and International Cooperation in Combating Violations of [Multi-Environmental] Agreements": "Implementation governs all laws, regulations, "Policies and other measures and measures that member states adopt and/or adapt to fulfill their obligations under multilateral environmental agreements and their annexes." As can be seen, the importance of cooperation in implementing, as one of the three pillars of global environmental governance, is such that the Board of Governors of the United Nations Environment Program has made international cooperation the subject of the guidelines mentioned above.

ISSUES AND CHALLENGES OF GLOBAL ENVIRONMENTAL GOVERNANCE

Extensive diversity in international conventions, agreements, and protocols; Diversity of international structures and institutions; Lack of adequacy of traditional theories in the face of deep, complex, and often cross-border environmental problems that do not pay much attention to the cooperation and participation of other actors, including civil society and the private sector, and only consider governments as the only major players in international and global arenas, The lack of democratic decision-making systems in decision-making and policymaking structures to manage complex environmental issues and sustainable development with stakeholders ranging from ordinary people to civil society institutions and governments on the world stage are just some of the issues in environmental governance. We are dealing with them globally. Fragmentation and multiplicity of documents; Structural fragmentation for fulfilling international obligations and managing severe and complex environmental problems that have been isolated by formal and informal procedures, separately and without coordination, cooperation and communication with each other in international and global mechanisms, structures and institutions, and the incompatibility of these existing fragmented structures. Compared to the existing environmental problems, sustainable development is one of the most critical challenges of global environmental governance. Also, as far as the institutions and structures of global environmental governance are concerned, the environment is associated with various elements in many cases. Thus in some way, actors, institutions, and international organizations find a role concerning it. In such a situation, one cannot expect an institution within the scope of UNEP - which does not even have the title of an organization and is only a program - to be able to coordinate the environmental activities of all organizations operating in the international system.

Some experts have specifically addressed the issue of inconsistencies in the legislative process and the creation of environmental documents. For example, Philip Sands, in his book Principles of International Environmental Law, states that although the principles and rules of international environmental law are contained in thousands of national, bilateral, subregional, regional, and global laws, the principles and rules that are commonly used, Have not been compiled in any single document. Despite the efforts of the International Union for Protection of the Environment's Commission on Environmental Rights in the 1990s, this does not appear to be the case shortly. The absence of a central legislature, or a coherent set of international legislative arrangements, has led to the legislative process and the body of rules being ad hoc, gradual, and fragmented (Sands & Peel, 2012). Sands believes there is a real need for a coherent framework for harmonizing existing rules and creating new ones. The years following the Rio Conference show that there is no political will to do so. According to him, progress in environmental protection is not possible only through the adoption of large bodies of regulations (Sands & Peel, 2012). He addresses the issue of "implementation" and believes that without the widespread use of fair and efficient economic tools, one can not expect this situation to improve. Referring to the "joint implementation" system under the 1992 Convention on Climate Change, he warns of the complexities of applying economic theories (Sands & Peel, 2012). In addition to the above, Najam has listed the six main challenges of effective global environmental governance(Jam & Blake, 2017):

Fragmentation of the Global Environmental Governance System

The rapid growth of multilateral environmental agreements, related documents, and geographically dispersed institutions has disrupted global environmental governance. Incompatibility of rules and norms can lead to depletion of government coffers, especially in developing countries. All this has caused the global environmental governance system to not function at the desired level, and this system's agreements, institutions, and resources cannot achieve full synergy. The ability of the global environmental governance system to deal with complex and intertwined environmental threats has become increasingly challenged as the asymmetric regime of solutions becomes even more complex than the problems it is supposed to solve(Griffin, 2003).

Lack of Cooperation and Coordination Among International Organizations

Coordination has always been one of the essential goals of environmental governance. The United Nations Environment Program (UNEP) was created to be at the top of the world's environmental coordinators pyramid. Still, it has never reached the more important institutions it was supposed to coordinate. Governments also never really wanted to provide the necessary political centrality or resources to play the coordinating role they had devoted to the United Nations Environment Program. The rapid increase in the number of international actors influencing the environmental governance system today has increased the importance of coordination and, at the same time, made it more difficult.

The creation of the Global Environment Facility as the principal financial mechanism, the Secretariats of the Multilateral Environmental Agreements, and the Committee on Sustainable Development, which later fell outside the scope of UNEP, led to further disruption and inconsistency within the global environmental institutions. Such an atmosphere of distrust among these institutions, the unbalanced power, and the vague (and sometimes conflicting) competencies of the member states have not facilitated the institutions' cooperation and coordination. Policy coordination and implementation lie at the heart of the global environmental governance crisis. This crisis has created a rift between international politics, fragmented efforts, and sometimes competing or incoherent decision-making structures.

Lack of Implementation, Compliance, Implementation, and Efficiency

Although many multilateral environmental agreements have been signed and ratified in the global environmental governance system, with a few exceptions, compliance with these agreements, and making fundamental changes in terms of the quality of the environment and those living in the environment, there is misery in vocabulary distinction between implementation (actions taken by members to integrate a treaty into their national legal system), compliance (compliance with treaty rules), enforcement (existing practices for forcing states to comply with MEAs), and efficiency (effect of a treaty as a The whole has more of an academic aspect in achieving its goals. Although such a distinction can be instrumental conceptually, from a practical point of view, the issue is the implementation of international environmental documents, and implementation is defined as the sum of implementation, compliance, enforcement, and efficiency. Ultimately, the global environmental governance system is supposed to guarantee the fulfillment of the environmental commitments of international actors, in which case the actual quality of the environment must be maintained and improved (efficiency), which is the result of the implementation of international instruments (implementation, compliance, implementation). The complexity of the issue lies in the fact that the global environmental governance system is so concerned about the conclusion of new agreements that it has paid almost no attention to the implementation of these agreements. The temptation to conclude new agreements, regardless of their implementation, has led to the depletion of the best human resources, especially in developing countries, and the lack of resources to implement agreed decisions(Thomas, 2000).

Inefficient Use of Resources

There are many sources of financing related to the global environment: Multilateral financial resources of various organizations; Multilateral environmental agreements and multilateral financial arrangements; Donations; Mobility of private capital; Non-traditional sources of financing; Financing through the non-governmental sector, and sources of domestic capital. However, the lack of financial resources and credit in the global environmental governance system is a significant obstacle to compliance, especially in developing countries. Another problem is the inefficiency of the use of money, which leads to various harms: It injects money into the system, damages the credibility of institutions, and disrupts the payment of money to those who need it most. Reducing inefficiencies through synergistic financial management can benefit many (Buchanan & Keohane, 2006).

Decision Making Outside the System of Global Environmental

Many important and influential decisions on the environment, including investment, development and trade, war, health, international trade, and the international financial and investment system, outside the complex network of international agreements and organizations that formally govern global environmental governance form, are taken. Decisions related to investment, development, and trade affect natural resource use, production, and consumption patterns(Stubbs, 2008). They are likely to have a more significant impact than multilateral environmental agreements in these areas. Environment and security are also related in two ways: first, the possibility of conflict due to lack of natural resources, and second, the destruction of the environment in war zones. Environmental issues have also spread to health, as health risks from environmental degradation have become increasingly common. The international trade system, consisting of the rules of the WTO trade regime and regional free trade agreements, each of which provides for separate mechanisms for resolving disputes, has a profound effect on global environmental governance. The international financial and investment system affects national development by financing projects and the national economic policies of developing countries. In addition, there is a sharp imbalance between trade and financial institutions, and environmental institutions. The United Nations Environment Program and the Global Environment Facility are downplayed compared to the financial resources and political influence of multilateral development banks and development agencies. Not surprisingly, economic priorities often outweigh environmental concerns. One of the critical issues facing global environmental governance is incorporating environmental concerns into the mainstream of economic decision-making and other non-environmental areas(Keohane, 2011).

Non-Governmental Actors in the State-Based System

In environmental governance, the debate is primarily focused on reforming the state-centered system of international organizations, multilateral treaties, and national implementation. The traditional approach to reforming the system often ignores the massive participation and increasing involvement of civil society and private sector actors in policymaking, capacity-building, and implementation. NGOs play an essential and growing role in global environmental governance, not only as stakeholders but also as the "driving force" of international environmental policymaking by formulating agendas, drafting treaties, providing scientific information, and implementing monitoring activities. Nor should we forget the role of local and international NGOs in implementation and capacity building. In addition to civil society, the influence of the private sector in the protection and development of the environment is increasing.

SOLUTIONS FOR CHALLENGES IN GLOBAL ENVIRONMENTAL GOVERNANCE

The crisis caused by the impact of human activities on the environment and its possible irreversible effects requires an appropriate response from governments and citizens. Nature does not place social and political barriers and the global dimensions of the crisis, the effects of any unilateral action by governments or individual institutions. They are also powerful; it neutralizes. Climate change, air and ocean pollution, nuclear or genetic hazards, depletion of natural resources, and biodiversity loss have all led to global misconduct and foreshadowing irreversible effects. On the other hand, over the past 30 years, the difficulties of implementing many multilateral environmental agreements and treaties, the ultimate goal of protecting the environment, have led to the proposal to establish a global environmental organization that may help the process of environmental degradation. Take control of life. The International Institute for Sustainable Development has also set a reform agenda for global environmental governance. These goals require the cooperation and coordination of international actors and environmental solid leadership and policy based on knowledge, solidarity and effective coordination, good management of the institutions of the environmental governance system, and the expansion of environmental concerns and actions to other areas of international policy and action.

Global environmental governance is one of the most significant challenges of the present age. The question is, how can unequal and evolving societies come together to protect their environment, which is often shared and deeply interdependent? Traditional approaches to solving global environmental problems (especially international treaties and conventions) are evolving and, in some cases, being replaced by innovative arrangements. In addition, the presence of non-governmental actors such as transnational corporations, NGOs, networks of scientists, and international organizations working to address global environmental problems is a hallmark of global environmental governance. The United Nations Institute for Advanced Study, in collaboration with the University of Kitakyushu in Japan and with the support of the Japan Foundation's Global Partnership Center, has researched the global environmental governance system, identifying issues and problems and providing solutions (Charnovitz, 2002): Integration of Multilateral Environmental Agreements; Strengthen UNEP, expand the role of the World Assembly of Environment Ministers, reform existing United Nations institutions, strengthen mechanisms and financial resources, build environmental capacity in the World Trade Organization, possible models for the World Environment Organization, reform the United Nations Board of Trustees, expand tasks UN Security Council and Establishment of the World Environmental Court.

DISCUSSION

The system of global environmental governance through organizations and institutions, decisions and policies, financing mechanisms, rules, rituals, and norms seeks to protect the environment and sustainable development, none of which will be achieved without coordination and cooperation among international actors. This system through the process of producing binding documents, and creating structures consisting of soft documents (declarations, declarations, etc.) and complicated documents (conventions, treaties, protocols, etc.) and the necessary institutions to interpret and manage the tools of governance (Consisting of formal and informal institutions, both private and public), must ultimately ensure that the requirements and commitments made as described above by international actors are met to protect the environment worldwide. But the experience of the last 50 years has shown that the components of this system, namely processes, structures, and implementation, are usually not interconnected and in perfect harmony with each other.

Inflation of multilateral environmental agreements and other international documents, and non-implementation of the provisions of existing documents, fragmentation of international structures and institutions, and inconsistency and lack of cooperation of international actors with each other separately and without coordination and communication in international mechanisms, structures, and institutions, Formal and informal practices see the isolated world. The incompatibility of these fragmented structures with environmental problems and sustainable development, along with a body of case-by-case, gradual, and incoherent rules, is one of the most critical challenges to global environmental governance. All the factors affecting the environment, especially the private sector and civil society institutions in policymaking, capacity building, and implementation, are the result of undemocratic decision-making systems for managing complex issues and deep problems arising from environmental requirements and sustainable development; Making effective environmental decisions and policies outside the global environmental governance system, especially in the areas of investment, development and trade, war, health, international trade, and the international financial and investment system; Paying too much attention to the process of approval and creation of documents and neglecting their implementation and implementation; Another reason is the failure of the global environmental governance system(Stone, 2008).

Reforming the global environmental governance system and stopping the process of environmental degradation and its improvement requires following the documents, treaties, and multilateral environmental agreements and avoiding limited financial resources in approving and creating new documents and implementing the same existing documents; Ensure the participation of all actors involved in environmental protection, including civil society and the private sector; Adopt all decisions and policies affecting the environment, within the global environmental governance system; Optimal allocation of resources and balancing the process of document production and implementation of commitments and document content. Achieving these requires strengthening UNEP as a coordinating body; Expanding the role of the World Assembly of Environment Ministers and cooperation between governments through this; Reform current UN institutions; Strengthen mechanisms and financial resources; Building environmental capacity in the World Trade Organization; Propose and review possible models for forming a global environmental organization; Reform of the United Nations Board of Trustees; Extending the mandate of the United Nations Security Council, and the establishment of the World Environmental Court and also, creating a global environmental organization, environmental solid leadership and policy based on knowledge, practical solidarity and coordination, good management of the constituent institutions of the environmental governance system, and extending environmental concerns and actions to other areas of international policy and facing the challenges of the global environmental governance system(Mol, 2008).

CONCLUSION

The global environmental governance system has not been very successful in achieving its primary goal: to protect the global environment and achieve sustainable development. Among the reasons for this is the incoherence and lack of cooperation of international actors; Inflation of multilateral environmental agreements, inappropriate international structures, and institutions; Failure to follow the documents and non-implementation of the obligations contained in them; Inefficient allocation of resources; Unpopular way of making decisions; Policy-making and decision-making outside the system of global environmental governance and the absence of non-governmental actors in a state-centered system. This article aims to identify the causes of the inefficiency of the global environmental governance system and examine the necessary measures to address the shortcomings of this system.

The leading article, using library research method and articles and books of researchers who have previously researched global environmental governance, tries to examine the current state of the global environmental governance system. Given the widespread extent of global environmental degradation, this paper examines some of the causes of the inefficiency of the existing global environmental governance system. It concludes by discussing several proposed solutions for reforming the global environmental governance system. In the global environmental governance system, through creating documents and institutions, a structure consisting of soft and hard documents and the necessary institutions for the implementation of these documents and the fulfillment of their obligations and requirements are created. But without the coordination and cooperation of all international actors at an acceptable level, even if governments have the political will to protect the environment, one cannot expect the phenomenon of border crossings to be controlled by inconsistent responses from governments. Existing institutions, including UNEP, have not been effective enough. Despite many declarations, conventions, and other international instruments, many environmental problems include climate change, air, soil, ocean pollution, nuclear hazards, or manipulation. Genetics, the depletion of natural resources, and the destruction of biodiversity and landscapes continue to plague parts of the world. Therefore, the provisions of existing instruments, especially multilateral environmental agreements, need to be implemented using existing structures, with the help of coordinating bodies, including UNEP, and closer and more effective cooperation of international actors, including governments, civil society organizations, and NGOs, and the private sector will be possible. Conclusion: Reforming the global environmental governance system in the first place requires cooperation and coordination between governments, civil society institutions, and the private sector. Also, adherence to multilateral environmental documents, treaties, and agreements strengthening UNEP as a coordinating body, expanding the role of the World Assembly of Environment Ministers, reforming existing UN institutions; Strengthening mechanisms and financial resources, building environmental capacity in the World Trade Organization, possible models for the World Environment Organization, reforming the UN Security Council, expanding the mandate of the UN Security Council, and establishing an international environmental tribunal are other requirements of the global environmental governance system.

REFERENCES

Alexandroff, A. S., & Cooper, A. F. (Eds.). (2010). *Rising states, rising institutions: Challenges for global governance*. Brookings Institution Press.

Birnie, P. (2002). Salman MA Salman and Kishor Uprety, Conflict and Cooperation on South Asia's International Rivers. The Hague & Law International.

Boyle, A., & Birnie, P. (1995). *Basic documents on international law and the environment*. Clarendon Press.

Buchanan, A., & Keohane, R. O. (2006). The legitimacy of global governance institutions. *Ethics & International Affairs*, *20*(4), 405–437. doi:10.1111/j.1747-7093.2006.00043.x

Charnovitz, S. (2002). A World Environmental Organization. *Colum. J. Envtl. L.*, *27*, 323.

Griffin, K. (2003). Economic globalization and institutions of global governance. *Development and Change*, *34*(5), 789–808. doi:10.1111/j.1467-7660.2003.00329.x

Jam, F., & Blake, J. (2017). Global environmental governance system: Challenges and solutions. *Environmental Sciences*, *15*(1), 141–156.

Karns, M. P., & Mingst, K. A. (2010). International Organizations: The Politics and Process.

Keohane, R. O. (2011). Global governance and legitimacy. *Review of International Political Economy*, *18*(1), 99–109. doi:10.1080/09692290.2011.545222

Koh, K. L. (2008). Regional and state level environmental governance. ASEAN's environment governance: An evaluation. In *UNITAR/Yale Conference on Environmental Governance and Democracy*, New Haven, CT.

Meganck, R. A., & Saunier, R. E. (2012). *Dictionary and introduction to global environmental governance*. Routledge. doi:10.4324/9781849771009

Miller, C. A. (2007). Democratization, international knowledge institutions, and global governance. *Governance: An International Journal of Policy, Administration and Institutions*, *20*(2), 325–357. doi:10.1111/j.1468-0491.2007.00359.x

Mol, A. P. (2008). *Environmental Reform in the Information Age. The Contours of Informational Governance*. Cambridge University Press. doi:10.1017/CBO9780511491030

Najam, A. (2003). The case against a new international environmental organization. *Global Governance*, *9*(3), 367–384. doi:10.1163/19426720-00903008

O'brien, R., Goetz, A. M., Scholte, J. A., & Williams, M. (2000). *Contesting global governance: Multilateral economic institutions and global social movements* (Vol. 71). Cambridge University Press. doi:10.1017/CBO9780511491603

Podder, S. (2014). Mainstreaming the non-state in bottom-up state-building: Linkages between rebel governance and post-conflict legitimacy. *Conflict Security and Development*, *14*(2), 213–243. doi:10.1080/14678802.2014.889878

Riazati, S. (2006). A closer look: Professor seeks stronger UN. *The Daily Bruin, 21*. dailybruin.com/2006/10/17/a-closer-look-professor-seeks.

Risse, T. (2004). Global governance and communicative action. *Government and Opposition*, *39*(2), 288–313. doi:10.1111/j.1477-7053.2004.00124.x

Sands, P., & Peel, J. (2012). *Principles of international environmental law*. Cambridge University Press. doi:10.1017/CBO9781139019842

Stone, D. (2008). Global public policy, transnational policy communities, and their networks. *Policy Studies Journal: the Journal of the Policy Studies Organization*, *36*(1), 19–38. doi:10.1111/j.1541-0072.2007.00251.x

Stubbs, R. (2008). The ASEAN alternative? Ideas, institutions and the challenge to 'global' governance. *The Pacific Review*, *21*(4), 451–468. doi:10.1080/09512740802294713

Thomas, C. (2000). *Global governance, development and human security: the challenge of poverty and inequality*. Pluto.

Thomas, C. (2001). Global governance, development and human security: Exploring the links. *Third World Quarterly*, *22*(2), 159–175. doi:10.1080/01436590120037018

Weiss, T. G., & Thakur, R. (2010). *Global governance and the UN: An unfinished journey*. Indiana University Press.

ADDITIONAL READING

O'brien, R., Goetz, A. M., Scholte, J. A., & Williams, M. (2000). *Contesting global governance: Multilateral economic institutions and global social movements* (Vol. 71). Cambridge University Press. doi:10.1017/CBO9780511491603

Podder, S. (2014). Mainstreaming the non-state in bottom-up state-building: Linkages between rebel governance and post-conflict legitimacy. *Conflict Security and Development*, *14*(2), 213–243. doi:10.1080/14678802.2014.889878

Riazati, S. (2006). A closer look: Professor seeks stronger UN. *The Daily Bruin, 21*. dailybruin.com/2006/10/17/a-closer-look-professor-seeks.

Risse, T. (2004). Global governance and communicative action. *Government and Opposition*, *39*(2), 288–313. doi:10.1111/j.1477-7053.2004.00124.x

Sands, P., & Peel, J. (2012). *Principles of international environmental law*. Cambridge University Press. doi:10.1017/CBO9781139019842

Stone, D. (2008). Global public policy, transnational policy communities, and their networks. *Policy Studies Journal: the Journal of the Policy Studies Organization*, *36*(1), 19–38. doi:10.1111/j.1541-0072.2007.00251.x

Stubbs, R. (2008). The ASEAN alternative? Ideas, institutions and the challenge to 'global' governance. *The Pacific Review*, *21*(4), 451–468. doi:10.1080/09512740802294713

Thomas, C. (2000). *Global governance, development and human security: the challenge of poverty and inequality*. Pluto.

Thomas, C. (2001). Global governance, development and human security: Exploring the links. *Third World Quarterly*, *22*(2), 159–175. doi:10.1080/01436590120037018

Weiss, T. G., & Thakur, R. (2010). *Global governance and the UN: An unfinished journey*. Indiana University Press.

KEY TERMS AND DEFINITIONS

Equity: Defined by UNEP to include intergenerational equity - "the right of future generations to enjoy a fair level of the common patrimony" - and intragenerational equity - "the right of all people within the current generation to fair access to the current generation's entitlement to the Earth's natural resources" - environmental equity considers the present generation under an obligation to account for long-term impacts of activities and to act to sustain the global environment and resource base for future generations. Pollution control and resource management laws may be assessed against this principle.

Polluter pays principle: The polluter pays principle stands for the idea that "the environmental costs of economic activities, including the cost of preventing potential harm, should be internalized rather than imposed upon society at large." All issues related to responsibility for environmental remediation costs and compliance with pollution control regulations involve this principle.

Precautionary principle: One of the most commonly encountered and controversial principles of environmental law, the Rio Declaration formulated the precautionary principle: To protect the environment, the precautionary approach shall be widely applied by States according to their capabilities. Where there are threats of serious or irreversible damage, lack of complete scientific certainty shall not be used as a reason for postponing cost-effective measures to prevent environmental degradation. The principle may play a role in any debate over the need for environmental regulation.

Prevention: The concept of prevention can perhaps better be considered an overarching aim that gives rise to a multitude of legal mechanisms, including prior assessment of environmental harm, licensing or authorization that set out the conditions for operation and the consequences for violation of the conditions, as well as the adoption of strategies and policies. Emission limits and other product or process standards, the use of best available techniques, and similar techniques can all be seen as applications of the concept of prevention.

Public participation and transparency: identified as necessary conditions for "accountable governments,... industrial concerns," and organizations generally, public participation and transparency are presented by UNEP as requiring "effective protection of the human right to hold and express opinions and to seek, receive and impart ideas,... a right of access to appropriate, comprehensible and timely information held by governments and industrial concerns on economic and social policies regarding the sustainable use of natural resources and the protection of the environment, without imposing undue financial burdens upon the applicants and with adequate protection of privacy and business confidentiality," and "effective judicial and administrative proceedings." These principles are present in environmental impact assessment, laws requiring publication and access to relevant environmental data, and administrative procedures.

Sustainable development: Defined by the United Nations Environment Programme as "development that meets the needs of the present without compromising the ability of future generations to meet their own needs," sustainable development may be considered together with the concepts of "integration" (development cannot be considered in isolation from sustainability) and "interdependence" (social and economic development, and environmental protection, are interdependent). Laws mandating environmental impact assessment and requiring or encouraging development to minimize environmental impacts may be assessed against this principle.

Transboundary responsibility: Defined in the international law context as an obligation to protect one's environment and prevent damage to neighboring environments, UNEP considers transboundary responsibility at the international level as a potential limitation on the sovereign state's rights. Laws that limit externalities imposed upon human health and the environment may be assessed against this principle.

Chapter 3
Principles and Concepts of Environmental Law

ABSTRACT

Environmental law has developed in response to emerging awareness of and concern over issues impacting the entire world. While laws have developed piecemeal and for various reasons, some effort has gone into identifying key concepts and guiding principles common to environmental law as a whole. The principles discussed below are not exhaustive and are not universally recognized or accepted. Nonetheless, they represent essential principles for understanding environmental law around the world. This chapter describes the fundamental principles of environmental law on which the discipline is based. Also, it explains the basic concepts of environmental law and the relationship between principles and concepts.

INTRODUCTION

Environmental law is a collective term encompassing aspects of the law that protect the environment. A related but distinct set of regulatory regimes, now strongly influenced by environmental legal principles, focus on managing specific natural resources, such as forests, minerals, or fisheries. Other areas, such as environmental impact assessment, may not fit neatly into either category but are essential environmental law components.

Early examples of legal enactments designed to consciously preserve the environment, for its own sake or human enjoyment, are found throughout history. In the common law, the primary protection was found in the law of nuisance, but this only allowed private actions for damages or injunctions if there was harm to the land. Thus, smells emanating from pigsties, strict liability against dumping rubbish, or damage from exploding dams. Private enforcement, however, was limited and found to be woefully inadequate to deal with major environmental threats, particularly threats to shared resources. During the "Great Stink" of 1858, the dumping of sewerage into the River Thames began to smell so ghastly in the summer heat that Parliament had to be evacuated. Ironically, the Metropolitan Commission of Sewers Act 1848 had allowed the Metropolitan Commission for Sewers to close cesspits around the city in an attempt to "clean up," but this led people to pollute the river. In 19 days, Parliament passed a further Act

DOI: 10.4018/978-1-6684-4158-9.ch003

to build the London sewerage system. London also suffered from terrible air pollution, culminating in the "Great Smog" of 1952, which triggered its legislative response: the Clean Air Act 1956. The primary regulatory structure was to set limits on emissions for households and businesses (mainly burning of coal) while an inspectorate would enforce compliance(Norouzi, 2022).

For a better understanding and roots of any argument, one should refer to its theoretical, intellectual, and philosophical foundations. The basis of environmental law, human rights, is the intrinsic value of the environment itself, embodied in environmentalism and Gaia theory and ecofeminisms. The inherent value of the environment means that the environment is not a living and dynamic being and the only bedrock for human exploitation. Therefore, all living things, including animals and ecosystems, are valuable, but not like human beings' human dignity.

Gaia's theory "considers the whole biosphere as a quasi-living, superstructural system called the climate of which the human race is only a part." It has been said that man should never manipulate the earth's natural order or change any part of it(Norouzi & Ataei, 2021).

The concepts are more general, and the principles are more objective and specific. Applying principles in environmental and international environmental law is the same as principles used in determining legislative and executive strategies.

Principles are the fundamental doctrines on which other principles or rules of conduct are based, and the concepts of ideas or themes are the leading integrators. Principles can serve as a guide in shaping and interpreting the legal norms of the field and filling the gaps in the subject law. Principles and concepts appear in constitutions and internal regulations.

"Principles of international environmental law may come from various proposals, problems or aspirations or threats to the environment, the development of "science and technology, various economic interests, intergovernmental relations, pressure from NGOs, scientific research, expert work, diplomatic negotiations(Shohani et al., 2021)."

The principles of international environmental law can be divided into common principles and specific principles. Common principles refer to those principles common to other traditional areas of international law, such as sovereignty, consent, fidelity, cooperation, good faith, and good neighborliness, but specific regulations refer to those principles that govern the field. Their application is in international environmental law, which includes the principles of joint but distinct liability, payment by the polluter, non-harmful use of the land, and reporting of environmental events(Wiener, 2000).

In formulating and developing international environmental law, some principles have been formed like other branches of jurisprudence, which can be mentioned as the main pillars and fundamental principles of international environmental law. These principles have played an essential role in establishing international environmental rules and implementing and interpreting environmental treaties, and resolving disputes between governments on environmental issues. These principles include the principle of prevention or prevention of environmental damage and the principle of payment by the polluter, which, if realized, many crises and environmental pollution will not occur.

PRINCIPLES OF ENVIRONMENTAL LAW

Environmental law has developed in response to emerging awareness of and concern over issues impacting the entire world. While laws have developed piecemeal and for various reasons, some effort has gone into identifying key concepts and guiding principles common to environmental law as a whole.

The principles discussed below are not exhaustive and universally recognized or accepted. Nonetheless, they represent essential principles for understanding

Principle of Sovereignty

According to the principle of sovereignty, states have exclusive competence in their territory. This principle is one of the basic principles of environmental law. Sovereignty is the main element of the government that has undergone various changes, so that in the second half of the twentieth century and the early years of the twenty-first century witnessed significant changes such as the establishment of international organizations such as the United Nations and the spread of concepts such as human rights, humanitarian intervention, doctrine. We supported and raised environmental issues and problems that took the concept of "sovereignty" out of absoluteness.

In international environmental law, one of the most important legal sources of the 1947 ruling principle is the dispute between Canada and the United States over the Trail Smelter case, which states: Governments should not use their territory in a way that harms other countries.

The principle of sovereignty in the environment means that no country can provide a means of harming others due to the exercise of sovereignty.

"Sovereignty is not a completely exclusive concept, but states are allowed to act according to their policies, provided they use their land in a way that does not conflict with the rights of others. The sovereignty of states is recognized, except in cases where it causes harm to other countries. This principle is one of the basic principles of international environmental law. "In any case, the evolution and development of this principle has led to good governance and reflects the development of this principle in the context of international environmental law."

Article 21 of the 1972 Stockholm Declaration, which marks the turning point of the principle of sovereignty in international environmental law, states that States must use their territory in a manner that does not cause harm to the environment or other States. The sovereignty of states has two bases; (1) Recognizing the sovereignty of States over the use of their natural resources; and (2) not causing damage to territories under the control of other States or to areas such as the high seas that are not under the control of States(Tarlock, 1993).

Principle of Cooperation

There is a core environment in international law called cooperation that emphasizes the cooperation of governments in all areas of the environment. There is also an independent principle of cooperation and assistance in emergencies, which indicates the cooperation and assistance of governments in a situation where the environment has been damaged or is about to collapse. "Although states have independent sovereignty, international co-existence will inevitably lead to international cooperation." This principle is based on a task and seems to be based on international interactions and global participation. Therefore, the principle of cooperation is rooted in public international law but is particularly important in international environmental law. This principle is one of the essential principles of international environmental law, the necessity of which has been emphasized in international documents of soft and complicated rules. According to this principle, governments are obliged to cooperate in all circumstances and in good faith to protect the environment. Commitment to cooperation includes a wide range of collaborations, from providing the necessary resources and technology and holding training courses to exchanging information and providing assistance during environmental emergencies. Based on the principle of cooperation and assistance in emergencies, in cases of natural disasters or other emergencies that appear to have adverse effects on the environment, governments should promptly inform other countries and provide appropriate assistance and cooperation following That incident. This principle has been developed as a duty in customary law. As a general rule, it has also explicitly appeared in numerous international treaties. In the Strait of Corfu, the International Court of Justice ruled in favor of the right of all countries to enjoy maritime privileges, which the Albanian government was obliged to inform the captain of British warships of the existence of a minefield in its waters. This requirement, on which the court relies, derives from the right established for the public and from the well-known principles of fundamental human rights, which must be observed. The most important international instrument in this regard is the United Nations Convention on the Law of the Sea, which summarizes Article 198, which has already addressed the issue of marine pollution in various treaties, at general levels, and in particular regional seas (Salzman & Thompson Jr, 2010).

Principle of Environmental Protection

This principle implies that within countries, the duty of all public, governmental, non-governmental, and private sectors is to pay attention to and protect the environment. That is, any activity in the country should be done taking into account the environmental aspects and checking whether the activity in question causes damage to the environment or not? If damage to the environment is estimated, it should be avoided, and otherwise, the activity can be carried out. For this purpose, an "environmental impact assessment" must first be performed. Before considering the economic, social, or any other dimension of the project, we must examine whether it is environmentally justifiable to do so; in fact, the negative and positive consequences of the project to be accurately estimated. In the Islamic Republic of Iran, the task of evaluation is the responsibility of the Environment Organization. The concept of environmental impact assessment is to estimate the activity's environmental impact before starting and deciding whether to start it. In connection with the environmental assessment in 1991, the Transboundary Environmental Assessment Treaty was signed in Espoo, Finland, under the United Nations Economic Commission for Europe (ESPO) and entered into force in 1997. Under the Convention, States Parties undertake to assess the environmental impact of certain activities at each stage of the project. The Convention also sets out general obligations for States to inform and consult with the other Party on most projects that may have transboundary environmental effects. It should be noted that from the author's point of view, environmental impact assessment is considered in the framework of the principle of environmental protection and not as an independent principle(Sands & Peel, 2012).

Principle of Transparency

The starting point for this principle can be seen in the ruling of the International Court of Justice in the 1949 Strait of Corfu. In this case, the International Court of Justice ruled that the Albanian government was obliged to inform the captain of the British ship of the presence of a mine in Albanian territorial waters. According to this principle, if a state becomes aware of a danger that could cause other states to be in a state of emergency, it must report the danger. In other words, the principle of informing about the dangers that may affect the country's environment has been assigned to governments as a duty(Sands & Peel, 2012).

In this regard, informing about accidents and accidents that have a devastating impact on the environment is more critical. Therefore, the principle of information in nuclear accidents has an essential place. The importance of this principle became more transparent following the negligence of former Soviet officials in reporting on the April 26, 1986 incident at the Chernobyl nuclear power plant. This disregard provoked a global backlash and led to the ratification of a particular treaty on prior warning of nuclear accidents or the dangers posed by nuclear material, which was concluded only five months after the incident, and under an unprecedented procedure. The record came into force about a month later.

In addition to emergencies, governments should also identify the effects of activities carried out within their jurisdiction on the environment and notify other governments, even if non-governmental organizations carry them out.

Article 18 of the Rio 1992 Declaration states: Governments must promptly inform other States of natural disasters or other emergencies which appear to have immediate effects on their environment. The international community must make every effort to help the affected government.

One of the manifestations of this principle is the 1998 Aras Convention, which is the first binding document in the international arena on public access to environmental information, public participation in decision-making, and public access to environmental information. This treaty is an important document to ensure the observance of democratic principles in adopting and implementing government decisions. The treaty is based on two principles: Public access to environmental information; Possibility of public participation in environmental decisions.

The Convention stipulates that"the requested information must be provided at least (at least) up to one month after the request, which, if the matter is complicated, may be extended for up to two months."

Article 3 of this treaty sets out the general principles. In addition to the government's duty to provide information at each individual's request, the relevant agencies and government should inform and assist individuals preliminarily on accessing and seeking. The provisions of the Convention encourage government agencies to develop and implement an information policy on citizens' access to environmental information(Sands & Peel, 2012).

Article 9 of the Convention provides for the possibility of referring to independent judicial and administrative authorities specified by law in the event of a refusal to submit environmental information. In addition, handling should be fast and low cost out of turn.

As a matter of necessity, we should refer again to Article 198 of the 1982 Convention on the Law of the Sea, which provides that damage will be reported immediately to other countries likely to be affected by the damage and to competent international authorities.

Article 13 of the 1982 Basel Convention on the Control of Transboundary Movements of Hazardous Wastes and their Disposal stipulates that the Contracting States shall, in the event of an accident occurring during the cross-border transfer of hazardous waste or other types of waste which may pose a risk to human health or the environment in other countries, to ensure prompt information to those countries.

The Basel Convention on the Control of Transboundary Movements of Hazardous Wastes and their Disposal; A fundamental legal response of the international community to the problems caused by the annual global production of 400 million tons of toxic, toxic, explosive, erosive, flammable, and infectious wastes. Improper disposal of these wastes causes contamination of soil, groundwater, and habitat destruction for fish and animals and is also associated with cancer and congenital disabilities(Palmer, 1992).

Principle of Prevention

Experience and expertise have shown that avoiding environmental hazards is a golden rule. For example, hazards such as the extinction of a plant or animal species, soil erosion, and the leakage of resistant pollutants into the sea create irreversible conditions. Even when the damage is reparable, the costs of returning it to its former state are often very high. Commitment to preventing environmental damage is a commitment to avoid posing a severe threat to the transboundary environment and a preventive approach that seeks to avoid danger regardless of its various dimensions in transboundary impact or the issue of international responsibility. The principle of prevention in international environmental law is to prevent environmental damage before it occurs.

In Iranian domestic law, the principle of prevention can be considered a kind of pre-delegated authority to allow it to intervene in a project, either to prevent its development or implementation is designed to meet the relevant criteria. Eliminate restrictions and other factors that alone or collectively cause damage to the environment. The principle of prevention has two basic foundations.

First: evaluation of activities before and underway; Second: monitoring the various environmental conditions and situations;

Environmental damage assessment methods are based on prevention. Such methods are designed to encourage project owners to reduce environmental damage. If the project does not comply with the specified requirements, the competent authority can prevent it from continuing or require it to follow the relevant rules and restrictions.

It is also important to note that the public and the public are legally invited to intervene in such cases to express their concerns or objections to specific initiatives. Also, conducting scientific experiments on new products before entering the market is considered a prevention principle. This type of legal regulation aims to prevent damage or control and limit its effects on the environment. This principle is enshrined in many international instruments of a different legal nature.

One of the main reasons for applying this principle is assessing and estimating environmental damage and, consequently, appropriate compensation. It should be noted that the possibility of accurately calculating "environmental damage" before initiating projects is low.

In 1972, the UN General Assembly recognized the importance of activities in the national jurisdiction to prevent significant damage to the environment of neighboring areas. The final decision of the European Conference on Security and Cooperation in Helsinki in 1975 stated that the best way to prevent damage to the environment was to use preventive measures(Palmer, 1992).

Principle of Payment by the Polluter

This principle is known in public international law as compensation for improper damages. In international environmental law, it is also known as the principle of "payment by the polluter" or "the polluter is the payer." According to this principle, the agent of an activity that has caused damage to the environment must compensate for the damage.

This principle had entered the field of international environmental law since the introduction of civil liability in international treaties on nuclear and oil accidents. Individuals have civil liability for activities that lead to environmental degradation. However, before, It was also used in the rulings of international lunatics. At the same time, it can be said that this principle is at the intersection of the path of prevention and compensation.

According to this principle, the polluter must pay the cost of eliminating environmental pollution. On the one hand, this principle recognizes the right of others to enjoy a healthy environment and, on the other hand, is a preventive measure to prevent environmental degradation. As defined in Article 1992 of the Convention for the Protection of the Marine Environment in the Northeast Atlantic, the polluter's payment is "the costs of preventing and controlling pollution and enforcing the regulations to reduce the pollution borne by the polluter." According to the definition of the "European Economic Community" November 7, 1974, a "polluter" is someone who directly or indirectly harms the environment or is the one who causes significant damage to the environment.

The emergence of liability has several effects, including compensation and compensation through compensation methods, so the principle of payment by the polluter is deeply related to the principle of sovereignty of states. This principle has a long history. The principle of payment by the polluter, which was approved by the Organization for Economic Co-operation and Development in 1972 as an economic principle, is one of the most critical environmental principles because all environmental policies in developed countries are based on it. This principle aims to impose the costs of pollution prevention and control decisions on the polluter. The principle leads to the rational use of limited environmental resources by avoiding international trade and investment deformation.

Article 16 of the Rio 1992 Declaration states that governments must be committed to the principle that the polluter must pay for its remediation. The principle of "payment by the polluter" is used in some treaties to hold private enterprises or industry sectors accountable for environmental damage. Only a national legal system can hold a company legally liable for damages to the environment, and how this happens depends on the rules relating to offshore damages. For example, Article 3 of the Nordic Environmental Protection Treaty between Denmark, Finland, Norway, and Sweden states: "Anyone affected by the disturbance caused by harmful environmental activities in another Member State is entitled to the right to such activities." Include the criteria for damage prevention in a court or competent administrative authority of that country and appeal against the decision of that court or administrative authority with the same scope.

One of the most critical points of contention between a country and a transnational corporation over environmental damage, which led to a public outcry in protest of the detrimental performance, was the so-called "jungle oil evacuation" case in 2011, owned by Chorun. An Ecuadorian court then sued Texaco for $ 9.5 billion in toxic lake discharges from its 1964-1992 operation in the Amazon(Palmer, 1992).

Principle of Joint but Different Liability

This principle is rooted in international custom. The concept of shared but differentiated responsibility is that whenever several countries, both developed and developing, are to establish specific conditions, measures, and constraints in the form of an international treaty to solve a global environmental challenge, depending on the extent of development Finding themselves and their role in creating problems arising from the specific environmental issue that is the subject of the treaty, to varying degrees and in proportion to the role of each in the occurrence of the said issue, as well as the amount of technical and financial facilities at their disposal And the obligations subject to that international treaty are the responsibility. This is the framework that takes a severe shape of environmental diplomacy. The principle is to strengthen the diplomatic process during international conferences. Despite the environmental risks and differences between North and South countries and the requirements of treaties and countries' adherence to international obligations, it seems familiar but different responsibility.

On soft rights, Article 7 of the Rio 1992 Declaration states: Governments must work together in a spirit of global partnership to preserve, protect, and restore the health and well-being of the planet. Due to the different roles of countries in the destruction of the world environment, different countries have common but different responsibilities. Of the strict rights, for example, Article 4 of the 1992 United Nations Convention on Climate Change refers to the principle of joint but different liability and is binding on the Contracting Parties(Orts, 1994).

CONCEPTS AND AIMS

Concepts play a fundamental role in understanding environmental law. The concepts in this field are general. By studying the concepts, one can understand environmental law's principles, perspectives, ideas, and goals.

Common Heritage of Humanity

As the name implies, the "common heritage of humanity" means anything significant to the world, not just a nation and a state, and every effort must be made to preserve it. The concept of the common heritage of humanity has a more general and broader meaning than common property. "The concepts of the common heritage of humanity and the common concern of humanity reflect the growing awareness of the interdependence of the biosphere and its disturbing environmental problems and the vital importance of those problems."

The common heritage of humanity is either those objects and property that are in possession of a state but are valuable to the world because of the value of the environment, culture, and civilization, or are an area or property that is not ruled by a state and a nation like marine resources but It is essential for the planet. In 1956, with the construction of the Aswan high dam on the Nile River in Egypt, there was a danger that the Nubia monuments would be permanently submerged. The international community realized that the flooding of these monuments is irreparable damage to Egypt and Sudan, and all of humanity. "So the idea of a common heritage and a shared responsibility to protect them was formed."

Following Article 1 of Resolution 2749, dated 1970, the General Assembly; The common heritage of humanity was proposed for the first time, and the general principles governing the resources of the seabed and subsoil were formulated, in which these resources were mentioned as the common heritage of humanity.

World Heritage is listed in the preamble to the 1972 UNESCO World Heritage Convention for the Protection and Preservation of Cultural and Natural Heritage. The reference of existing international treaties, resolutions, and recommendations to cultural and natural property indicates the importance of preserving this unique and irreplaceable property to any nation to which it belongs. It should be borne in mind that some cultural and natural heritage has benefits. They are exceptional and should be preserved as a common heritage of humanity.

Introduction to the International Trade Convention on Endangered Species of Wild Fauna and Flora, Washington 1973 Introduces these irreplaceable species and the importance of preserving them for present and future generations, and calls on all countries of the world to cooperate. To preserve these resources and prevent improper exploitation. Convention on the Conservation of Wild Migratory Species, Bonn 1979; It refers to the irreplaceability of these species and the emphasis on their preservation and protection. In the 1982 Convention on the Law of the Sea, Monte Gobi referred to the resources of the seabed and the seabed as a common heritage of humankind.

Introduction, The 1992 Convention on Biological Diversity calls on member states to make biodiversity conservation a priority in their agenda, citing biodiversity as an issue for all human beings and their shared heritage. The International Court of Justice's 1974 adversarial decision on the Irish-British fisheries case stipulates that it is necessary to emphasize the preservation of the natural environment, which must be taken into account in the extraction of these resources as they are not present for generation only. The future must be preserved, and no government can monopolize the use of natural resources and the environment and ignore the common heritage of humanity(Nanda, 1995).

Rights of Future Generations

"The rights of future generations" includes the concept of protecting human rights and the environment against climate change to protect human beings in the future exhaustively. "These rights seek to establish 'generational justice.'" The concept of the rights of future generations entered into international environmental law at the 1972 Stockholm Conference. "This concept is based on two pillars. First: human life and human dependence on the earth's natural resources and consequently the inseparable dependence of all generations on environmental conditions; Second: Man is constantly making changes in the environment. As a result, it has been argued that the people of the present generation have a special obligation to protect the planet earth from guaranteeing the rights of future generations. Thus, future generations' rights and use of this valuable heritage limit man from indiscriminately using the environment. This limitation is that what we have gained from the environment will not be improperly passed on to future generations. On the other hand, we can owe the current technology, wealth, and progress to some extent to the environment, so this progress should not destroy the environment itself. The current environment and progress should be appropriately transmitted to the next generation. As a result, the current generation must avoid irreversible environmental degradation as much as possible to preserve this valuable trust for future generations. The concept of the rights of future generations derives from the concept of sustainable development. The current generation is obliged to use the environment wisely and rationally. There are three instances of the concept of the rights of future generations. First, every generation must protect various environmental resources, which means that that generation must use its environment rationally and desirably so that future generations can also benefit from this valuable heritage. Second: Use the environment in a way that does not cause irreparable damage. "Third, cultural and natural heritage must always be protected for future generations(Lazarus, 2008)."

Sustainable Development

This concept has been a core concept in the United Nations global strategy since the 1980s. According to Brandt Land at the 1983 Development and Environment Summit, the concept of sustainable development is an advanced process that meets the needs of the current generation without reducing the ability of future generations to meet their own needs—in other words, providing an opportunity for everyone to live on the planet forever. Today, sustainable development is one of the most fundamental issues of international environmental law. Without a doubt, it can be named the primary goal of developing international environmental law. On the other hand, environmental law is essential for monitoring and managing sustainable development. The importance of sustainable development in international environmental law is that the world is witnessing the transformation of this legal branch into international sustainable development law. Sustainable development means the process of development that meets the needs of the present generation while protecting the environment and taking into account the needs of future generations. Several international documents and treaties on the environment have addressed attention to sustainable development. The Stockholm and Rio Declarations state that the pursuit of sustainable development requires the protection of the environment and that Agenda 21 for the development of these two concepts is inseparable. Agenda 21, with particular attention to this goal and proposing the establishment of the Commission for Sustainable Development, has called for the orientation of international environmental law in line with the concept of sustainable development. In drafting new international documents in this field, environmental protection will be considered as part of the process of sustainable development, and the relationship between the environment and other areas of human life and international relations such as trade and investment will also be considered. The preamble to the 1992 Climate Change Convention states that all countries, especially "developing countries," need access to the resources needed for sustainable development. It also emphasizes the cooperation of all countries to protect the climate system for future generations. Under Article 2 of the 1997 Kyoto Protocol, to fulfill the commitment to reduce greenhouse gas emissions and promote sustainable development, Contracting Governments must implement their national policies and measures to increase energy efficiency in various sectors and use new and renewable types. Use energy and work with other members on these issues and develop sustainable forms of agriculture. Article 2 of the 1992 Convention on Biological Diversity defines "sustainable exploitation" as "the use of the components of" biodiversity "to such an extent that it is possible to exploit different and diverse species to meet the needs of present and future generations." The Rio 1992 Declaration, while emphasizing the 1972 Stockholm Declaration in general, addresses the issue of sustainable development. The first principle of the Rio Declaration states that human beings are at the center of sustainable development. The right to health and a day-to-day drinking life is in harmony with nature(Kiss et al., 1991).

The second principle recognizes the right of States to act following the Charter of the United Nations and international law to exploit their resources following their own environmental and development measures. Principle 3 of the Rio Declaration states that the right to development must be exercised in such a way as to equally meet the needs of the present and future generations in the field of development and the preservation of the environment. From the point of view of international jurisprudence, the Gabchikovo Nagimaros case was the first binding case to address issues related to international environmental law directly. "According to the majority of the judges of the Court, the verdict, in this case, has had an impact on the development of international environmental law." "The International Court of Justice has referred to the concept of sustainable development for the first time following this case." The fourth principle states that environmental protection must become an integral part of the development process to achieve sustainable development. The pursuit of development should not be considered something separate from environmental protection. And the last principle, which is the twenty-seventh, states that governments must work together in good faith and in a spirit of cooperation to develop international law in the field of sustainable development. The Climate Change Treaty states that "all members have the right to development and the obligation to promote it." The Convention on Biological Diversity also mentions the concept of sustainable exploitation in several sections (Kiss & Shelton, 1997).

CONCLUSION

Concepts play a fundamental role in understanding environmental law. The concepts in this field are general. By studying the concepts, one can understand environmental law's principles, perspectives, ideas, and goals. This chapter refers to the basic principles and concepts of environmental law concerning international treaties and regulations. The differences between the principles and concepts and their role in environmental law were briefly described.

REFERENCES

Kiss, A., & Shelton, D. (1997). *Manual of European environmental law*. Cambridge University Press.

Kiss, A. C., Shelton, D., & Shelton, D. (1991). *International environmental law* (Vol. 3). Transnational Publishers.

Lazarus, R. J. (2008). *The making of environmental law*. University of Chicago Press.

Nanda, V. (1995). *International environmental law & policy*. Brill Nijhoff.

Norouzi, N. (2022). Regulating Sustainable Economics: A Legal and Policy Analysis in the Light of the United Nations Sustainable Development Goals. In Handbook of Research on Changing Dynamics in Responsible and Sustainable Business in the Post-COVID-19 Era (pp. 266-287). IGI Global. doi:10.4018/978-1-6684-2523-7.ch013

Norouzi, N., & Ataei, E. (2021). Environmental Protection Regulations in the Light of Public Law and Social Obligations. *Research Journal of Ecology and Environmental Sciences*, *1*(1), 1–16.

Orts, E. W. (1994). Reflexive environmental law. *Nw. UL Rev.*, *89*, 1227.

Palmer, G. (1992). New ways to make international environmental law. *The American Journal of International Law*, *86*(2), 259–283. doi:10.2307/2203234

Salzman, J., & Thompson, B. H. Jr. (2010). *Environmental Law and Policy: Concepts and Insights*. Foundation Press.

Sands, P., & Peel, J. (2012). *Principles of international environmental law*. Cambridge University Press. doi:10.1017/CBO9781139019842

Shohani, A., Ataei, E., & Norouzi, N. (2021). Prevention and Suppression of Environmental Crimes in the Light of the Actions of Non-Governmental Organizations in the Iranian Legal System. *Research Journal of Ecology and Environmental Sciences*, *1*(1), 57–70.

Tarlock, A. D. (1993). The nonequilibrium paradigm in ecology and the partial unraveling of environmental law. *Loy. LAL Rev.*, *27*, 1121.

Wiener, J. B. (2000). Something borrowed for something blue: Legal transplants and the evolution of global environmental law. *Ecology Law Quarterly*, *27*, 1295.

ADDITIONAL READING

Kiss, A., & Shelton, D. (1997). *Manual of European environmental law*. Cambridge University Press.

Kiss, A. C., Shelton, D., & Shelton, D. (1991). *International environmental law* (Vol. 3). Transnational Publishers.

Lazarus, R. J. (2008). *The making of environmental law*. University of Chicago Press.

Nanda, V. (1995). *International environmental law & policy*. Brill Nijhoff.

Orts, E. W. (1994). Reflexive environmental law. *Nw. UL Rev.*, *89*, 1227.

Palmer, G. (1992). New ways to make international environmental law. *The American Journal of International Law*, *86*(2), 259–283. doi:10.2307/2203234

Salzman, J., & Thompson, B. H. Jr. (2010). *Environmental Law and Policy: Concepts and Insights*. Foundation Press.

Sands, P., & Peel, J. (2012). *Principles of international environmental law*. Cambridge University Press. doi:10.1017/CBO9781139019842

Tarlock, A. D. (1993). The nonequilibrium paradigm in ecology and the partial unraveling of environmental law. *Loy. LAL Rev.*, *27*, 1121.

Wiener, J. B. (2000). Something borrowed for something blue: Legal transplants and the evolution of global environmental law. *Ecology Law Quarterly*, *27*, 1295.

KEY TERMS AND DEFINITIONS

Equity: Defined by UNEP to include intergenerational equity - "the right of future generations to enjoy a fair level of the common patrimony" - and intragenerational equity - "the right of all people within the current generation to fair access to the current generation's entitlement to the Earth's natural resources" - environmental equity considers the present generation under an obligation to account for long-term impacts of activities and to act to sustain the global environment and resource base for future generations. Pollution control and resource management laws may be assessed against this principle.

Polluter pays principle: The polluter pays principle stands for the idea that "the environmental costs of economic activities, including the cost of preventing potential harm, should be internalized rather than imposed upon society at large." All issues related to responsibility for environmental remediation costs and compliance with pollution control regulations involve this principle.

Precautionary principle: One of the most commonly encountered and controversial principles of environmental law, the Rio Declaration formulated the precautionary principle: To protect the environment, the precautionary approach shall be widely applied by States according to their capabilities. Where there are threats of serious or irreversible damage, lack of complete scientific certainty shall not be used as a reason for postponing cost-effective measures to prevent environmental degradation. The principle may play a role in any debate over the need for environmental regulation.

Prevention: The concept of prevention can perhaps better be considered an overarching aim that gives rise to a multitude of legal mechanisms, including prior assessment of environmental harm, licensing or authorization that set out the conditions for operation and the consequences for violation of the conditions, as well as the adoption of strategies and policies. Emission limits and other product or process standards, the use of best available techniques, and similar techniques can all be seen as applications of the concept of prevention.

Public participation and transparency: identified as necessary conditions for "accountable governments,... industrial concerns," and organizations generally, public participation and transparency are presented by UNEP as requiring "effective protection of the human right to hold and express opinions and to seek, receive and impart ideas,... a right of access to appropriate, comprehensible and timely information held by governments and industrial concerns on economic and social policies regarding the sustainable use of natural resources and the protection of the environment, without imposing undue financial burdens upon the applicants and with adequate protection of privacy and business confidentiality," and "effective judicial and administrative proceedings." These principles are present in environmental impact assessment, laws requiring publication and access to relevant environmental data, and administrative procedures.

Sustainable development: Defined by the United Nations Environment Programme as "development that meets the needs of the present without compromising the ability of future generations to meet their own needs," sustainable development may be considered together with the concepts of "integration" (development cannot be considered in isolation from sustainability) and "interdependence" (social and economic development, and environmental protection, are interdependent). Laws mandating environmental impact assessment and requiring or encouraging development to minimize environmental impacts may be assessed against this principle.

Transboundary responsibility: Defined in the international law context as an obligation to protect one's environment and prevent damage to neighboring environments, UNEP considers transboundary responsibility at the international level as a potential limitation on the sovereign state's rights. Laws that limit externalities imposed upon human health and the environment may be assessed against this principle.

Chapter 4
Systems of Environmental Protection Responsibility

ABSTRACT

Environmental protection protects the natural environment by individuals, organizations, and governments. Its objectives are to conserve natural resources and the existing natural environment and, where possible, to repair damage and reverse trends. Due to the pressures of overconsumption, population growth, and technology, the biophysical environment is being degraded, sometimes permanently. This has been recognized, and governments have begun restricting activities that cause environmental degradation. Since the 1960s, environmental movements have created more awareness of the multiple environmental problems. The purpose of this chapter is to explain that it is essential that all actors on earth, including humans, governments, institutions, transnational corporations, international organizations, and other actors, be held accountable for any action that leads to environmental degradation. Responsible for existing cracks that result in damage to ecosystems.

INTRODUCTION

Environmental degradation is the deterioration of the environment through depletion of resources such as quality of air, water, and soil; the destruction of ecosystems; habitat destruction; the extinction of wildlife; and pollution. It is defined as any change or disturbance to the environment perceived as harmful or undesirable.

Environmental degradation is one of the ten threats officially cautioned by the high-level panel on Threats, Challenges, and Change of the United Nations. The United Nations International Strategy for Disaster Reduction defines environmental degradation as "the reduction of the capacity of the environment to meet social and ecological objectives and needs." Environmental degradation comes in many types. When natural habitats are destroyed, or natural resources are depleted, the environment is degraded. Efforts to counteract this problem include environmental protection and environmental resources management (Campbell-Lendrum & Corvalán, 2007).

DOI: 10.4018/978-1-6684-4158-9.ch004

The destruction of the environment did not exist in its current form until the advent of the Industrial Revolution (1750 AD). Until then, the environment has been damaged by natural disasters such as earthquakes, floods, volcanoes, and the like, especially by internal and international armed conflicts. The discovery of oil and coal, the establishment of industrial units, and the rapid population growth, especially in cities for housing and the use of more natural resources (renewable and non-renewable) to feed and obtain energy from the causes of environmental degradation. Therefore, efforts should be made to balance and preserve the environment and replace it with environmental resources. Of course, some people try to destroy the environment for their short-term material gain. After a while, those set fire to their own or others' farmland to turn it into a home and promenade. Among the forms and situations that follow the destruction of the environment, we must mention the following. Deforestation takes place after the felling of trees to create more places and agricultural lands for human life, which causes soil erosion and desertification. Also, the growth of industries and pollution caused by industrial plant effluents, some of which have lasting effects, and the production of greenhouse gases, which themselves cause the depletion of the ozone layer and cause profound climate change, irreparable environmental damage such as the extinction of animal species and It causes a plant. Different groups do environmental degradation. One of the reasons for the destruction of the environment by non-urban residents is environmental resources to provide the required energy. Due to a lack of knowledge and familiarity with the correct principles of agriculture and large amounts of chemical fertilizers, the soil is eroded. Plowing sloping lands, burning forests for agriculture, grazing livestock before the emergence of grasses, and destroying pastures are examples of environmental degradation. In many cases, urban dwellers' leading cause of environmental degradation is the low cultural level and lack of environmental ethics, including releasing waste, waste, and sewage into the environment. This section refers to environmental degradation factors such as pollution and dust(Dunlap & Brulle, 2015).

GENERAL RESPONSIBILITY FOR ENVIRONMENTAL PROTECTION

Everyone is responsible for protecting the environment. No person should shift his responsibility to another natural or legal person. It is not suitable for people to ignore their civic obligations to clean the environment and blame the government and institutions responsible for environmental crises. The government only deals with public policy, and it is the citizens who should think more about protection than any other actor is from the environment.

Of course, the government should also prioritize environmental issues and programs, allocate appropriate budgets and agile organizations to this matter, and use the relevant managers and experts.

Perhaps it can be said that the only part of the various affairs of society that people can influence more than other areas is the environment. Because at the general level of every human being and the national level of every citizen, the possibility of protecting the environment and influencing others by keeping the environment clean, managing water consumption (both safe drinking water and agriculture), waste separation, less meat consumption In general, consumption management can fulfill its citizenship obligations.

The protection of the environment is the responsibility of human beings (whether in the role of a citizen who is a citizen of a state or an individual outside their own country), companies, production units, international organizations, and governments(Floyd, 2008).

Western culture has seldom considered the man-nature relationship in the moral and legal dimension. Environmental ethics has been developed by and for man, without having offered other spaces.

The principle of intergenerational solidarity on which the theory of sustainable development is based is a timid advance because it only includes human beings, not other living beings, and because the duties in relation to our around.

As an ethical foundation of environmental law, it does not make sense to continue nurturing an exclusively anthropocentric relationship. We must nourish ourselves with new visions of a biocentric and holistic nature to support the values, principles and norms of the environment. Only from an ethic of life can we build an environmental law for the new century and for all beings on the planet.

The current of environmental thought, called deep ecology, led by Bill Devally Arne Naess raises some statements that dispute traditional tendencies and classic legal visions that must be carefully examined, because in the construction of environmental law they are of enormous importance; among them we mention the following:

- The well-being and flourishing of human and non-human life on Earth are values in themselves. These values are independent of the utility of the non-human world for human purposes.
- The richness and diversity of life forms contribute to the realization of these values and are therefore also values in themselves.
- Humans have no right to reduce this richness and diversity, unless it is to satisfy vital needs.
- Human intervention in the non-human world is currently excessive and the situation is rapidly deteriorating.
- and. Therefore, we have to drastically change our political orientations in terms of economic, technological and ideological structures. The result of the operation will be profoundly different from the current state.
- Ideological change consists mainly in valuing the quality of life, rather than endlessly trying to achieve a higher standard of living.

International Responsibility of Governments for Environmental Protection

Governments have a greater responsibility for the environment than any other actor. Because the presence, exploration and research, and protection of areas of the environment are beyond individuals' capacity and environmental civic institutions. Also, suppose the government fails to protect and manage the environment. In that case, it is possible to cause irreparable damage to the environment, such as the extinction of plant or animal species or severe pollution of cities, which leads to disease and death of citizens, which can not be compensated at any cost. Or the drying up of wetlands and lakes and the pollution of rivers and soil erosion can only be repaired at a very high cost.

Today, governments can not be indifferent to the destruction of the environment within their country and consider it a matter of national sovereignty. The reason is the globalization of environmental culture, the efforts of many individuals and environmental institutions, the positive approach of most governments to protect the environment, and various treaties in various fields. The same should be said about the damage to a country's environment. Suppose the damage causes damage to a neighboring country or other ecosystems. In that case, the international responsibility of the government of the source of the damage will be established, in which case it will face pressure from world public opinion and other governments due to the need to protect the global environment. Suppose the damage does not affect neighboring countries or other regions whose existence is essential for the entire world. In that case, the government is responsible for the damage to its environment due to the violation of its citizens' right to a healthy environment. In this regard, of course, it must be said that any damage, however small and imperceptible, affects the whole world.

In the international dimension, the international responsibility of states is realized when infringing acts, including acts, omissions, or a combination of both, are carried out in conflict with either or a set of mutual obligations, international custom, mandatory rules, and obligations the general public. Of course, if there are no factors to remove the offensive description of the verb.

According to the rules of international responsibility of states, a country to which an act or omission contrary to international law and regulation is attributable to it must compensate the country or countries affected by it and the global environment violating any obligation under international law provided It is responsible for assigning it to one or more governments.

"The 2001 draft of the International Law Commission, entitled 'Drafting the Liability of States for International Violations,' considers only damages to be incurred as a result of breaches of international obligations." Article 19 In the 2001 Plan, the International Responsibility of States, which reflected the International Responsibility of Governments for the Environment, was removed. Regarding the need to compensate for the damage caused by environmental degradation, it can be said that the international responsibility of governments in the event of damage to the environment is realized.

A noteworthy point about international environmental law is its various dimensions and elements and the breadth of its issues, especially that part of transnational environmental issues, such as border pollution and international destruction. Researchers in international law has established rules governing governments' responsibilities and obligations, especially in responding to border demolition and pollution. Through this system of responsibility, governments have developed international agreements to prevent the activities of Harmful to the regulated environment. On the other hand, the establishment of the International Criminal Court can be considered one of the significant international measures in the open criminal protection of the environment (Gienapp et al., 2008).

International environmental law, as it is currently analyzed, has deficiencies that need to be urgently overcome. Among these weaknesses we highlight: The normative dispersion. There is a profuse number of international legal instruments that apparently create the feeling of great protection and security, but in reality show enormous difficulties in their application and internal contradictions that allow multiple interpretations and, ultimately, make the rules ineffective. "There are more than 4,000 conventions, treaties and instruments with provisions to deal with the protection of the environment, most of them bilateral or with no claim to universality. Of a strictly international nature, with a claim to universality and globality, there are around 152." The non-existence of an international environmental jurisdictional system. It is a reality that international environmental law lacks its own jurisdictional bodies that settle disputes and apply sanctions. In environmental matters, access to other already existing regional or equivalent bodies (generally human rights) is practically nil. The system proposed in most of the rules of this system, to resolve conflicts, is arbitration and its practice leaves much to be desired. Lack of procedures, mechanisms or instruments, so that, as in the international human rights system, individuals and organizations can access international organizations to present cases of violation of the right to the environment.

Faced with such a weak international panorama, we must look at the internal systems of environmental regulation and jurisdiction. We observe that the trend in comparative environmental law is directed towards the search for new procedural alternatives to resolve environmental conflicts and collective conflicts. Among these instruments, the development that has been taking place (slowly, but surely) in the figure of popular actions or collective actions, as some scholars have recently called them, stands out.

In the international context and comparative law, we can see that in many countries the issue of popular actions has been gradually incorporated into the constitutions and laws of the world. It is clear that in the United States, Canada, Brazil, Portugal, France, and more recently in Argentina, this institution is one of the greatest procedural revolutions of all time, since it has proven to be an effective means of resolving many of the tensions and conflicts derived from industrialization and massification. Popular actions in these countries are an element of agglutination and democratic participation in the administration of justice.

Through the exercise of collective actions, those activities that cause damage to large sectors of the community, as is the case of the inadequate exploitation of natural resources, defective medical products, unforeseen construction of public works or unforeseen in the construction of a private work, the excessive charging of goods or services, the alteration in the quality of food, the very recurring phenomenon of misleading advertising, fraud in the financial sector, will have popular actions, as a way different but very effective legal system to resolve such conflicts.

These actions have the merit of strengthening human groups as a whole, by allowing vulnerable sectors or those who live in circumstances of greater vulnerability, greater risk, those who are in a situation of economic disadvantage to be placed in a condition of equality and can viably legally confront, with the possibility of success, those most powerful sectors.

A second aspect refers to procedural developments regarding the issue of legitimacy to file these actions. Law 472 of 1998 in its article 48 paragraph states: "In the group action the actor or whoever acts as plaintiff, represents the other people who have been individually affected by the infringing facts, without the need for each of the interested parties to exercises its own action separately, nor has it granted power".

From this brief tour of the environmental protection procedural instruments, we can conclude that it is necessary to strengthen these mechanisms both internationally and internally. This is a clear challenge of environmental law. How should these new instruments be interpreted, applied and exercised? How to train new judges with an environmental vision who can hear and resolve the conflicts of the next century around water, air, biodiversity, ecosystem protection, and the protection of indigenous and ethnic communities?

Elements Constituting the International Responsibility of States

The international responsibility of governments is the same whether in environmental matters or in other areas. The existence of a state with political sovereignty is essential for the realization of international responsibility because states are formed on the basis of a political and sovereign element and have independent legal personality and enjoy rights and obligations within the framework of international rules and regulations. From the twentieth century onwards, the international community considers legal and even natural persons to have rights and obligations. International organizations have a degree of international character in order to carry out their responsibilities within the framework of the objectives set out in their statutes, but their competence is not as complete as that of governments because their competence is determined by their statutes, procedures, and practices. Who have an international responsibility for their wrongdoing.

A government or international organization is liable and liable for damages when it commits an act or omission contrary to its legal obligations under the rules of international law (treaty, custom, general principles of law, or bilateral agreement) because of a breach. Each of them gives rise to the responsibility of the offending function.

Environmental damage is one of the components of fulfilling international responsibility, but in general, there is no need to cause damage to be held liable. It becomes.

In international law, damages are sometimes material and compensation in the form of compensation to the state, sometimes moral, which is compensation through consent, and sometimes material and spiritual. Loss, on the other hand, is divided into two categories in terms of the relationship between the damage and the harmful act:

First - direct or immediate damage; It is harm that there is a causal relationship between it and the harmful act.

Second - indirect or mediated loss; Damage that does not have an actual causal relationship between it and the harmful act.

One of the most critical and challenging issues of assigning international responsibility to the state is finding the causal relationship between the actions of states and environmental damage. To invoke government liability for transboundary pollution, the plaintiffs must first prove that the government attributed the harmful act. The aggrieved country must prove a breach of an international obligation and the causal relationship between the act performed and the breach of the obligation. And to receive compensation for material damage must also be proven.

Proving the causal relationship between the harmful action of the state and the harm to another is one of the most critical issues of international responsibility, which is doubly difficult in the field of environmental activities because the harm to human beings may not be known and the cause can not be easily identified, or Its effects took years to unfold; These difficulties have led to the failure of many environmental claims.

An essential point in the discussion of international responsibility in the field of environment is that, unlike other obligations in other fields of international law, the indirect effects and long-term damage to the environment are much more significant and can even be said impossible in some cases (O'Connor et al., 1999; Qian & Zhu, 2001; Raleigh & Urdal, 2007; Schlosberg & Collins, 2014; Shepardson et al., 2012).

New Systems Environmental Responsibility of States

The issue of civil liability and even more, the responsibility of the State in environmental matters is in full evolution and the reason is very clear: with old schemes we cannot solve new and transcendental problems vital for the survival of the planet and its species. Traditionally, the issue has been examined from an ethical and not a legal point of view, frequently pointing out that we are all responsible: "The preservation and conservation of the environment is a responsibility that commits the joint action of the State and individuals. The development of productive work, as well as free private initiative, within a framework of legality, cannot be considered in absolute terms., since it is clear that the preservation of a healthy environment, in addition to being an unalterable and unconditional duty, is perennial, since it falls on something necessary: the dignity of human life".

This trend that we could call "joint responsibility" cannot avoid the other vision of the problem: the exact definition of legal responsibilities for the purposes of prevention and repair of environmental damage.

Three regulatory systems can be clearly differentiated on the subject of environmental responsibility:
Subjective liability systems are:

In them, environmental liability is interpreted within civil liability schemes based on fault and fraud. These are the prevailing systems throughout the world, both in continental and Anglo-Saxon systems.

Anglo-Saxon systems: United Kingdom: there is an environmental protection regulation called the Environmental Protection Act of 1990, modified in 1995. In matters of civil liability, the prevailing rule is liability for fault; the exception is strict liability. The evolution towards strict liability has been taking place in very specific areas, such as contamination with waste or dangerous substances; In these cases, according to current environmental regulations, it is only necessary to prove the causal link between the deposit of hazardous substances and the damage; the fraud or fault of the agent subject is irrelevant. Australia and New Zealand: in these countries, in the same way, progress is being made, from subjective liability systems, towards objective liability.

Civil systems: Holland: in this country the concern for environmental issues is very broad. Regulations such as the Public Damage Law, the Law on General Regulation of Environmental Protection, the Air Pollution Law and the Law on Substances Dangerous to the Environment stand out. In terms of liability, its system is based on the classic scheme of the agent's fault, but jurisprudence is beginning to evolve towards risk-based liability. In Italy its liability system is based on the norms of the civil code, where, for example, article 844 stands out, which states: Art. 844. "The owner of a farm cannot prevent the emission of smoke or heat, the exhalation, the noise, the vibrations and the like derived from the neighboring farm, if they do not exceed the normal tolerability, and also taking into account the condition of the place". In applying this rule, courts must balance the demand for production with the right to property. You can take into account the priority of a certain use. The concept of tolerability that is introduced and the harmonization of property rights and economic freedom that the article proposes is very interesting. Liability for the exercise of a dangerous activity is also established in what writers call civil liability for damage to the environment, a quasi-objective system. In Japan as a country hit by multiple environmental tragedies. They have marked the need for evolution of their legislation and jurisprudence. The case of Minamata stands out, where the courts established the compensation of the injured parties for the physical and mental damages suffered, lost profits and expenses of advisors, as well as the obligation of the defendant companies to prevent future contamination. All this was achieved based on an interpretation of the right to enjoy a healthy environment. In Japan there are many environmental laws; among others we can highlight: Water Pollution Control Law, Maritime Pollution Control Law, Waste Disposal Law, Environmental Crimes Law and Conflict Resolution Law related to pollution. The main experience that this country contributes is its system of compensation funds related to the environment, an alternative that is beginning to make its way in the world. In Latin American Systems there is a great similarity in the way in which Latin American systems regulate the issue of responsibility, Colombia, among them. These are subjective systems with some rules that allow them to evolve towards quasi-objective systems, especially those referring to liability for dangerous activities (Art 2341 of the Colombian Civil Code).

In strict liability systems which is globally practiced by three countries with independent and autonomous environmental responsibility systems stand out: Germany, the United States and Canada.

Germany: This country is the pioneer of strict liability in environmental matters in continental systems. Outstanding regulations in this matter are: the Wasser-haushaltsgesetz (Water Law) which establishes strict liability for damages to people and property as a result of any change in the physical, chemical or biological composition of the water. In the same way, specific strict liability systems are established in the following laws: Bundesberggesetz (mining facilities), Atomgesetz (nuclear reactors), Haftpflichtgesetz (power generation facilities), Bundes. Immissionsschutzgesetz (emissions dangerous to the environment). As a general rule on environmental responsibility, Germany has the UmweltHG. The main characteristics of this standard are: A general system of strict liability is established "If someone suffers death, personal injury or property damage due to an environmental impact emitted by one of the facilities contained in appendix 1 of the law, the owner of the facility will be liable against to the victim for the damages caused by that". (Art 1 of the law). The basic requirements are: a) that the defendant carry out his activity in one of the facilities defined in the annex to the law; b) that the environmental impact has been emitted from the defendant's facilities; c) that there is a causal link between the environmental impact and the damage for which reparation is claimed. The list of facilities that generate environmental impact and for which said law operates is legally defined. "There are ninety-six activities divided into ten groups, among which are: mining and energy production; production of stones, sand and other construction materials; production of steel, iron and other metals; production of chemicals, pharmaceuticals or petroleum derivatives; treatment with organic matter and production of artificial materials; wood and pulp processing; production of food, feed, and agricultural products; waste treatment; storage and disposal of certain materials; (and others)". Environmental impact is defined as materials, vibrations, noise, pressure, lightning, gases, vapors, heat or other phenomena emitted into the ground, air or water. A presumption of causal link is established. The following are indicated as exemptions from liability: a) force majeure; b) non-substantial damages or reasonable damages according to local conditions; c) any polluting activity before the entry into force of the law (January 1, 1991); d) Any liability that gives rise to compensation greater than the maximum amount for which one can respond, which is estimated at DM 320,000,000. So-called development or damage risks arising from those substances, which at the time of the damage, were not recognized as dangerous are not excluded. Extraterritoriality. This characteristic allows the plaintiff to bring his action in Germany, even if the damages had occurred in another country, as long as the person responsible is a German natural or legal person.

United States: "There are three salient features of US environmental law: the general imposition of strict liability, the broad powers of authorities to inspect and obtain information, and the important role of the public in enforcing compliance with the law." law". In this country, the main powers in environmental matters belong to the Federal Government and are exercised through the EPA (Environmental Protection Agency). Among the environmental regulations, the National Environmental Policy Act (1969), the Clean Air Act (1955), the Clean Water Act (1972) and the Comprehensive Environmental Response, Compensation and Liability Act CERCLA (1980) stand out. In terms of liability, this last rule is the most important. CERCLA has its central aspect in the duty to carry out the decontamination before the Environmental Protection Agency, directly by the obligated or indemnifying the Agency for the expenses incurred to decontaminate. As some elements that stand out from this law we can point out the following: It defines who is responsible for contamination of places containing dangerous substances. This is called a "potentially responsible party." The places that need decontamination are defined. This is a list of national priorities. Strict liability is imposed on the owners of the place in which the damage or threat due to the emission of dangerous substances occurs and on the producers, transporters and even on the creditors (in relation to the damages of their debtor). The cases in which these alleged perpetrators exempt themselves from responsibility are indicated: force majeure, war and action or omission of a third party who is not an employee or agent of the possible person responsible. In these cases, you must prove the diligence and precautions taken. There is a fund system for pollution damage repairs. "In the United States, they have the Superfund, intended to clean up contaminated sites and which is financed by contributions from the oil and chemical industry. The Environmental Protection Agency (EPA) acts trying to recover the expenses of the landowners, the causes pollution (carriers, products, deposited) 27,000 contaminated sites have been identified with an average unit cost of recovery of 26 million dollars

The amount of expenses involved in litigation is greater than that of remediation and there have already been cases companies that have had to close".

Canada: This country is also a leader in environmental protection. In its legislation, the Canadian Environmental Protection Act stands out, a norm in which obligations are established for the federal government to take preventive and remedial measures. Canadian jurisprudence has also been a pioneer in providing solutions of enormous interest in environmental matters, such as that related to the Brosseau v. The Alberta Securities Commission where the Supreme Court of Canada "accepted and applied the principle that a sanction can be imposed on a person in relation to an event that occurred in the past, provided that the objective is not penalty but to protect the public."

In the European Union, three main actions have been developed around the search for unification of a regime of responsibilities:

1. Convention of the Council of Europe for civil liability for damage caused by activities dangerous to the environment. (1993)
2. Draft directive on civil liability for environmental damage caused by waste.
3. Green Paper of the Commission to repair environmental damage. (1993)

The international initiatives and those of the European Union "are focused on the damage caused by cross-border pollution. They are based on the Tripartite Convention on Nuclear Liabilities (1960), the International Convention on Liability for Oil Spill Damage (1969) and the analogous one for damage to the environment by dangerous activities, of the Council of Europe (1993) In the Convention of the Council of Europe reference is made to the responsibilities for damage to the environment through dangerous activities.

It is the most important Convention for this continent and can serve as the basis for regional regulations. Its main regulated aspects are:

*   The person responsible is the operator and while the activity lasts.
*   Partial liability cases are accepted.
*   Retroactivity is prohibited, except in the case of landfills.
*   A system of free access to administration information is established by any citizen, within two months of the request (there are exceptions).
*   For actions, a role of environmental protection organizations is defined.
*   The prescription is limited to 30 years and the objection is created due to general ignorance of the matter.

In the Draft of the directive on responsibilities for damages caused by waste, proposed since 1991, reference is made to all waste generated in industries, homes and transport. A homogeneous system is proposed, and the responsibility may lie with the receiver. Responsibility is also shared and cascading for those who have handled or received the product. Force majeure and having been deceived are excuses, as well as the actions of the victim. Each State will decide who has the right to initiate the actions and the applicable remedies. The prescription is limited to 30 years. It also proposes the imposition of limits to the legal actions of the green pressure groups and the obligatory system of insurance policies for producers and handlers of waste. In the so-called Green Book, the policy of application of the figure of civil liability in damages to the environment, as a direct object of protection, begins. The Green Paper sets out the legal problems of civil liability legal systems when applied to environmental problems. Said document states that it is necessary to redefine the basic conceptual framework of traditional liability in order to apply it in environmental matters. Here the basic legal issues are:

- Establish the type of violations.
- Establish who is individually or collectively responsible.
- Establish the cause (violation) / effect (damage) relationship.
- Test the reality of the damage and quantify it.
- The effective repair of the damages caused.

States can resort to the two civil liability systems by extension in environmental matters, being able to choose between:

a)   Fault liability system, applicable in cases of negligence or recklessness that have been the cause of environmental damage. This option would complement the basic regulation proposed by the document.

b)   Strict liability system (no fault). In this case, the existence of an infringement does not have to be demonstrated. The effort is presented in the prevention position, avoiding blame, but entering into the need to define damages, quantify civil responsibilities and designate the affected activities.

The Green Paper raises the problems arising from specific situations such as: cases of chronic pollution, emission levels authorized by public authorities, damage originated in the past, "adequate repair" and the problem of damage insurance.

As regards the compensation system to be established when the damage has already been caused and cannot be attributed to a specific person or entity, the Green Paper calls for the introduction of community compensation mechanisms (joint compensation), whose costs would be distributed among the various economic sectors involved in one way or another in environmental problems.

In short, the Green Paper claims to be based on a system of incentives and penalties. Some of the notable aspects are:

Strict liability: if the person responsible is not found, there will be a compensation system.

- Establishment of responsibilities: between the issuer and the administration.
- Shared responsibility. Attempts to determine a distribution rule.
- Chronic and/or diffuse pollution: in these cases it is intended that the distribution be among the community.
- Limits of liability: the solution could be to set the limits of liability and guarantee very high, so that it is difficult to circumvent the principle "polluter pays".
- Environmental damage: there is no single definition of damage or when it starts.
- Who can act?: the damaged, and possibly the pressure groups with limitations.
- Insurability: the amount and obligation of the policy, the repair of the damages caused by the insured and the indefinite limits of liability, must be better specified. States could participate in these aspects.
- Compensation system: proposes creating a special fund fed by rates paid by polluting sectors.

In this journey through the main trends in comparative law, we can clearly see the urgency of a regulation on environmental responsibility. Our Constitution allows the development of a strict liability system for collective rights in its article 88, but the legislator has not assumed the challenge of regulating it.

LIABILITY FOR UNAUTHORIZED ACTS IN THE ENVIRONMENT

When an act (current act or omission) is committed, that does not violate any obligation but is followed by an act without fault or fault, damages to one of the states or the environment of a stateless region such as the high seas. It is formed without prohibition.

So that it is enough to establish a relationship between the act of harm and the government that caused the damage to create international responsibility, it can be said that this type of responsibility is the same as "pure responsibility," which is also known as objective responsibility.

"Pure responsibility has a dual function. "This theory seeks to ensure immediate and effective reparation by not only exempting the victim of environmental damage caused by hazardous activities from proving the guilt of the perpetrators of such activities but also seeks to encourage such perpetrators to prevent environmental damage."

"Several major environmental incidents, such as the crude oil spill from the Tory Canyon off the coast of Britain in 1967 and the sinking of the Cosmos reconnaissance satellite on Canadian soil in 1979, have drawn the attention of governments and the International Law Commission to the need to extend unfettered liability." "It attracted international attention."

The difference between the theory of international responsibility of states in permissible and legitimate actions and the theory of liability due to error is that international obligations must be violated to compensate for the damage in the theory of error. Also, in risk theory, the beneficiary who created the hazardous environment is required to compensate for the damage caused by it. Still, in liability for unrestricted actions, compensation is an initial obligation, and there is no need to act contrary to international obligations to prove it. The perpetrator is responsible for any unauthorized or prohibited action.

Another advantage of this approach over other views is that it removes the burden of proving the error element that the victim has traditionally been born. This view is primarily used in terms of damage to the environment because if a government causes significant damage to human society, especially the environment, by carrying out a legitimate activity that is not prohibited by international law, the country that caused the damage or the state of origin It is responsible and based on the legitimacy of its action, it cannot present itself as irresponsible and harm the environment of other countries. But to fulfill this kind of responsibility, it is necessary to inform the government that causes the dangerous activity, the significant damages, the legitimacy of the activity, and the causal relationship between the harmful activity and the damage caused(Trombetta, 2008).

Necessity for Total Compensation for Damage to the Environment

Once the international responsibility has been established, the responsible government or international organization is obliged to compensate for the damage. Therefore, liability's main result and effect is the commitment to total compensation. In the case of the Kurzo plant, the Permanent Court of International Justice has ruled in favor of the party responsible for the breach of any international obligation.

Estimating damage to figures is significant in international law. In environmental damage, this is problematic because, in many cases, environmental elements can hardly be assessed with money.

Undoubtedly, with the increasing importance of insurance and financial guarantees in the environment and its interrelationship with economic activities, we expect this legal field to witness fundamental developments shortly.

In principle, the determination of the method or methods and how to compensate is primarily in the competence of the countries or international organizations that are parties to the dispute and is based on their agreement. Otherwise, the matter is within the jurisdiction of the judiciary or international arbitration, which, of course, also involves the subsequent or prior agreement of the states.

In general, compensation methods are the restoration of the status quo ante to restore the condition, provided that it is not fundamentally impossible. According to Article 34 of the Government's International Liability Plan, restoring the status quo is the first form of compensation available to the state affected by an international offense. Article 35 of the draft stipulates that the government responsible for the offense must restore the status quo ante. However, it is almost impossible to restore the status quo due to environmental damage due to its conditions and characteristics.

The method of compensation is the most common method of compensation. Compensation must be accurate, correct, and as far as possible following the damage. Of course, if the damage has a material aspect, the possible damage should be considered in determining the amount of compensation. Compensation, therefore, includes any damages that are financially assessable and can include benefits and even, if necessary, non-profit conditions, but indirect damages are not claimable in any way. In addition, the precise determination of environmental damage caused by the destruction of the human or natural environment to receive compensation has its complexities.

Loss of consent is another way to compensate. When the damage is caused directly to a country or international organization, he has the right to ask the offending country or international organization to compensate for the damage, especially if the damage is moral, by taking action. These actions take many forms: formal regrets and apologies, which may be given orally or in writing by a competent official or even the head of state, symbolic acts such as the military tribute to the flag by sending special envoys to apologize; Thus, the satisfaction of the third type of compensation is that the responsible government can, in fulfilling its obligation, provide total compensation for the damage caused by the international offense.

The most common method of obtaining consent, which is used in moral or non-material damage to a state, is to declare the act offensive by a competent court or tribunal. The International Court of Justice has pointed to the usefulness of this method of obtaining consent to compensate for non-financial losses incurred by a government.

But in international environmental law, the question remains, given that the environment is considered a common human heritage and a part of the rights of present and future generations, how can the satisfied consent be obtained? In general, in the discussion of international liability of states, it seems that customary law of international liability of states to compensate for environmental damage is not very effective, and contract law for the effectiveness of environmental compensation is far from environmental problems and events in the world.

The new trend of international environmental law towards pure liability, according to which no damages remain without compensation, has opened a new perspective on the international responsibility of states. Establish environmental insurance, establishment of joint national and international funds for environmental assistance, etc. It is considered an example of the new trend of international environmental law in compensation for environmental damages. On the other hand, according to customary international law, governments must cooperate internationally in environmental protection. Due to international responsibility's problems, governments must protect their environment with a precautionary approach. In this context, the need to apply the principles of cooperation and prevention in international environmental law is doubled.

Hence, along with the classical legal systems based on subjective and objective responsibility, the tendency towards pure responsibility is also progressing and developing and increasing the conclusion of international treaties and agreements on compensation for environmental damage in various issues such as marine pollution, such as the 1992 Civil Liability Agreement on oil pollution at sea and nuclear, air, and other issues. The establishment of various mechanisms for compensating for environmental damage through the establishment of funds and promoting "insurance" contracts indicates the development of the theory of pure liability in international environmental law(Tranter, 2011; Warner et al., 2010).

COLOMBIAN NEW SYSTEM

The new regime of environmental responsibility is basically framed in five principles of enormous importance in environmental matters, equally valid in matters of responsibility of the State or of individuals:

Precautionary Principle

The importance of the precautionary principle, established in the Rio Declaration of 1992 and included in Colombian environmental regulations by Law 99 of 1993, requires a rethinking of the activity of the State and civil society in the face of environmental problems.

It is no longer a matter of waiting for the damage to occur, or for the authorities (judges, officials from the environmental sector, mayors, etc., to sit at their desks to demand that damage be scientifically and technically proven to impose a precautionary measure or initiate preventive action The spirit of the principle of prevention or precaution requires acting before the damage occurs, taking all possible measures, at the slightest evidence of damage to health, the environment or the life of people or the living beings that it has the institutional and ethical mission to protect.

"In terms of civil liability, we have to move from the category of right to damage repair and structure a risk law. ...In criminal matters we cannot be satisfied with criminal figures of result. The criminal figures of modern law are figures of abstract danger; with this the consummative moment of the crime is anticipated and it is not necessary for a concrete result to occur".

In terms of State responsibility, the situation is one of greater demand in its preventive task and the derivation of responsibility for omission (in compliance with this principle) or for ignorance of it. A clear example can be seen in the processes of granting environmental licenses when standards and control requirements established by the Colombian Constitution and the laws are ignored and which can cause irreparable damage to the environment. The new popular actions law is illuminated by the precautionary principle, as can be seen when in article 5 it establishes the obligation of unofficial promotion of the action by the Judge and the obligation to adopt the necessary measures to make the action viable and promptly order precautionary measures.

Principle of Legal Certainty

Given the advances in science and technology, contemporary man and the legal systems that have evolved are reluctant to bear catastrophes and calamities without repair. It becomes a necessity, not only legal but also social, to search not only for those responsible for the damage, but also for those who have put communities, their members or nature at risk with their behavior. "On the other hand, there has been a profound change in the mentality of man. Today there is a tendency of the spirits to demand security. For this reason, in the face of any damage, a person is sought who is responsible to whom they can charge the obligation to repair it. Wherever In the past, the damage caused was borne by bowing before the disastrous chance, today we are trying to find the author of the damage".

The need to broadly develop the concept of environmental insurance as a quick and easy instrument for reparation to victims or the community is observed here. The international trend confirms the urgency of this tool in Colombian law. Likewise, alternative funds and compensation mechanisms deserve to be studied in depth.

Regarding the determination of those responsible, Colombian new class action law has interesting advances. Article 14 states that "if there is a violation or threat and those responsible are unknown, it will be up to the Judge to determine them." In this way, a broader and more protective judicial work of the environment is allowed and the limits for the actor are eliminated in the sense that the person responsible does not have to be necessarily identified, but rather the damage to the environment or to collective rights.

It is clear that the potential for environmental damage is increasing. This requires taking clear and effective measures to point out the responsibilities that concern, not only those who develop or execute the projects, but also the environmental or state authorities that grant the licenses for their implementation.

Principles of Full Reparation for Damage and Protection of Victims

The paradigm of contemporary civil and environmental responsibility is the full repair of the damage. For this assumption to be fulfilled, new demands arise, previously unthinkable from traditional law. One of these advances is the consideration of the obligation of reparation for lawful acts. It is estimated that there are not only unfairly caused damages, but unfairly suffered.

"Even illegality is not a temperament here, since at least in relation to the affected individuals, as established by intercontinental jurisprudence, compliance with the established conditions cannot be invoked for the exoneration of responsibilities. Even for the authority that created the situation, the allegation of respect for the initial clauses may be indifferent, if the best possible technology is not subsequently applied". This advance is very interesting, because, in environmental matters, it allows us to understand, for example, within a new context the so-called social, economic and cultural impacts of the projects. Aspects that are neglected, on many occasions, by those who prepare Environmental Impact Studies, and that require a very deep look at the implications that carrying out a project has for a community or a region.

Let's examine just one example: the construction of the Urra Hydroelectric Plant in the department of Córdoba. Faced with political and economic interests in building said dam, arguing that it would constitute an energy salvation for the Atlantic Coast, the communities and social, ethnic and environmental organizations of the region and the country repeatedly pointed out the enormous damages that bring. The authorities have remained deaf to the serious impacts on indigenous cultures (Emberá and Zenú), on the fishing communities of the Sinú River, on marine ecosystems, especially mangroves, on endangered species that would be lost due to the flooding of a large part of the Parque del Paramillo, in the species of the river, since the traditional bocachico cannot cross the dikes established by the project and dies in the attempt to reach the spawning areas. Today the region is going through a huge crisis: the victims are not only the indigenous, but also the fishermen, the peasants and in general the nature that waits (with great despair) for reparation, a solution that perhaps the law can offer.

The vision of responsibility must pass from the author of the damage (traditional vision) to the victim of the damage (new vision). With this transformation, the law is humanized and even more so it is extended to previously unthinkable spheres such as considering nature as a victim that also requires reparation. "Thus, the law reacts to any harm unjustly suffered, looks at the victim and from his angle judges the justice or injustice of the damage. It does not look for a person responsible to whom to make a judgment of reproach, it looks for damage to compensate."

Principles of Breaking the Classic Axiom of Guilt

In the traditional view of responsibility, guilt was an insurmountable shell, built to protect those who cause harm, freeing them from the obligation to respond. In this scheme the victim is abandoned to her fate. For the new law, the axiom that stated that "there is no responsibility without fault" must be changed to say that "there is no responsibility without damage".

"The right to reparation for damages thus becomes more of a right to compensation than a right to sanction faults, reaching to a certain extent a depersonalization of responsibility on the part of the person responsible".

Modern authors then agree on the need to overcome this classic concept of guilt in environmental matters. "The volatilization of the idea of fault when it comes to determining the obligation of compensation and its replacement by risk, is an already assimilated consequence of the expansion of industrial civilization."

In the environmental criminal field, interesting advances are beginning to be glimpsed, although enormously difficult given the rigidity of classic criminal schemes. In Europe, doctrinal and normative elaboration is directed towards the criminal liability of legal persons (a field traditionally limited by this branch of law), and towards objective criminal liability. Likewise, a theory is beginning to be built that extends criminal responsibility to the State authorities that intervene in environmental decision-making and that by omission or corruption promote or allow the execution of works with serious impacts on the environment and communities.

In recent jurisprudence that analyzed the objections of unconstitutionality to the Bill of Law 235 of 1996 Senate - 154 of 1996 Chamber "By which the ecological insurance is established, the Criminal Code is modified and other provisions are dictated" partially objected by the President of the Republic, the Constitutional Court adheres to this necessary step that must be taken in our country: "The imputation of criminal responsibility to the legal person in relation to the crimes to which mention has been made, does not violate the Political Constitution. On the other hand, In the case of legal persons and de facto companies, the presumption of responsibility, supported by evidence of the clandestine performance of the punishable act or without having obtained the corresponding permit, does not imply a breach of the Political Constitution either. described, fully authorize the legislator to qualify the responsibility of a subject based on certain facts".

Investment in the Burden of Proof Principle

The principle enshrined in the Civil Code (Art. 1757) and in the Code of Civil Procedure, article 177 according to which "it is incumbent on the parties to prove the assumption of fact of the norms that enshrine the legal effect that they pursue" it is necessary to invert it and reformulate it in environmental matters. The scheme that is imposed is called by Professor Bermúdez Muñoz22 as dynamic burden of proof and according to him "it is the judge who in each specific case must determine which of the parties must bear the consequences of the lack of proof of a certain fact, by virtue of the fact that it is easier for it to supply it". There are precedents in Colombian jurisprudence, especially referring to the issue of medical liability of official entities, in which the Council of State has inverted the traditional rule and has imposed on the defendants the obligation to prove their diligence.

CONCLUSION

Environmental law is a system under construction, full of challenges and proposals. Its importance is vital and its trends constitute new alternatives for traditional legal systems and for the structuring of principles of responsibility that allow us to preserve our planet. Environmental responsibility as one of those new paradigms requires us to reflect from a preventive approach, and forces us to think that we should not wait for new industrial or nuclear accidents to occur, or for more hazardous waste dumps to emerge in the world or for the destruction of our ecological heritage so that the main actors in this process take action on the matter and adequately regulate a matter that is of the greatest interest to present and future generations of living beings on the planet. From the tour carried out by the different regulatory systems of environmental responsibility, we find new elements of assessment for the basic concepts of responsibility. Important proposals regarding the regulation of strict liability and procedural matters. Finally, it is necessary to study alternative proposals for reparation For the victims of environmental damage and for the community, when this damage is not individualized, the experiences that exist in the world on two issues are important: funds and environmental liability insurance. While a new law on environmental responsibility and collective rights is being developed, it will be in the hands of the judges to illuminate with constitutional and international principles the interpretation of the recently issued procedural norms that provide Colombians with the possibility of defending themselves effective environment and collective rights. Environmental authorities must be rigorous in fulfilling their task of environmental control and due implementation of tools to demand prevention as a general rule or repair damage in those cases in which situations of environmental deterioration are verified. Social organizations, especially environmental groups, also have the challenge of appropriating these new instruments and socializing their exercise. We cannot continue to think that the classic conservationist attitude is enough, when we have in front of our eyes, permanent violations of the right to a healthy environment. We also believe that the productive sectors and, in general, those who develop projects that imply serious environmental impacts, must assume the urgent task of self-control and development of compliance environmental audit schemes.

REFERENCES

Campbell-Lendrum, D., & Corvalán, C. (2007). Climate change and developing-country cities: Implications for environmental health and equity. *Journal of Urban Health*, *84*(1), 109–117. doi:10.100711524-007-9170-x PMID:17393341

Dunlap, R. E., & Brulle, R. J. (Eds.). (2015). *Climate change and society: Sociological perspectives*. Oxford University Press. doi:10.1093/acprof:oso/9780199356102.001.0001

Floyd, R. (2008). The environmental security debate and its significance for climate change. *The International Spectator*, *43*(3), 51–65. doi:10.1080/03932720802280602

Gienapp, P., Teplitsky, C., Alho, J. S., Mills, J. A., & Merilä, J. (2008). Climate change and evolution: Disentangling environmental and genetic responses. *Molecular Ecology*, *17*(1), 167–178. doi:10.1111/j.1365-294X.2007.03413.x PMID:18173499

O'Connor, R. E., Bard, R. J., & Fisher, A. (1999). Risk perceptions, general environmental beliefs, and willingness to address climate change. *Risk Analysis*, *19*(3), 461–471. doi:10.1111/j.1539-6924.1999.tb00421.x

Qian, W., & Zhu, Y. (2001). Climate change in China from 1880 to 1998 and its impact on the environmental condition. *Climatic Change*, *50*(4), 419–444. doi:10.1023/A:1010673212131

Raleigh, C., & Urdal, H. (2007). Climate change, environmental degradation and armed conflict. *Political Geography*, *26*(6), 674–694. doi:10.1016/j.polgeo.2007.06.005

Schlosberg, D., & Collins, L. B. (2014). From environmental to climate justice: Climate change and the discourse of environmental justice. *Wiley Interdisciplinary Reviews: Climate Change*, *5*(3), 359–374. doi:10.1002/wcc.275

Shepardson, D. P., Niyogi, D., Roychoudhury, A., & Hirsch, A. (2012). Conceptualizing climate change in the context of a climate system: Implications for climate and environmental education. *Environmental Education Research*, *18*(3), 323–352. doi:10.1080/13504622.2011.622839

Tranter, B. (2011). Political divisions over climate change and environmental issues in Australia. *Environmental Politics*, *20*(1), 78–96. doi:10.1080/09644016.2011.538167

Trombetta, M. J. (2008). Environmental security and climate change: Analysing the discourse. *Cambridge Review of International Affairs*, *21*(4), 585–602. doi:10.1080/09557570802452920

Warner, K., Hamza, M., Oliver-Smith, A., Renaud, F., & Julca, A. (2010). Climate change, environmental degradation and migration. *Natural Hazards*, *55*(3), 689–715. doi:10.100711069-009-9419-7

ADDITIONAL READING

Campbell-Lendrum, D., & Corvalán, C. (2007). Climate change and developing-country cities: Implications for environmental health and equity. *Journal of Urban Health*, *84*(1), 109–117. doi:10.100711524-007-9170-x PMID:17393341

Dunlap, R. E., & Brulle, R. J. (Eds.). (2015). *Climate change and society: Sociological perspectives.* Oxford University Press. doi:10.1093/acprof:oso/9780199356102.001.0001

Floyd, R. (2008). The environmental security debate and its significance for climate change. *The International Spectator*, *43*(3), 51–65. doi:10.1080/03932720802280602

Gienapp, P., Teplitsky, C., Alho, J. S., Mills, J. A., & Merilä, J. (2008). Climate change and evolution: Disentangling environmental and genetic responses. *Molecular Ecology*, *17*(1), 167–178. doi:10.1111/j.1365-294X.2007.03413.x PMID:18173499

O'Connor, R. E., Bard, R. J., & Fisher, A. (1999). Risk perceptions, general environmental beliefs, and willingness to address climate change. *Risk Analysis*, *19*(3), 461–471. doi:10.1111/j.1539-6924.1999.tb00421.x

Qian, W., & Zhu, Y. (2001). Climate change in China from 1880 to 1998 and its impact on the environmental condition. *Climatic Change*, *50*(4), 419–444. doi:10.1023/A:1010673212131

Raleigh, C., & Urdal, H. (2007). Climate change, environmental degradation and armed conflict. *Political Geography*, *26*(6), 674–694. doi:10.1016/j.polgeo.2007.06.005

Schlosberg, D., & Collins, L. B. (2014). From environmental to climate justice: Climate change and the discourse of environmental justice. *Wiley Interdisciplinary Reviews: Climate Change*, *5*(3), 359–374. doi:10.1002/wcc.275

Shepardson, D. P., Niyogi, D., Roychoudhury, A., & Hirsch, A. (2012). Conceptualizing climate change in the context of a climate system: Implications for climate and environmental education. *Environmental Education Research*, *18*(3), 323–352. doi:10.1080/13504622.2011.622839

Tranter, B. (2011). Political divisions over climate change and environmental issues in Australia. *Environmental Politics*, *20*(1), 78–96. doi:10.1080/09644016.2011.538167

Trombetta, M. J. (2008). Environmental security and climate change: Analysing the discourse. *Cambridge Review of International Affairs*, *21*(4), 585–602. doi:10.1080/09557570802452920

Warner, K., Hamza, M., Oliver-Smith, A., Renaud, F., & Julca, A. (2010). Climate change, environmental degradation and migration. *Natural Hazards*, *55*(3), 689–715. doi:10.100711069-009-9419-7

KEY TERMS AND DEFINITIONS

Equity: Defined by UNEP to include intergenerational equity - "the right of future generations to enjoy a fair level of the common patrimony" - and intragenerational equity - "the right of all people within the current generation to fair access to the current generation's entitlement to the Earth's natural resources" - environmental equity considers the present generation under an obligation to account for long-term impacts of activities and to act to sustain the global environment and resource base for future generations.

Polluter pays principle: The polluter pays principle stands for the idea that "the environmental costs of economic activities, including the cost of preventing potential harm, should be internalized rather than imposed upon society at large." All issues related to responsibility for environmental remediation costs and compliance with pollution control regulations involve this principle.

Precautionary principle: One of the most commonly encountered and controversial principles of environmental law, the Rio Declaration formulated the precautionary principle: To protect the environment, the precautionary approach shall be widely applied by States according to their capabilities. Where there are threats of serious or irreversible damage, lack of complete scientific certainty shall not be used as a reason for postponing cost-effective measures to prevent environmental degradation.

Prevention: The concept of prevention can perhaps better be considered an overarching aim that gives rise to a multitude of legal mechanisms, including prior assessment of environmental harm, licensing or authorization that set out the conditions for operation and the consequences for violation of the conditions, as well as the adoption of strategies and policies.

Sustainable development: Defined by the United Nations Environment Programme as "development that meets the needs of the present without compromising the ability of future generations to meet their own needs," sustainable development may be considered together with the concepts of "integration" (development cannot be considered in isolation from sustainability) and "interdependence" (social and economic development, and environmental protection, are interdependent).

Transboundary responsibility: Defined in the international law context as an obligation to protect one's environment and prevent damage to neighboring environments, UNEP considers transboundary responsibility at the international level as a potential limitation on the sovereign state's rights. Laws that limit externalities imposed upon human health and the environment may be assessed against this principle.

Chapter 5
Legal Authorities of Environmental Law

ABSTRACT

The environment is more vulnerable to damage than any other area. Human beings of all ages and situations— childhood and old age, poverty and wealth, peace and happiness, and conflict and sorrow—can cause damage to the environment. However, most of the damage is done by governments with their wrong policies, and increased individual damages are also due to those policies. Therefore, the environment should be protected by the judiciary and arbitration. In case of intentional damage or negligence, the perpetrators will be dealt with legally, and the ways of compensation will be determined. This chapter describes domestic and international legal authorities related to the environment and can influence it with their decisions and actions.

INTRODUCTION

Environmental law is a collective term encompassing aspects of the law that protect the environment. A related but distinct set of regulatory regimes, now strongly influenced by environmental legal principles, focus on managing specific natural resources, such as forests, minerals, or fisheries. Other areas, such as environmental impact assessment, may not fit neatly into either category but are essential components of environmental law(Campbell-Lendrum & Corvalán, 2007).

Early examples of legal enactments designed to consciously preserve the environment, for its own sake or human enjoyment, are found throughout history. In the common law, the primary protection was found in the law of nuisance, but this only allowed private actions for damages or injunctions if there was harm to the land. Thus, smells emanating from pigsties, strict liability against dumping rubbish, or damage from exploding dams. Private enforcement, however, was limited and found to be woefully inadequate to deal with major environmental threats, particularly threats to shared resources. During the "Great Stink" of 1858, the dumping of sewerage into the River Thames began to smell so ghastly in the summer heat that Parliament had to be evacuated. Ironically, the Metropolitan Commission of Sewers Act 1848 had allowed the Metropolitan Commission for Sewers to close cesspits around the city in an

DOI: 10.4018/978-1-6684-4158-9.ch005

attempt to "clean up," but this led people to pollute the river. In 19 days, Parliament passed a further Act to build the London sewerage system. London also suffered from terrible air pollution, culminating in the "Great Smog" of 1952, which triggered its legislative response: the Clean Air Act 1956. The primary regulatory structure was to set limits on emissions for households and businesses (mainly burning coal) while an inspectorate would enforce compliance (Campbell-Lendrum & Corvalán, 2007; Dunlap & Brulle, 2015).

THEORY OF ENVIRONMENTAL LAW

Global and regional environmental issues are increasingly the subjects of international law. Debates over environmental concerns implicate core principles of international law and have been the subject of numerous international agreements and declarations.

Customary international law is an essential source of international environmental law. These are the norms and rules that countries follow as a matter of custom, and they are so prevalent that they bind all states in the world. When a principle becomes customary, the law is not clear-cut, and many arguments are put forward by states not wishing to be bound. Examples of customary international law relevant to the environment include the duty to warn other states promptly about icons of an environmental nature and environmental damages to which another state or states may be exposed, and Principle 21 of the Stockholm Declaration ('good neighborliness' or sic utere)(Floyd, 2008).

Given that customary international law is not static. Still, ever-evolving and the continued increase of air pollution (Carbon Dioxide) causing climate changes have led to discussions on whether basic customary principles of international law, such as the jus cogens (peremptory norms) erga omnes principles, could be applicable for enforcing international environmental law.

Numerous legally binding international agreements encompass various issues, from terrestrial, marine, and atmospheric pollution to wildlife and biodiversity protection. International environmental agreements are generally multilateral (or sometimes bilateral) treaties (a.k.a. convention, agreement, protocol, etc.). Protocols are subsidiary agreements built from a primary treaty. They exist in many areas of international law but are especially useful in the environmental field, where they may be used to incorporate current scientific knowledge regularly. They also permit countries to reach an agreement on a framework that would be contentious if every detail were to be agreed upon in advance. The most widely known protocol in international environmental law is the Kyoto Protocol, followed by the United Nations Framework Convention on Climate Change.

While the bodies that proposed, argued, agreed upon, and ultimately adopted existing international agreements vary according to each agreement, specific conferences, including 1972's United Nations Conference on the Human Environment, 1983's World Commission on Environment and Development, 1992's United Nations Conference on Environment and Development and 2002's World Summit on Sustainable Development have been significant. Multilateral environmental agreements sometimes create an International Organization, Institution, or Body responsible for implementing the agreement. Significant examples are the Convention on International Trade in Endangered Species of Wild Fauna and Flora (CITES) and the International Union for Conservation of Nature (IUCN).

International environmental law also includes the opinions of international courts and tribunals. While there are few and they have limited authority, the decisions carry much weight with legal commentators and are pretty influential on the development of international environmental law. One of the biggest challenges in international decisions is determining adequate compensation for environmental damages. The courts include the International Court of Justice (ICJ), the International Tribunal for the Law of the Sea (ITLOS), the European Court of Justice, the European Court of Human Rights, and other regional treaty tribunals(Gienapp et al., 2008).

DOMESTIC AND REGIONAL ENVIRONMENTAL LEGAL SYSTEMS

African Environmental Law

According to the International Network for Environmental Compliance and Enforcement (INECE), the major environmental issues in Africa are "drought and flooding, air pollution, deforestation, loss of biodiversity, freshwater availability, degradation of soil and vegetation, and widespread poverty." The US Environmental Protection Agency (EPA) is focused on the "growing urban and industrial pollution, water quality, electronic waste and indoor air from cookstoves." They hope to provide enough aid on concerns regarding pollution before their impacts contaminate the African environment and the global environment. By doing so, they intend to "protect human health, particularly vulnerable populations such as children and the poor." To accomplish these goals in Africa, EPA programs are focused on strengthening the ability to enforce environmental laws and public compliance. Other programs develop stronger environmental laws, regulations, and standards(Brunch et al., 2001).

Asian Environmental Law

The Asian Environmental Compliance and Enforcement Network (AECEN) is an agreement between 16 Asian countries dedicated to improving cooperation with environmental laws in Asia. These countries include Cambodia, China, Indonesia, India, Maldives, Japan, Korea, Malaysia, Nepal, Philippines, Pakistan, Singapore, Sri Lanka, Thailand, Vietnam, and Lao PDR(Boer, 1998).

EU Environmental Law

The European Union issues secondary legislation on environmental issues that are valid throughout the EU (so-called regulations) and many directives that must be implemented into national legislation from the 28 member states (national states). Examples are the Regulation (EC) No. 338/97 on the implementation of CITES; or the Natura 2000 network the centerpiece for nature & biodiversity policy, encompassing the bird Directive (79/409/EEC/ changed to 2009/147/EC)and the habitats directive (92/43/EEC), which are made up of multiple SACs (Special Areas of Conservation, linked to the habitats directive) & SPAs (Special Protected Areas, linked to the birds directive) throughout Europe(Van Calster & Reins, 2017).

EU legislation is ruled in Article 249 Treaty for the Functioning of the European Union (TFEU). Topics for common EU legislation are:

* Climate change
* Air pollution
* Water protection and management
* Waste management

- Soil protection
- Protection of nature, species, and biodiversity
- Noise pollution
- Cooperation for the environment with third countries (other than EU member states)
- Civil protection

MENA Environmental Law

Environmental law is rapidly growing in the Middle East. The US Environmental Protection Agency works with countries in the Middle East to improve "environmental governance, water pollution, water security, clean fuels and vehicles, public participation, and pollution prevention(Samimi et al., 2012)."

Oceanian Environmental Law

The main concerns about environmental issues in Oceania are "illegal releases of air and water pollutants, illegal logging/timber trade, illegal shipment of hazardous wastes, including e-waste and ships slated for destruction, and insufficient institutional structure/lack of enforcement capacity." The Secretariat of the Pacific Regional Environmental Programme (SPREP) is an international organization between Australia, the Cook Islands, FMS, Fiji, France, Kiribati, Marshall Islands, Nauru, New Zealand, Niue, Palau, PNG, Samoa, Solomon Island, Tonga, Tuvalu, USA, and Vanuatu. The SPREP was established to improve and protect the environment and assure sustainable development for future generations(Clarke, 2008).

Australian Environmental Law

Commonwealth v Tasmania (1983), also known as the "Tasmanian Dam Case," was a highly significant case in Australian environmental law. The Environment Protection and Biodiversity Conservation Act 1999 is the centerpiece of environmental legislation in Australia. It sets up the "legal framework to protect and manage nationally and internationally important flora, fauna, ecological communities and heritage places" and focuses on protecting world heritage properties, national heritage properties, wetlands of international importance, nationally threatened species and ecological communities, migratory species, Commonwealth marine areas, Great Barrier Reef Marine Park, and the environment surrounding nuclear activities. However, it has been subject to numerous reviews examining its shortcomings, the latest in mid-2020. The interim report of this review concluded that the laws created to protect unique species and habitats are ineffective(Fisher, 2014).

Brazilian Environmental Law

The Brazilian government created the Ministry of Environment in 1992 to develop better strategies for protecting the environment, using natural resources sustainably, and enforcing public environmental policies. The Ministry of Environment has authority over policies involving the environment, water resources, preservation, and environmental programs involving the Amazon(de Aguiar Patriota, 2008).

Canadian Environmental Law

The Department of the Environment Act establishes the Department of the Environment in the Canadian government and the Minister of the Environment. Their duties include "the preservation and enhancement of the quality of the natural environment, including water, air, and soil quality; renewable resources, including migratory birds and other non-domestic flora and fauna; water; meteorology;" The Environmental Protection Act is the main piece of Canadian environmental legislation that was put into place 31 March 2000. The Act focuses on "respecting pollution prevention and the protection of the environment and human health to contribute to sustainable development." Other federal statutes include the Canadian Environmental Assessment Act and the Species at Risk Act. When local and federal legislation conflict, federal legislation takes precedence. Individual provinces can have legislation, such as Ontario's Environmental Bill of Rights and Clean Water Act(Wood et al., 2010).

Chinese Environmental Law

According to the US Environmental Protection Agency, "China has been working with great determination in recent years to develop, implement, and enforce a solid environmental law framework. Chinese officials face critical challenges in effectively implementing the laws, clarifying the roles of their national and provincial governments, and strengthening the operation of their legal system." Explosive economic and industrial growth in China has led to significant environmental degradation, and China is currently developing more stringent legal controls. The harmonization of Chinese society and the natural environment is billed as a rising policy priority(Van Rooij, 2006).

Congolese Environmental Law

In the Republic of Congo, inspired by the African models of the 1990s, the constitutionalization of environmental law appeared in 1992, which completed a historical development of environmental law and policy dating back to the years of independence and even long before the colonization. It gives a constitutional basis to environmental protection, which traditionally was part of the legal framework. The two Constitutions of 15 March 1992 and 20 January 2002 concretize this paradigm by stating a legal obligation of a clean environment by establishing a compensation principle and a criminal nature foundation. By this phenomenon, Congolese environmental law is situated between non-regression and the search for efficiency(Kennedy, 2012)."

Ecuadorian Environmental Law

With the enactment of the 2008 Constitution, Ecuador became the first country to codify the Rights of Nature. The constitution, specifically Articles 10 and 71–74, recognizes the inalienable rights of ecosystems to exist and flourish, gives people the authority to petition on behalf of ecosystems, and requires the government to remedy violations of these rights. The rights approach is a break away from traditional environmental regulatory systems, which regard nature as property and legalize and manage degradation of the environment rather than prevent it. The Rights of Nature articles in Ecuador's constitution are part of a reaction to a combination of political, economic, and social phenomena. Ecuador's abusive past with the oil industry, most famously the class-action litigation against Chevron, and the failure of an extraction-based economy and neoliberal reforms to bring economic prosperity to the region has resulted in the election of a New Leftist regime, led by President Rafael Correa, and sparked a demand for new approaches to development. In conjunction with this need, the principle of "Buen Vivir," or good living—focused on social, environmental, and spiritual wealth versus material wealth—gained popularity among citizens and was incorporated into the new constitution. The influence of indigenous groups, from whom the concept of "Buen Vivir" originates, informing the constitutional ideals, also facilitated the incorporation of the Rights of Nature as a basic tenet of their culture and conceptualization of "Buen Vivir(Kimerling, 1990; Shohani et al., 2021a)."

Egyptian Environmental Law

The Environmental Protection Law outlines the responsibilities of the Egyptian government to "preparation of draft legislation and decrees pertinent to environmental management, collection of data both nationally and internationally on the state of the environment, preparation of periodical reports and studies on the state of the environment, formulation of the national plan and its projects, preparation of environmental profiles for new and urban areas, and setting of standards to be used in planning for their development, and preparation of an annual report on the state of the environment to be prepared to the President(Frihy, 2001)."

Indian Environmental Law

In India, Environmental law is governed by the Environment Protection Act, 1986. This act is enforced by the Central Pollution Control Board and the numerous State Pollution Control Boards. Apart from this, there are also individual legislations enacted explicitly for the protection of Water, Air, Wildlife, etc. Such legislation includes(Nomani & Hussain, 2020):

- The Water (Prevention and Control of Pollution) Act, 1974
- The Water (Prevention and Control of Pollution) Cess Act, 1977
- The Forest (Conservation) Act, 1980
- The Air (Prevention and Control of Pollution) Act, 1981
- Air (Prevention and Control of Pollution) (Union Territories) Rules, 1983
- The Biological Diversity Act, 2002 and the Wild Life Protection Act, 1972
- Batteries (Management and Handling) Rules, 2001
- Recycled Plastics, Plastics Manufacture and Usage Rules, 1999
- The National Green Tribunal established under the National Green Tribunal Act of 2010 has jurisdiction over all environmental cases dealing with a substantial environmental question and acts covered under the Water (Prevention and Control of Pollution) Act, 1974.
- Water (Prevention and Control of Pollution) Cess Rules, 1978
- Ganga Action Plan, 1986
- The Forest (Conservation) Act, 1980
- Wildlife protection Act, 1972
- The Public Liability Insurance Act, 1991 and the Biological Diversity Act, 2002. The acts covered under Indian Wild Life Protection Act 1972 do not fall within the jurisdiction of the National Green Tribunal. Appeals can be filed in the Hon'ble Supreme Court of India.

- Basel Convention on Control of TransboundaryMovements on Hazardous Wastes and Their Disposal, 1989 and Its Protocols
- Hazardous Wastes (Management and Handling) Amendment Rules, 2003

Japanese Environmental Law

The Basic Environmental Law is the basic structure of Japan's environmental policies replacing the Basic Law for Environmental Pollution Control and the Nature Conservation Law. The updated law aims to address "global environmental problems, urban pollution by everyday life, loss of accessible natural environment in urban areas and degrading environmental protection capacity in forests and farmlands." The three basic environmental principles that the Basic Environmental Law follows are "the blessings of the environment should be enjoyed by the present generation and succeeded to the future generations, a sustainable society should be created where environmental loads by human activities are minimized, and Japan should contribute actively to global environmental conservation through international cooperation." From these principles, the Japanese government have established policies such as "environmental consideration in policy formulation, establishment of the Basic Environment Plan which describes the directions of long-term environmental policy, environmental impact assessment for development projects, economic measures to encourage activities for reducing environmental load, improvement of social infrastructure such as sewerage system, transport facilities etc., promotion of environmental activities by corporations, citizens and NGOs, environmental education, and provision of information, promotion of science and technology(Kumamoto, 1989)."

New Zealand Environmental Law

The Ministry for the Environment and Office of the Parliamentary Commissioner for the Environment was established by the Environment Act 1986. These positions are responsible for advising the Minister on all areas of environmental legislation. A common theme of New Zealand's environmental legislation is sustainably managing natural and physical resources, fisheries, and forests. The Resource Management Act 1991 is the main piece of environmental legislation that outlines the government's strategy to managing the "environment, including air, water soil, biodiversity, the coastal environment, noise, subdivision, and land use planning in general(Salmon & Grinlinton, 2015)."

Russian Environmental Law

The Ministry of Natural Resources and Environment of the Russian Federation makes regulations regarding "conservation of natural resources, including the subsoil, water bodies, forests located in designated conservation areas, fauna and their habitat, in the field of hunting, hydrometeorology and related areas, environmental monitoring and pollution control, including radiation monitoring and control, and functions of public environmental policymaking and implementation and statutory regulation(Sofronova et al., 2014)."

Singaporean Environmental Law

Singapore is a signatory of the Convention on Biological Diversity. Most of its CBD obligations are overseen by the National Biodiversity Reference Centre, its National Parks Board (NParks) division. Singapore is also a signatory of the Convention on International Trade in Endangered Animals, with its obligations under that treaty being overseen by NParks. The Parliament of Singapore has enacted numerous legislation to fulfill its obligations under these treaties, such as the Parks and Trees Act, Endangered Species (Import and Export) Act, and Wildlife Act. The new Wildlife (Protected Wildlife Species) Rules 2020 marks the first instance in Singapore's history that direct legal protection has been offered for specifically named species, as listed in Parts 1-5 of the Rules' schedule(Fowler, 2020).

South African Environmental Law

South African environmental law describes the legal rules in South Africa relating to the social, economic, philosophical, and jurisprudential issues raised by attempts to protect and conserve the environment in South Africa. South African environmental law encompasses natural resource conservation and utilization and land-use planning and development. Issues of enforcement are also considered, together with the international dimension, which has shaped much of the direction of environmental law in South Africa. The role of the country's constitution, crucial to any understanding of environmental law application, is also examined. The National Environmental Management Act (NEMA) provides the underlying framework for environmental law(Henderson, 2001).

UK Environmental Law

United Kingdom environmental law concerns the protection of the environment in the United Kingdom. Environmental law is increasingly a European and international issue due to the cross-border issues of air and water pollution and manufactured climate change.

In the common law, the primary protection was found in the tort of nuisance, but this only allowed private actions for damages or injunctions if there was harm to the land. Thus smells emanating from pigsties, strict liability against dumping rubbish, or damage from exploding dams. Private enforcement, however, was limited and found to be woefully inadequate to deal with major environmental threats, particularly threats to common resources.

- 1306, Edward I briefly banned coal fires in London.
- John Evelyn, Fumifugium (1661) argued for burning fragrant wood instead of mineral coal, which he believed would reduce coughing.
- Ballad of Gresham College (1661) describes how the smoke "does our lungs and spirits choke, Our hanging spoil, and rust our iron."
- In 1800, one million tons of coal were burned in London, and 15 million across the UK.
- Smoke Nuisance Abatement (Metropolis) Act 1853
- John Snow, in 1854, discovered that the water pump on Broad Street, Soho was responsible for 616 cholera deaths because it was contaminated by an old cesspit leaking fecal bacteria. The germ theory of disease began to replace the miasma theory that had lingered since the Black Death.

During the "Great Stink" of 1858, the dumping of sewerage into the River Thames began to smell so ghastly in the summer heat that Parliament had to be evacuated. The Metropolitan Commission of Sewers Act 1848 had allowed the Metropolitan Commission for Sewers to close cesspits around the city in an attempt to "clean up," but this simply led people to pollute the river. In 19 days, Parliament passed a further Act to build the London sewerage system.

- Alkali Act 1863 and Alkali Act 1874, amended 1906
- WS Jevons, The Coal Question; An Inquiry Concerning the Progress of the Nation, and the Probable Exhaustion of Our Coal Mines (1865)
- Ground Game Act 1880, Night Poaching Act 1828, Game Act 1831, game preservation
- James Johnston (socialist politician), president of the Smoke Abatement League, an international conference in 1911.

London also suffered from terrible air pollution, culminating in the "Great Smog" of 1952, which triggered a legislative response: the Clean Air Act 1956. The primary regulatory structure was to set limits on emissions for households and businesses (mainly burning coal) while an inspectorate would enforce compliance. It required zones for smokeless fuel to be burned and relocated power stations(Reid, 2016).

- Clean Air Act 1968 required tall chimneys to disperse pollution.

United States Environmental Law

United States environmental law concerns legal standards to protect human health and improve the natural environment of the United States. While subject to criticism at home and abroad on protection, enforcement, and over-regulation, the country remains an important source of environmental legal expertise and experience. The United States Congress has enacted federal statutes intended to address pollution control and remediation, including, for example, the Clean Air Act (air pollution), the Clean Water Act (water pollution), and the Comprehensive Environmental Response, Compensation, and Liability Act (CERCLA, or Superfund) (contaminated site cleanup). There are also federal laws governing natural resources use and biodiversity which are strongly influenced by environmental principles, including the Endangered Species Act, National Forest Management Act, and Coastal Zone Management Act. The National Environmental Policy Act, governing environmental impact review in actions undertaken or approved by the US federal government, may implicate all of these areas. Federalism in the United States has played a role in national environmental legislation. Many federal environmental laws employ cooperative federalism mechanisms - many federal regulatory programs are administered in coordination with the US states. Furthermore, the states have enacted laws to cover areas not preempted by federal law. This includes areas where Congress had acted in a limited fashion (e.g., state site cleanup laws to handle sites outside Superfund) and where Congress has left regulation primarily to the states (e.g., water resources law)(Lazarus, 2001).

Vietnamese Environmental Law

Vietnam is currently working with the US Environmental Protection Agency on dioxin remediation and technical assistance to lower methane emissions. In March 2002, the US and Vietnam signed the U.S.-

Vietnam Memorandum of Understanding on Research on Human Health and the Environmental Effects of Agent Orange/Dioxin(Dung et al., 2021).

Iranian Environmental Law

The only legislative authority in Iran is the Islamic Consultative Assembly, whose resolutions can be implemented after the approval of the Guardian Council. Of course, in the event of a dispute between these two institutions, the opinion of the Expediency Council regarding the submitted resolution is correct. But some authorities implement the law to issue regulations (instructions) in the field of environment.

It should be noted that these authorities, which are involved in environmental protection and public culture, contribute to the formation of Iranian environmental law. These organizations and centers include the Ministry of Energy, the Environmental Protection Organization, the Land Affairs Organization (Agricultural Jihad), the Forests and Rangelands Organization, the Municipal Environment Department, the Ports, and the Maritime Organization, the Shipping Organization, and the National Oceanographic Center.

Considering that the introduction of the Islamic Consultative Assembly is in constitutional law, other legal authorities are introduced.

Its tasks include protection of the country's natural ecosystems and repair of past adverse effects on the environment, prevention, and prevention of environmental degradation and pollution, assessment of the tolerable capacity of the environment for reasonable and continuous efficiency of environmental resources, monitoring Continuous use of environmental resources, active response to critical environmental issues, including pollution beyond the tolerable capacity of the environment.

This organization conducts studies and research in environmental protection in the following cases.

- Environmental pollutants in water, air, soil, waste, pesticides, chemical fertilizers, noise, and the like.
- Areas with special and unique ecological characteristics (ecosystem) determine their boundaries.
- How to establish the country's development phenomena such as industrial units, power plants, dams, agricultural complexes, development, human settlements, and the like.
- Use of environmentally friendly technology.
- Environmental education to disseminate and improve the level of environmental knowledge and insight of the community to create interest, sense of responsibility, and public participation of the people in environmental protection with the cooperation of educational and research centers and mass media, NGOs, and all Facilities within the country and internationally.
- Economic valuation of natural resources and environmental costs of development.
- Study and layout to achieve sustainable development.

The Environment Organization has no role in providing services such as energy supply, including oil, gasoline, water, electricity, but according to the monitoring tools at its disposal can achieve the right to a healthy environment for citizens and Iranians, and all human beings to enjoy a healthy environment. Play a significant role. Of course, the essential tool in achieving this goal is to create a culture and educate officials and people and remind them of environmental crises so that everyone can take the most crucial problem in the world, namely disregard for the environment, and meet environmental requirements.

Because it provides water and electricity to the country (the most crucial energy consumption of the people), the Ministry of Energy will introduce it in the institutions and organizations related to the environment. This ministry can play a significant role in achieving environmental goals: maintaining a healthy environment for current and future generations by adequately implementing laws and adopting policies in line with environmental protection and sustainable development (Norouzi et al., 2021a; Shohani et al., 2021b; Norouzi et al., 2021b; Norouzi, 2022; Norouzi & Ataei, 2021).

INTERNATIONAL ENVIRONMENTAL LAW

International Court of Justice

The International Court of Justice is one of the pillars of the Charter of the United Nations, which operates under its statute, which is part of the Charter. The court was established in 1946 as the successor to the Permanent Court of International Justice, although unlike the Permanent Court, which was outside the system of the United Nations, the International Court of Justice is the primary judicial body of the United Nations. In general, the court's task is to resolve disputes between states following international law and to issue advisory opinions (opinions) on legal matters at the request of the General Assembly, the Security Council, and other organs of the United Nations and specialized agencies.

It is important to note that the United Nations members accept the Statute of the Court, but this does not mean that States are required to refer to the court to resolve their disputes. Instead, they must first, following Article 36 of the Statute of the Court, the "condition of arbitration" or the agreement to refer to the court, declare their subsequent or prior consent to refer the dispute to the court.

"The International Court of Justice has taken important steps, directly and indirectly, in the development of international environmental law in some interlocutory and consultative cases." The court is the principal judicial organ of the United Nations, which, like other organs of the organization, participates in establishing and maintaining international peace and security. For this purpose, it is obliged to apply the existing legal rules following Article 38 of its Statute(Gienapp et al., 2008)."

Environmental Cases From the International Court of Justice

In this section, among the various advocacy and advisory cases dealt with by the International Court of Justice, those that refer directly to environmental protection or those related to one of the environmental issues are mentioned.

Nuclear Tests (New Zealand and Australia vs. France) 1947

Following the French nuclear test on 20 December 1974, the governments of Australia and New Zealand each filed a complaint against the country with the International Court of Justice over the release of radioactive material from the French nuclear test into their territories. These experiments were performed. France, on the other hand, objected to the court's jurisdiction. The court issued an interim injunction on 22 June 1973, urging France to refrain from further testing. However, the court did not enter into a substantive hearing due to France's unilateral announcement that the tests had been suspended. The court ruled with nine votes that Australia's claim was irrelevant and therefore did not need to be decided. Australia's ultimate goal was to halt France's nuclear tests in the South Pacific and France to suspend its intentions. Such experiments have been announced. According to the ruling of the Arbitral Tribunal in

Smelter's case, no country has the right to use its territory in any way or allow another to cause unusual damage to a neighboring country. In addition, the International Court of Justice's rulings, such as the Corfu Channel rulings and the nuclear test claims, are set out in Article 21 of the Stockholm Declaration on the Human Environment(O'Connor et al., 1999).

Mills on the Uruguay River (Argentina and Uruguay) 2010

4 May 2006 Argentina has filed a lawsuit against Uruguay in the International Court of Justice. Argentina accused the Uruguayan government of violating its obligations under the Uruguay River Charter in its lawsuit. In this case, Argentina claims that Uruguay violated the Uruguay River Charter on 26 February 1975 by building two mills on the banks of the Uruguay River at the entrance to a city in Argentina, causing damage to the environment of the Uruguay River and its coastal region. This action has caused damage to his country. Following a formal and substantive examination of the case, the court ruled on 20 April 2010: the obligation of the parties to cooperate internationally, to negotiate for the protection of the environment; Although the Court condemned the Uruguayan government for violating formalities for informing the Argentine government, it did not, in essence, rule in favor of the Uruguayan government for failing to cause environmental damage(Schlosberg & Collins, 2014).

Aerial spraying (Ecuador vs. Colombia) 2013

The Ecuadorian government filed a lawsuit with the Colombian government, alleging that it violated its obligations under international environmental law. The Ecuadorian government has claimed that Colombia, by aerial spraying of farms in border areas, has caused damage to people, crops, animals, and the natural environment and posed a more significant threat to future generations of citizens.

In its petition, the Ecuadorian government claims that the indigenous communities where the toxins are distributed live according to ancient traditions and are deeply rooted in the environment and extreme poverty of agricultural materials such as corn, coffee, and other products for their survival. They need to be dependent, which deepens their connection to the earth. On the other hand, medical care in these areas is primary. Formal education is low due to the Colombian government's measures to disinfect this vulnerable area, posing significant risks to the region's human and natural environment. The Ecuadorian government also invoked the "American Treaty for the Peaceful Settlement of International Disputes" signed in 1948 by the governments of Ecuador and Colombia to prove the jurisdiction of the International Court of Justice. Ecuador's arguments, therefore, were based on the court's assertion that it violated the rules of customary and contractual international law, which oblige states not to use the land in a way that does not harm other states. Finally, the International Court of Justice (ICJ) dismissed the case on 13 September 2013 after the two governments agreed on 9 September 2013 and announced that the Colombian government would not continue air spraying to affect Ecuadorian territory.

Whaling (Australia vs. Japan) 2014

"This is one of the issues that has somehow established the sovereignty of the environment and can be a basis for further development of international environmental law and another opportunity for the participation of dispute resolution authorities in the evolution of international environmental law." The Government of Australia has filed a petition with the International Court of Justice on behalf of the Government of Japan according to paragraphs 1 and 2 of Articles 36, 38, and 40 of the Statute of the Court. The Australian government claims that the Government of Japan, in continuation of the long-term whaling program related to the second phase of the Japan Research Program under the Antarctic Special Permit, fulfills its obligations under the International Covenant on Whaling Protection of Marine Mammals and the environment violated marine life. In its petition, the Australian government stated that in 1982, the International Whaling Commission approved the suspension of whaling for commercial purposes according to Article 5 paragraph 1 (e) of the convention. The Australian government has repeatedly asked Japan to abandon the program or to do a substantial review. In response to these requests, Japan stated that it was fully aware of the international community's strong response to its whaling science program, particularly the "humpback whale," but that its purpose was to research appropriate and appropriate whaling management tools. It has been related to international treaties. Japan has stated that while it will not change its research plans, it will postpone hunting back whales until the "normal" process at the International Whaling Commission progresses. The Australian government has applied to the court for compensation by the Government of Japan. The court ruled that Japan had "violated" several articles of the Whaling Convention and should refrain from issuing a license to its research center to kill whales.

Coastal Road Construction (Nicaragua vs. Costa Rica) 2015

The Nicaraguan government has filed a lawsuit against the Costa Rican government in the International Court of Justice. In this case, the Nicaraguan government stated that the Costa Rican government was preparing to destroy the region's environment by constructing a road parallel and very close to the southern shore of the San Juan border river. The implementation of this project has led to the discharge and release of a significant volume of sediments such as uprooted vegetation soil and cut down trees into the river, which these sediments have been created as a result of clearing and leveling the land. In addition, cutting down trees and harvesting topsoil and vegetation near the riverbank facilitates erosion and leaching of more significant amounts of sediment into the river. River sediments have also posed an imminent threat to the quality of aquatic life and rare species and the diversity of animals and plants in the border river. The Nicaraguan government's legal arguments are based on the principle that under customary international law, states should not cause harm to neighboring countries and, under common international law, should commit to activities that risk harming the territory of a neighboring country. Should inform and consult with them and commit to environmental impact assessments where these activities are likely to have significant cross-border side effects. Nicaragua, therefore, believes that the Government of Costa Rica's customary international law and bilateral and multilateral treaties, including the Ramsar Convention 1971, the UNESCO World Heritage Convention 1972, the United Nations Conference on the Human Environment Stockholm Convention 1972 Ecology And violated the 1992 Convention on Biological Diversity and Conservation of Central Wildlife Areas in Central America, and is seeking compensation from the guilty government. The International Court of Justice (ICJ) ruled on 16 December 2015 that Costa Rica posed a severe threat to the region's environment by launching a road construction project along the Nicaraguan border. Costa Rica has therefore not complied with its obligations under public international law regarding the need for environmental assessment of projects. The court also concluded that Nicaragua's claim for compensation from Costa Rica was a declaration of Costa Rica's acceptance of the conduct concerning the environmental assessment to obtain Nicaragua's consent (compensation).

Arbitration Case of Trail Smelter (the USA and Canada) 1941

Trail Smalter's arbitration case is considered one of the most critical and influential votes in the macro-environment. "In terms of the reputation of the case, it is enough that, according to Pro Taylor, it can be considered a proverb of international environmental law." This issue was also mentioned in the principle of sovereignty that the territory of countries should not be a factor in causing damage against other countries. In the 1920s, parts of the United States were affected by transboundary pollution, in the case of the Trail smelter. A zinc smelter in Trail, Canada, 10 miles from the border between British Columbia and Washington, DC, emitted "sulfur dioxide" into the United States, damaging crops and property. Under that law, US citizens and private entrepreneurs tried to sue the smelter (a joint venture between Canadian mining and smelting limited liability companies). Still, both US and Canadian legal systems made it impossible at the time. So they left the matter to the United States to turn it into an international dispute against Canada to provide diplomatic support to its citizens. In this regard, both countries initially resorted to their sovereignty. Canada, under its sovereignty, had authorized legal activity in its territory, and the United States, citing its right to territorial integrity, declared that it should not tolerate harmful Canadian interference(Shepardson et al., 2012).

They first referred the dispute to the Dispute Settlement Commission under the US-Canada Border Water Treaty, which in 1931 ordered the payment of $350,000. Both sides rejected the ruling and agreed to set up an arbitral tribunal to resolve the dispute. The Arbitral Tribunal issued its interim ruling in 1938 and its final ruling in 1941.

The arbitrators found that international law could not resolve the dispute in its due time. Of course, the right of sovereignty protects both the right of a country to pursue legal activities in its territory and the territorial integrity of the state against pollution from other countries. In this case, both countries had resorted to the principle of sovereignty. But the judges applied an ancient Roman principle that they thought was also applicable in intergovernmental disputes: use your property but do not damage the property of others.

Therefore, the arbitral tribunal found that according to international law and American law principles, no country has the right to use its territory in any way when this situation has dangerous consequences. The resulting damage is proven by clear and convincing evidence of damage to another country's land or property and persons due to smoke and dust(Qian & Zhu, 2001).

The arbitral tribunal held the Canadian government liable for the conduct of the Trail Smelter plant, following the principles of international law to comply with its obligations under international law. This reflects the principle of liability of States, which in the field of international environmental law is explicitly enshrined in the 1993 Convention on Civil Liability for Environmental Hazardous Activities (Lugano), which contains absolute liability (whether with or without fault)(Raleigh & Urdal, 2007).

It is also necessary to mention the Court of Justice of the European Union, which is active in the environment. By promoting a culture of human rights, victims seek recourse to mechanisms and human rights dispute resolution bodies to address the effects of climate change. Petitions have been filed with several international commissions regarding the devastating effects of climate change on indigenous peoples or protected areas. In December 2005, for example, the head of the Inuit ethnic group filed a lawsuit against the US government with the Inter-American Commission on Human Rights, the governing body of the Organization of American States, on behalf of himself and other members of that community in the United States and Canada. The lawsuit alleges that the United States violated human rights by destroying the North Pole, primarily due to its failure to limit its greenhouse gas emissions, because the United States did not accept the court's jurisdiction. He could not process the lawsuit. However, he invited the plaintiffs to attend a public hearing in which at least (theoretically) the joint responsibility of the states was raised. Human rights proceedings in cases involving climate change have only recently begun, and this trend is likely to continue (Trombetta, 2008; Tranter, 2011; Warner et al., 2010).

CONCLUSION

This chapter expressed environmental-related opinions and rulings in the domestic and international spheres resulting from arbitration and justice. Several well-known and critical environmental issues were explained to understand the role and position of the environment in the issued verdicts and judicial procedure. Each case has its environmental law model or concept, which helps resolve disputes. On the other hand, it should be said that the dispute resolution authority, depending on its position, the subject matter of the dispute, the scientific position of the judges or judges, plays a role in environmental law and creates judicial procedure. Like the important and very influential case of Smelter arbitration, which affected the principle of sovereignty to such an extent that it somehow limited it to the extent that states did not, as before, see themselves as absolute rulers with full authority in exploiting their territory to the detriment of others.

REFERENCES

Boer, B. (1998). The Rise of Environmental Law in the Asian Region. *U. Rich. L. Rev.*, *32*, 1503.

Brunch, C., Coker, W., & VanArsdale, C. (2001). Constitutional environmental law: Giving force to fundamental principles in Africa. *Colum. J. Envtl. L.*, *26*, 131.

Campbell-Lendrum, D., & Corvalán, C. (2007). Climate change and developing-country cities: Implications for environmental health and equity. *Journal of Urban Health*, *84*(1), 109–117. doi:10.100711524-007-9170-x PMID:17393341

Clarke, P. (2008). Supporting conservation and sustainable livelihoods in the Pacific: The IUCN Regional Environmental Law Programme for Oceania 2007-2008. *NATIONAL ENVIRONMENTAL LAW REVIEW*, (2), 49–55.

de Aguiar Patriota, A. (2008). Introduction to Brazilian Environmental Law. *Geo. Wash. Int'l L. Rev.*, *40*, 611.

Dung, M. T., Khoa, N. M., & Thi Thu Huong, P. (2021). Integrating Environmental Requirements into Vietnamese Sectoral Laws: Some Legal Issues. *Environmental Policy and Law*, *51*(5), 343–350. doi:10.3233/EPL-201010

Dunlap, R. E., & Brulle, R. J. (Eds.). (2015). *Climate change and society: Sociological perspectives*. Oxford University Press. doi:10.1093/acprof:oso/9780199356102.001.0001

Fisher, D. (2014). *Australian environmental law: norms, principles and rules*. Thomson Reuters Professional (Austalia) Limited.

Floyd, R. (2008). The environmental security debate and its significance for climate change. *The International Spectator*, *43*(3), 51–65. doi:10.1080/03932720802280602

Fowler, R. (2020). Environmental Law in Singapore, written by Joseph Chun and Lye Lin Heng. *Chinese Journal of Environmental Law*, *4*(1), 111–114. doi:10.1163/24686042-12340052

Frihy, O. E. (2001). The necessity of environmental impact assessment (EIA) in implementing coastal projects: Lessons learned from the Egyptian Mediterranean Coast. *Ocean and Coastal Management*, *44*(7-8), 489–516. doi:10.1016/S0964-5691(01)00062-X

Gienapp, P., Teplitsky, C., Alho, J. S., Mills, J. A., & Merilä, J. (2008). Climate change and evolution: Disentangling environmental and genetic responses. *Molecular Ecology*, *17*(1), 167–178. doi:10.1111/j.1365-294X.2007.03413.x PMID:18173499

Henderson, P. G. (2001). Some thoughts on distinctive principles of South African environmental law. *South African Journal of Environmental Law and Policy*, *8*(2), 139–184.

Kennedy, K. (2012). The environmental Law Framework of the Democratic Republic of the Congo and the Balancing of Interests. *The balancing of interests in environmental law in Africa*.

Kimerling, J. (1990). Disregarding environmental law: Petroleum development in protected natural areas and indigenous homelands in the Ecuadorian Amazon. *Hastings Int'l & Comp. L. Rev.*, *14*, 849.

Kumamoto, N. (1989). Japanese environmental law and ocean resources. *Ecology Law Quarterly*, *16*, 267.

Lazarus, R. J. (2001). The Greening of America and the Graying of United States Environmental Law: Reflections on Environmental Law's First Three Decades in the United States. *Virginia Environmental Law Journal*, 75-106.

Nomani, M. Z. M., & Hussain, Z. (2020). Innovation technology in health care management in the context of Indian environmental planning and sustainable development. *International journal on emerging technologies, 11*(2), 560-564.

Norouzi, N. (2022). A Practical and Analytic View on Legal Framework of Circular Economics as One of the Recent Economic Law Insights: A Comparative Legal Study. *Circular Economy and Sustainability,* 1-26.

Norouzi, N., & Ataei, E. (2021). Covid-19 Crisis and Environmental law: Opportunities and challenges. *Hasanuddin Law Review*, *7*(1), 46–60. doi:10.20956/halrev.v7i1.2772

Norouzi, N., Sheikhi, M., Jafari, M., Kalantari, S., Narani, S. V., & Shaebani, A. (2021a). Green Victimology View in Iranian Criminology System. *Research Journal of Ecology and Environmental Sciences*, *1*(2), 82–95.

Norouzi, N., Sheikhi, M., Jafari, M., Kalantari, S., Narani, S. V., & Shaebani, A. (2021b). Criminal Legislative Policy in the Protection of Water Resources with Regard to International Treaties: A case for Iranian Legal System. *Universal Journal of Social Sciences and Humanities*, *1*(1), 67–79.

O'Connor, R. E., Bard, R. J., & Fisher, A. (1999). Risk perceptions, general environmental beliefs, and willingness to address climate change. *Risk Analysis*, *19*(3), 461–471. doi:10.1111/j.1539-6924.1999.tb00421.x

Qian, W., & Zhu, Y. (2001). Climate change in China from 1880 to 1998 and its impact on the environmental condition. *Climatic Change*, *50*(4), 419–444. doi:10.1023/A:1010673212131

Raleigh, C., & Urdal, H. (2007). Climate change, environmental degradation and armed conflict. *Political Geography*, *26*(6), 674–694. doi:10.1016/j.polgeo.2007.06.005

Reid, C. T. (2016). Brexit and the future of UK environmental law. *Journal of Energy & Natural Resources Law*, *34*(4), 407–415. doi:10.1080/02646811.2016.1218133

Salmon, P., & Grinlinton, D. P. (Eds.). (2015). *Environmental Law in New Zealand*. Thomson Reuters.

Samimi, A. J., Ahmadpour, M., & Ghaderi, S. (2012). Governance and environmental degradation in MENA region. *Procedia: Social and Behavioral Sciences*, *62*, 503–507. doi:10.1016/j.sbspro.2012.09.082

Schlosberg, D., & Collins, L. B. (2014). From environmental to climate justice: Climate change and the discourse of environmental justice. *Wiley Interdisciplinary Reviews: Climate Change*, *5*(3), 359–374. doi:10.1002/wcc.275

Shepardson, D. P., Niyogi, D., Roychoudhury, A., & Hirsch, A. (2012). Conceptualizing climate change in the context of a climate system: Implications for climate and environmental education. *Environmental Education Research*, *18*(3), 323–352. doi:10.1080/13504622.2011.622839

Shohani, A., Ataei, E., & Norouzi, N. (2021a). Environmental Constitutionalism in Latin America. *Universal Journal of Social Sciences and Humanities*, *1*(1), 54–66.

Shohani, A., Ataei, E., & Norouzi, N. (2021b). The Extent of the Researcher's Liability for Environmental Damage Caused by Academic Research. *Research Journal of Ecology and Environmental Sciences*, *1*(2), 71–81.

Sofronova, E., Holley, C., & Nagarajan, V. (2014). Environmental non-governmental organizations and Russian environmental governance: Accountability, participation and collaboration. *Transnational Environmental Law*, *3*(2), 341–371. doi:10.1017/S2047102514000090

Tranter, B. (2011). Political divisions over climate change and environmental issues in Australia. *Environmental Politics*, *20*(1), 78–96. doi:10.1080/09644016.2011.538167

Trombetta, M. J. (2008). Environmental security and climate change: Analysing the discourse. *Cambridge Review of International Affairs*, *21*(4), 585–602. doi:10.1080/09557570802452920

Van Calster, G., & Reins, L. (2017). *EU environmental law*. Edward Elgar Publishing. doi:10.4337/9781782549185

Van Rooij, B. (2006). Implementation of Chinese environmental law: Regular enforcement and political campaigns. *Development and Change*, *37*(1), 57–74. doi:10.1111/j.0012-155X.2006.00469.x

Warner, K., Hamza, M., Oliver-Smith, A., Renaud, F., & Julca, A. (2010). Climate change, environmental degradation and migration. *Natural Hazards, 55*(3), 689–715. doi:10.100711069-009-9419-7

Wood, S., Tanner, G., & Richardson, B. J. (2010). What ever happened to Canadian environmental law. *Ecology Law Quarterly, 37*, 981.

ADDITIONAL READING

Campbell-Lendrum, D., & Corvalán, C. (2007). Climate change and developing-country cities: Implications for environmental health and equity. *Journal of Urban Health, 84*(1), 109–117. doi:10.100711524-007-9170-x PMID:17393341

Dunlap, R. E., & Brulle, R. J. (Eds.). (2015). *Climate change and society: Sociological perspectives.* Oxford University Press. doi:10.1093/acprof:oso/9780199356102.001.0001

Floyd, R. (2008). The environmental security debate and its significance for climate change. *The International Spectator, 43*(3), 51–65. doi:10.1080/03932720802280602

Gienapp, P., Teplitsky, C., Alho, J. S., Mills, J. A., & Merilä, J. (2008). Climate change and evolution: Disentangling environmental and genetic responses. *Molecular Ecology, 17*(1), 167–178. doi:10.1111/j.1365-294X.2007.03413.x PMID:18173499

O'Connor, R. E., Bard, R. J., & Fisher, A. (1999). Risk perceptions, general environmental beliefs, and willingness to address climate change. *Risk Analysis, 19*(3), 461–471. doi:10.1111/j.1539-6924.1999.tb00421.x

Qian, W., & Zhu, Y. (2001). Climate change in China from 1880 to 1998 and its impact on the environmental condition. *Climatic Change, 50*(4), 419–444. doi:10.1023/A:1010673212131

Raleigh, C., & Urdal, H. (2007). Climate change, environmental degradation and armed conflict. *Political Geography, 26*(6), 674–694. doi:10.1016/j.polgeo.2007.06.005

Schlosberg, D., & Collins, L. B. (2014). From environmental to climate justice: Climate change and the discourse of environmental justice. *Wiley Interdisciplinary Reviews: Climate Change, 5*(3), 359–374. doi:10.1002/wcc.275

Shepardson, D. P., Niyogi, D., Roychoudhury, A., & Hirsch, A. (2012). Conceptualizing climate change in the context of a climate system: Implications for climate and environmental education. *Environmental Education Research, 18*(3), 323–352. doi:10.1080/13504622.2011.622839

Tranter, B. (2011). Political divisions over climate change and environmental issues in Australia. *Environmental Politics, 20*(1), 78–96. doi:10.1080/09644016.2011.538167

Trombetta, M. J. (2008). Environmental security and climate change: Analysing the discourse. *Cambridge Review of International Affairs, 21*(4), 585–602. doi:10.1080/09557570802452920

Warner, K., Hamza, M., Oliver-Smith, A., Renaud, F., & Julca, A. (2010). Climate change, environmental degradation and migration. *Natural Hazards*, *55*(3), 689–715. doi:10.100711069-009-9419-7

KEY TERMS AND DEFINITIONS

Equity: Defined by UNEP to include intergenerational equity - "the right of future generations to enjoy a fair level of the common patrimony" - and intragenerational equity - "the right of all people within the current generation to fair access to the current generation's entitlement to the Earth's natural resources" - environmental equity considers the present generation under an obligation to account for long-term impacts of activities and to act to sustain the global environment and resource base for future generations. Pollution control and resource management laws may be assessed against this principle.

Polluter pays principle: The polluter pays principle stands for the idea that "the environmental costs of economic activities, including the cost of preventing potential harm, should be internalized rather than imposed upon society at large." All issues related to responsibility for environmental remediation costs and compliance with pollution control regulations involve this principle.

Precautionary principle: One of the most commonly encountered and controversial principles of environmental law, the Rio Declaration formulated the precautionary principle: To protect the environment, the precautionary approach shall be widely applied by States according to their capabilities. Where there are threats of serious or irreversible damage, lack of complete scientific certainty shall not be used as a reason for postponing cost-effective measures to prevent environmental degradation. The principle may play a role in any debate over the need for environmental regulation.

Prevention: The concept of prevention can perhaps better be considered an overarching aim that gives rise to a multitude of legal mechanisms, including prior assessment of environmental harm, licensing or authorization that set out the conditions for operation and the consequences for violation of the conditions, as well as the adoption of strategies and policies. Emission limits and other product or process standards, the use of best available techniques, and similar techniques can all be seen as applications of the concept of prevention.

Public participation and transparency: identified as necessary conditions for "accountable governments,... industrial concerns," and organizations generally, public participation and transparency are presented by UNEP as requiring "effective protection of the human right to hold and express opinions and to seek, receive and impart ideas,... a right of access to appropriate, comprehensible and timely information held by governments and industrial concerns on economic and social policies regarding the sustainable use of natural resources and the protection of the environment, without imposing undue financial burdens upon the applicants and with adequate protection of privacy and business confidentiality," and "effective judicial and administrative proceedings." These principles are present in environmental impact assessment, laws requiring publication and access to relevant environmental data, and administrative procedures.

Sustainable development: Defined by the United Nations Environment Programme as "development that meets the needs of the present without compromising the ability of future generations to meet their own needs," sustainable development may be considered together with the concepts of "integration" (development cannot be considered in isolation from sustainability) and "interdependence" (social and economic development, and environmental protection, are interdependent). Laws mandating environmental impact assessment and requiring or encouraging development to minimize environmental impacts may be assessed against this principle.

Transboundary responsibility: Defined in the international law context as an obligation to protect one's environment and prevent damage to neighboring environments, UNEP considers transboundary responsibility at the international level as a potential limitation on the sovereign state's rights. Laws that limit externalities imposed upon human health and the environment may be assessed against this principle.

Chapter 6
Environmental and International Trade Legislation

ABSTRACT

One of the fundamental questions that have always been raised in international trade law and international environmental law is to state the relationship between trade and the environment. Did the two interact, or are they in conflict? Are the goals of international trade law that have led to conventional development in developed countries consistent with the goals of environmental law? Hence, first, a view that shows the conflict between international trade law and international environmental law will be presented. Then another view that expresses the attention to the environment in international trade law will be presented.

INTRODUCTION

The interconnected nature of trade, environment, and development issues is reflected in a new approach called the "green economy." It is considered a framework for linking economic and environmental pillars of sustainable development. The United Nations Environment Program defines a green economy as an entity that promotes human well-being and social justice while significantly reducing environmental risks and deficiencies. Thus, this concept recognizes the inseparability of the three main pillars of sustainable development - social, economic, and environmental development - intending to cultivate these three conditions in which trade is an inevitable option. The interaction between trade and the transition to a green economy is complex and can be considered a two-way interaction. Trade can facilitate the transition to a green economy by supporting the exchange of environmentally friendly goods and services, increasing resource productivity by creating economic opportunities and employment, and helping to eradicate poverty. And the green economy also has the potential to create sustainable business opportunities. In this study, it tries to examine the measures resulting from the realization of the green economy on international trade law and also points out that to realize the concept of the green economy in today's world, the requirements arising from the agreements and conventions of the World Trade Organization should

DOI: 10.4018/978-1-6684-4158-9.ch006

be considered. Let them be because they pave the way for the realization and emergence of a green and sustainable economy as soon as possible.

One of the essential issues that have been raised in recent decades is the concern and concern of the international community in the face of the globalization of the economy and trade and has forced governments and international organizations, and other actors in the field of international relations to react harmful effects. It is destructive that economic activities can enter the environment and lead to crises that have developed simultaneously and rapidly. In the final document of the Rio + 20 conference, the idea and theme of the green economy in the face of threats to the environment Does is raised. In the international arena, a fundamental principle is essential. The concept of the green economy does not replace sustainable development. Today, the view that achieving full sustainability depends on creating an environmentally friendly economy has become very popular.

The right to a healthy environment is one of the recognized rights of human beings. According to the first principle of the Stockholm Declaration, "He is officially responsible for protecting and improving the environment for future and present generations." The right is to an environment where the truth belongs to all human beings. All subjects of international law have a shared responsibility to realize it may conflict with another right, which is the right to economic activity and the right to economic activity development. Given that the interconnected nature of trade, environment, and development is reflected in a new approach called the green economy, supporting trade in environmentally friendly goods and services can increase resource productivity and eradicate poverty., Analyzing the interrelationships of trade and achieving a sustainable and inclusive economy at the international level to help the global community and protect the environment against the many current crises is the motivation for choosing this issue.

TRADE AND GREEN ECONOMY

Leading research has been conducted using literature review and authoritative internet resources. The sources of this research are mainly articles and books of researchers who have previously researched global environmental governance. In this article, while studying the basic concepts such as governance and, more specifically, global environmental governance, the current state of the global environmental governance system has been examined. Given the overall rate of environmental degradation, both nationally and at the regional and global levels, some of the causes of the inefficiency of the existing system of global environmental governance have been studied. In the end, several proposed solutions have been discussed(Thomas, 2001).

International trade is a critical component of sustainable development and a green economy. This is recognized at both the Rio and Johannesburg conferences. Trade helps us achieve a more efficient allocation of scarce resources and makes it easier for countries. Both rich and developing countries can access environmental services, products, and technologies. Countries can use the tools and plans of development in the WTO multilateral trading system. This chapter intends to examine the role of trade and trade in environmental goods and services in achieving a green and inclusive economy.

World Trade Organization

"Successful transition to a comprehensive SAMBS economy would not happen automatically. An effective framework for continued support for green development will require reform at the national and international levels and national policies and actions. They are needed to stimulate a green economy. At the international level, organizational structures are needed to ensure that developing countries enjoy the benefits of engaging in a global movement(Salzman & Ruhl, 2000)."

"The WTO provides a strong supportive framework to support sustainable development and the concept of a green economy. Enabling the environment through supportive goals, institutions, and oversight, the enforcement mechanism of a growing set of rules and regulations in the field of the environment." It builds and strengthens (Salzman & Ruhl, 2000)."

"WTO rules seek to strike an essential balance. On the one hand, they defend the rights of the organization members to support measures to advance legitimate goals such as environmental protection. On the other hand, they ensure that many measures are not taken arbitrarily and are not achieved in protecting domestic production. Sustainable Development One of the goals of the Doha Round was the latest round of multilateral talks on global trade. Negotiations have been able to help eliminate environmentally unfavorable trade, that is, trade that is harmful to the environment and promotes greater access to environmental goods and services at lower prices.

According to the WTO Secretariat, the Rio + 20 conference should have emphasized the commitment to:

- A multilateral trading system based on open and fair laws that is both predictable and equitable and beneficial to all countries in achieving sustainable development and a green economy.
- Ensure that commercially practical actions for environmental purposes are not considered an arbitrary and unjustifiable means of discrimination or restriction in international trade.

Use WTO mechanisms to monitor and monitor national actions with trade implications, including green measures, to increase understanding and dialogue and prevent the risk of trade tensions.

Promote an international trade system that meets the needs of developing countries. For example, by ensuring that capacity-building business plans help developing countries reap trade benefits in the transition to a green economy.

- Support the successful conclusion of the Doha Summit as a significant contribution to the vision of sustainable development. Of course, it should also be noted that WTO agreements and the requirements arising from these agreements indirectly affect the progress and achievement of a green economy. And some of these agreements include exceptions that allow members of the organization to implement green economy measures when pursuing a set of legitimate goals. The following is a brief overview of these agreements:

General Tariff and Trade Agreement (GATT) 3 Article 20: GATT is the main agreement for goods trade. Article 20 GATT On General Exceptions, there are several examples in which members' business transactions may be exempt from GATT rules. Otherwise, GATT rules and principles should apply. This article seeks, among other things, to ensure that green economy measures are not implemented arbitrarily. And is not used as a support device.

Agreement on Technical Barriers to Trade (TBT) and agreement on the Application of Plant Health Measures (SPS): Rules such as the TBT Agreement, sales with technical regulations and production standards, and the SPS Agreement The sale of food, human, and animal safety provides the basis for WTO members to take regulatory and legal action to protect the environment and promote a green economy, while at the same time It also imposes rules to ensure that such measures do not constitute unnecessary restrictions on international trade.

Agreement on Subsidies and Countermeasures (SCM): The SCM Agreement seeks to prevent members from providing subsidies that would alter international trade. Some of these basic rules and principles have been respected. The agreement has left members with a political climate, among other issues, to support the establishment and spread of green technologies.

Agreement on Trade-Related Aspects of Intellectual Property Rights (TRIPS): The TRIPS Agreement is a framework for using the intellectual property system to expand access to, and dissemination of green technologies provides a political space to promote the public interest in areas vital to socio-economic and technological development and specific incentives for technology transfer. Deprivation of incompatible and destructive technologies is considered as protection of intellectual property.

Government Purchase and Sale Agreement (GPA): This agreement shall apply only to the World Trade Organization members who have ratified it. Its goal is to open markets to international competition based on transparency and non-discrimination. Under this agreement, the parties and operators may adopt or request technical specifications to promote green sales or trade.

Facilitate International Trade in the Process of Achieving a Green Economy

The Bali Conference (WTO Ministers) was held in 2013, and among other issues, a legally binding agreement was reached to "facilitate trade." The agreement sought to help economic growth and reduce poverty by reducing trade costs and inefficiency, increasing competition and exports, and thus increasing incomes and jobs. Trade facilitation meant that traded goods could be easily traded across borders. The World Trade Organization defines it as "the simplification and harmonization of international trade practices that are practiced through trade practices." Activities, practices, and rituals, including the collection, presentation, communication, and data processing required to transfer and exchange goods in international trade.

Before Tiafi's conclusion, trade facilitation was included in the GATT Agreement, Article 5 (freedom of transit), Article 7 (costs and formalities related to imports and exports), Article 10 (publication and management of trade laws and regulations). In 2004, WTO members formally launched talks to explain, review, and formalize the principles behind the material as part of the Doha Development Summit. In addition, in 2004, the Trade Facilitation Order included negotiations on a "specific and different deal" for developing countries, considering the need for technical assistance and capacity building. The Trade Facilitation Agreement consists of two main parts (Akinyemi et al., 2017):

- Basic requirements to facilitate trade
- SDT provisions for technical assistance to assist developing countries in implementing those commitments and requirements.

Obligations and requirements include the following:

- Timely information and transparency about customary legal requirements
- Limits on costs and charges for delivery of goods and services at the border
- Simplify border formalities through only one way

The impact of the trade facilitation agreement on the transition to a greener economy is unclear. Positive consequences, such as negative consequences, are expected after implementing the agreement. To the extent that this agreement (TFA) leads to more trade, it also imposes more pressure and impact on the environment, as greenhouse gas emissions from production and transportation increase and more resources are consumed. We are natural with risks to water supply, fertile land, and natural biodiversity, and in terms of positive outcomes, the TFA offers opportunities for transition to a green economy. By reducing costs and increasing productivity, we can reduce waste and adverse effects on the environment and take a step towards achieving a green economy.

In addition, the TFA may authorize additional trade in environmental goods, services, and technology. In particular, trade in these goods and services and technologies should be more straightforward in developing countries, leading, for example, to better use renewable energy technologies. Combined with the reduction of other trade barriers, the implementation of the TFA may well lead to a significant improvement in achieving a green and sustainable trade.

Sustainable or green trade plays a crucial role in the relationship between international trade and the transition to a green economy. This type of business broadly refers to a business that does not deplete natural resources, does not harm the environment, does not disturb and change conditions and social security, but promotes economic growth.

In general, sustainable trade can be associated with the following elements: Positive social, economic, and environmental consequences of international trade in goods and services, production of economic values and reduction of poverty and inequality, reduction of environmental impacts of trade-related economic activities, natural Resources Reconstruction (Akinyemi et al., 2017), positive social, economic and environmental consequences of international trade in goods and services, production of economic values and reduction of poverty and inequality, reduction of environmental effects of trade-related economic activities and reconstruction of natural resources.

"As a result, facilitating international trade can both contribute to the green economy and create problems for the world's environment and natural resources. But if we try to expand and facilitate business to achieve sustainable trade, and these trade exchanges develop in the field of environmental goods and services, it can itself be a stimulus and factor for faster access to a low-carbon and inclusive economy that the world and the global environment are waiting to achieve. It can indirectly create the legal requirements to achieve a new and green economy."

Trade of Environmentally Friendly Goods and Services

"Environmental goods and services" play an essential role in transforming the green economy. Such goods and services can cover as many areas as they want, including air pollution control, renewable energy, water, waste management, environmental monitoring, environmental assessment and consulting, cleaning services, cleaning technology, carbon capture, and storage (Bucher et al., 2014).

"Negotiations on trade liberalization in the field of environmental goods and technologies are part of the Doha meeting of the World Trade Organization. The purpose of paragraph 31 (3) of the Doha Declaration of the Ministers Is development, which, of course, is done by reducing or eliminating tariff and non-tariff barriers to environmental goods and services "(Norouzi, 2021a).

"This seems good from an environmental point of view, meaning that the easier it is to trade environmental goods and technologies, the better for the global environment (Norouzi, 2021a)."

"However, this directive does not define environmental goods and technologies, nor does it specify the speed and depth of trade liberalization of environmental goods and technologies." But as negotiators understand, defining environmental goods is challenging. Some examples of environmental goods are:

- Goods used in cleaning the environment, such as equipment for preventing oil and oil spills, preventing environmental damage in industrial production stages such as air pollution control, waste management, and energy conservation or environmental protection equipment.
- Environmentally friendly technologies and products such as electric cars and wind turbines and technology related to clean coal fuel(Norouzi & Fani, 2022).
- Products produced and made environmentally friendly, such as organic products and recycled paper.

"The importance of facilitating trade in environmental goods and services is recognized in several international agreements, such as the Rio + 20 Final Document, entitled 'The Future We Want." The final document states: "It is a stimulus for sustainable economic growth and development, and it also emphasizes the vital role that a multilateral, just and law-based trading system, as well as trade liberalization, can play in stimulating economic growth and development." As a result, trade plays a vital role in accelerating the dissemination and absorption of environmental goods and services. It can facilitate access to cost-effective environmental goods and services to enable a faster and less costly trend toward a green economy. At the same time, it creates new jobs and job opportunities (Norouzi, 2021a). "

Arguably the most essential and fundamental international agreement on the trade-in environmental goods to date has been the Declaration of the Westpack 10 in 2012, which agreed on a list of 54 environmental goods. This list includes, for example, machines for generating heat and energy based on renewable fuels, controlling air pollution from industrial activities, gas turbines for generating electrical energy from gas from waste recycling, and others. Until the World Trade Organization (WTO) Ministerial Declaration in Doha in 2001, however, regulations explicitly targeted the reduction or elimination of tariff and non-tariff barriers to environmental services and goods, without specifying which environmental goods and services should fall under this terminology.

"Trade liberalization in environmental services, if well managed, can have significant benefits for the private sector as well as the general public, albeit by creating opportunities such as increasing the market and improving health and environmental sustainability, especially in countries with development is taking place, investment and expertise from foreign companies can create jobs and skills, and facilitate technology transfer, and domestic companies, in areas such as engineering, construction, and tourism, gain knowledge from growing business in the field. Environmental services can benefit (Akinyemi et al., 2017). " It should be noted that significant potential is found in developing countries for the growth of trade in environmental goods. Many of them already understand these opportunities to invest in and support environmental infrastructure by creating supportive legal frameworks that create many opportunities for trade in environmental goods and services. Emerging economies have become essential players in the production and trade of various clean technologies, and this is due to the significant increase in investment in research and development.

Depending on what stage of development a country is in or what environmental pressures it faces, different environmental products may be of different levels of importance. Thus, developed countries prioritize energy efficiency, renewable energy, and reduced carbon emissions, while developing countries, and significantly less developed countries, prioritize investment in waste and waste management.

"While much of the trade liberalization debate has focused on environmental goods, the service sector has also been somewhat overshadowed in contrast to the close cooperation and relationship between environmental trade and environmental services, products limit, which in reality seems to be one. "It simply came to our notice then.

The classification of environmental services by the World Trade Organization took place during the negotiations in Uruguay in 1991. Since the rapid development of the environmental goods and services sector, many have expressed concern that the classification is too precise. Fortunately, no agreement has been reached on a revised version so far(Norouzi & Fani, 2021).

"Environmental services can be classified as infrastructure or non-infrastructure environmental services.

Environmental infrastructure services include collecting and disposing of waste and services that require significant investment, such as construction and maintenance. These services are usually defined as public goods, often managed or regulated by government agencies. And non-infrastructure environmental services include pollution prevention and remediation as a response to solutions to the environmental problems that occur naturally in modern industrial economies. Unlike the first category, non-infrastructure services comply with government pollution or environmental degradation regulations. The market is open and competitive for non-infrastructure services (Akinyemi et al., 2017). According to the data obtained, the growth of markets for environmental services and products is strongly influenced by environmental regulations, although the need for environmental goods and services is often quite clear, based on the environmental challenges at the level of We face it nationally and internationally, there is a gap between the need and demand and potential benefits for environmental services and goods in developing countries, and this is primarily due to the lack or non-implementation of environmental regulations (Bucher et al., 2014) ".

"One of the obvious ways in which trade policies contribute to the greening of the economy is by reducing tariff and non-tariff barriers on goods such as wind turbines and efficient light bulbs and services such as environmental engineering, and that environmental goods and services are environmental benefits. "It offers tangible importers, especially in developing countries, where access to renewable energy can be the key to tackling poverty. As a result, free trade in environmental goods and services can be a way to achieve a green economy in developing countries."

INTERNATIONAL TRADE LAW VS INTERNATIONAL ENVIRONMENTAL LAW

One of the fundamental questions that have always been raised in international trade law and international environmental law is to state the relationship between trade and the environment. Did the two interact, or are they in conflict? Are the goals of international trade law that have led to conventional development in developed countries consistent with the goals of environmental law? Hence, first, a view that shows the conflict between international trade law and international environmental law will be presented. Then another view that expresses the attention to the environment in international trade law will be presented.

Conflict Between International Trade Law and the Environment

What first comes to mind about the relationship between international trade and environmental law is their conflicting approach. For example, we can mention the performance of some large international companies that, to escape the strict environmental laws in developed countries, seeks to select countries such as developing or less developed countries to achieve their goals, i.e., maximum benefits. Economically, at a lower cost and without responding to competent government agencies in the field of environment, hazardous waste in the bed of nature. Unfortunately, several developing and underdeveloped countries welcome the entry of these companies to raise capital and resolve economic crises, such as the unemployment crisis, without the usual strictures on environmental protection(Norouzi, 2022a).

Another reason for these companies to be present in developing or less developed countries is that implementing environmental commitments requires a lot of costs that these companies do not consider cost-effective. Therefore, they have no choice but to act in standard ways and destroy the environment like their predecessors. For example, we can mention climate pollution as the work of international oil companies in oil-rich countries, which has endangered human health and the species of animals and plants. The "Race to the Bottom Hypothesis" theory illustrates the approach of developed and developing countries to large international corporations.

Stricter regulations in developed countries and more lenient or non-strict regulations in other countries, and the high cost of implementing environmental commitments are among the barriers that lead companies to environmental degradation. According to the "Pollution Haven Hypothesis" hypothesis, in some developing and less developed countries, contaminated industries have been transferred to these countries due to applying more lenient environmental laws or the lack of strict enforcement of these laws. And it has made these countries a haven for the world's polluting industries. In addition, trade liberalization and environmental austerity make developed countries specialize in sanitary goods and development. Less developed countries produce polluted goods and become a haven for polluting industries. Become. Hence, it is necessary to revise and amend environmental laws and policies in developing and less developed countries. An important point to keep in mind when enacting strict laws is that enacting them alone is not enough, as environmental legislation implies government interference in the market and is based on the objectives of international trade law. From that, it can be concluded that trade liberalization is contradictory(Norouzi, 2022b).

Many of these companies object to the current practice and give up due to the high cost of implementing environmental standards. Comparing these costs with economic benefits allows companies operating in industrial operations, such as large oil and gas projects, to implement the same old methods for transporting their waste and wastes, which result in nothing but environmental pollution. That is, if low-cost environmentally friendly technologies spread to industry and developing or less developed countries impose stricter environmental regulations, these companies will live up to their environmental commitments. Although there is usually a tendency to evade environmental obligations without strict regulations, a review of what has happened in developed countries shows that, while necessary to protect the environment, strict environmental regulations are not enough. Instead, companies must transfer their waste through environmentally friendly, low-cost technologies.

Interaction of International Trade Law and International Environmental Law

Another view states that international trade and the environment are not two different categories. It is necessary to pay attention to both because they are not in conflict and can be implemented in harmony. According to the previous view, environmental laws and regulations, which usually impose strict standards and obligations on companies, constitute a form of government interference in private markets, contrary to international trade law purposes, such as trade liberalization. On the contrary, those who believe in the interaction of these two categories state that with the entry of the government in international trade, as a powerful player in the current world that can deal with environmental crises, the actions that in the past in law International trade has taken place and has led to the destruction of the environment, he said. In other words, the presence of the government and the enactment of environmental laws and regulations can not only lead to the prevention of environmental crises or the intensification of current environmental crises; Rather, it can achieve the goals of international trade law correctly and in the form of achieving sustainable development. Following the principles of sustainable development, unilateral growth in international trade and achieving favorable economic development, although in the short term, will benefit governments. It will lead to significant crises in the future, including environmental crises. It becomes the environment that today's societies face. Therefore, according to this view, which emphasizes the presence of the government in the arena of interactions between international trade law and international environmental law, not only environmental goals; Rather, business goals will be aligned with the principles of sustainable development.

Experience has shown that if powerful actors in international relations, such as governments and international organizations, do not support the environment, companies operating in the field of international trade will not move towards environmental protection because the costs of They consider the observance of environmental commitments and standards high. They consider it an obstacle in achieving maximum economic benefits(Movahedian et al., 2021).

In addition, given the growing power of international organizations, it is no longer possible to focus solely on the role of governments as in the past. Therefore, international organizations can also support environmental goals. In addition to being the beginning of the enactment of laws and regulations aimed at preventing environmental crises, this legal protection can play a secondary role and guarantee the implementation of the law in the event of such crises. There are criminal, and civil liability rights pay. The first organization for which a constructive role can be considered is the International Criminal Court, which, if interpreted broadly, could deal with environmental crimes(Norouzi & Sheikhi, 2021).

Further, the International Criminal Court (ICC) started in September 2016 that it would give priority to crimes that have led to "environmental degradation," "exploitation of natural resources," and "illegal occupation of land." (Posner & Sykes, 2013) This legal protection can also be implemented by using the capacities of criminal law in the national territory. As mentioned in the chapter on the settlement of environmental disputes, in the event of criminalization of environmental degradation in national criminal law and the provision of a mechanism for filing lawsuits and damages in ceremonial laws such as procedural law can be compensated. Wardeh took the initiative. In the mentioned chapter, the method of applying criminal enforcement guarantees in the Iranian legal system is mentioned.

WTO Approaches to the Environment

The same chapter describes the developments of international trade law and its influential organizations such as the World Trade Organization. The WTO is one of the essential players in international trade law. It is necessary to study its developments from its formation until now, which is the period of its development and display of power. Therefore, when talking about different approaches to the interaction or conflict of international trade law and international environmental law, it is necessary to refer to the current views on how the WTO interacts or conflicts with the environment.

A.   The views of economists and proponents of free trade:

Some economists believe that the WTO's primary goal is free trade and facilitating trade between countries for greater market access. Therefore, they consider the current WTO approach to environmental issues to deviate from its mission and goals. One theorist argues that the WTO should separate trade from other issues and refocus on its core goal of free trade and market access. (Staiger, 2004)

One of the consequences of accepting this view will be the opposition to environmental measures by the World Trade Organization. By recognizing the primary business objectives and not paying attention to non-commercial issues such as environmental issues, the WTO will gradually neglect environmental measures.

B.   A view on the connection between trade and the environment:

This view states that the purpose of implementing a trading system is to liberalize trade and that environmental issues related to trade are addressed. In other words, if there is a connection, the interaction between trade and environmental issues can be considered. (Alvarez. 2002)

One of the criticisms of this view is that it sees trade and the environment as alternatives. That is why WTO member states are reluctant to consider environmental issues. They see trade and the environment as alternatives and consider paying attention to environmental issues as a risk and obstacle to profiting from trade issues.

C.   The view of environmentalists:

This view states that the rules of the World Trade Organization are part of international environmental law and can be considered an organization related to the environment. This view has been met with an opposing view by others because it is one-sided and not pragmatic. In other words, this view is somewhat idealistic and abstract. For example, the Secretary-General of the World Trade Organization stated that "one of the indicators of the debate on trade and the environment is that the World Trade Organization is not an environmental organization."

The Sutherland Commission report can also be cited as another example, which states in its report that "although trade is an important factor in achieving development goals, the WTO can not be considered a development organization" (Bown, 2017).

D.   A view on the multifaceted nature of the World Trade Organization:

It is not correct to say that the WTO is merely a trade organization or an environmental organization. Due to any intellectual property rights, environmental issues, and human rights. Also, the view that this organization was considered an environmental organization has faced many problems in its implementation and has been opposed by the World Trade Organization itself.

What reinforces this view is the Declaration of the World Trade Organization (WTO) Parliamentary Conference in 2003, which recognized the WTO as an institution that deals with various issues, not just trade issues. Therefore, by accepting this view, it can be stated that environmental goals are also among the goals of the World Trade Organization, and the possibility of resolving disputes in this regard is possible. Thus, accepting that the WTO is an organization with different functions and tasks will resolve some trade-related environmental crises.

ENVIRONMENTAL CRITERIA SET OUT IN WTO AGREEMENTS

The 24 agreements and several memoranda of understanding are considered WTO treaties, many of which reflect environmental regulations.

GATT Environmental Regulations

Paragraphs (b) and (g) of the GATT Article make an exception to its general rules and can be considered the inspiration for the environmental rules set out in most WTO agreements. Knowledge of this material can determine GATT's approach to the environment. Paragraph (b) sets out the measures necessary to protect humans, animals, and plants' health and life, and paragraph (g) sets out measures to protect limited and non-renewable natural resources. It should be noted that the application of these two clauses is subject to the implementation of Article 20. The preamble to this article states that the measures referred to in paragraphs (b) and (g) must be applied so as not to cause unjustifiable discrimination between countries with identical conditions and not to impose restrictions on international trade.

Essential Measures to Protect Health and Life Regulations

Paragraph (b) of Article 20 of the GATT allows the Member States to take measures to protect the health and life of humans, animals, and plants. No provision in this document shall be construed as implying the implementation of the measures referred to in this paragraph. Such measures must be necessary and must not lead to unjustified discrimination. Due to the wide range of concepts mentioned above, there were disagreements between governments over implementing the regulation. Therefore, the dispute resolution pillar of the World Trade Organization interpreted the concepts while dealing with the disputes raised by this institution. For example, the "Thai cigarette case" in which the primitive element explains the concept of necessity. "Import restrictions imposed by Thailand can be considered necessary if another alternative action that is not contrary to the General Agreement or is more in line with it and can be expected to be reasonably used to achieve Thailand's health goals. "Does not exist."

Over time, this interpretation has become a "less commercial constraint" and a pillar of revision in the case of "asbestos," the "proportionality test," which means assessing the proportionality between the goal and the implementation measures to achieve the goal to determine the need for action. Therefore, according to the mentioned criterion, the necessity of actions is determined by the importance of the interests and values pursued. "The goals that health measures have been taken to achieve are the protection of human health and life, both of which are of the highest importance and are vital."

Measures Related to the Protection of Invisible Natural Resources Regulations

This clause contains two conditions: (first) measures must be related to protecting non-renewable natural resources; (Second) measures must be accompanied by restrictions on domestic production or consumption. The first condition expresses the criterion of "communication" that was first raised in the case of "salmon - herring." The relevance of the action means that its primary purpose is to protect the invisible natural resources.

The second condition restricts the citation to paragraph "g" of Article 20 of the GATT. The dispute resolution pillar in the "new formulation gasoline" case referred to government measures that impose restrictions on the production or consumption of domestic natural resources. These restrictions should apply without discrimination to domestic products and not just to imported goods.

One of the criticisms is that this article does not meet the challenges of the present age and can not solve some of the environmental problems that have arisen today. Also, a narrow interpretation of Article 20 by the Dispute Settlement Body has led to the loss of part of its practical effectiveness. This narrow interpretation of the exceptions in Article 20 implies the creation of an implicit classification between the goal of trade liberalization to protect the environment (Hilf, 2001).

Technical Barriers to Trade Agreement (TBT)

The Technical Barriers to Trade Agreement was adopted in the "Round of Uruguay," part of the regulations dealing with trade environmental issues. The primary purpose of this document is to ensure that technical regulations, standards, and technical evaluation methods do not become unnecessary obstacles to trade. The preamble to the agreement states that "no country shall be prohibited from taking any measures necessary to ensure the quality of its exports or to protect the health and life of humans, animals, plants or the environment ...". In addition, the second paragraph of Article 2 states that members must restrict international trade beyond what is necessary for the implementation of lawful and legitimate purposes. Paragraph 4 of this article stipulates that "if there is a relevant international standard, the technical regulations of a State shall apply that standard as its technical rules; "Unless the international standard is effective or appropriate for enforcing the legal objectives."

Compared to the GATT and the GATS, one of the advantages of this agreement is that regulations on non-trade issues such as the environment can be seen as part of the substantive rules contained in that treaty, which show a significant change in the WTO approach in Especially environmental issues.

Sanitary and Phytosanitary Agreement (SPS) (1994)

The Plant Health and Consumer Protection Act address human, animal, and plant health protection measures that directly or indirectly affect international trade. This agreement seeks to establish a framework containing the rules and principles necessary for developing, adopting, and implementing health and plant health measures to reduce the harmful effects of these measures on trade.

Article 2 of the agreement states that members have the right to take the necessary health and plant health measures to protect the health and life of humans, animals, and plants, provided that they do not conflict with the agreement's provisions. The agreement's criticism includes the Committee on Health and Plant Health's refusal to grant oversight of environmental conventions. In addition, the lack of rules on liability and compensation for the transboundary movement of harmful organisms is one of the negative points of this agreement.

Agriculture Agreement (1994)

The Agricultural Agreement seeks to reform the trade in agricultural products to establish a fair and market-oriented trade system. Therefore, obligations such as reducing domestic subsidies in the agricultural sector and export subsidies are reflected in the agreement. The Introduction to the Agriculture Agreement states that the obligations contained in this agreement were made because of the need and need to protect the environment. Appendix 2 to this agreement contains a list of different types of subsidies, some of which do not have adverse effects on the environment. Some subsidies were also set up to implement environmental programs.

General Agreement on Trade of Services (GATS)

The General Service Trade Agreement sets out vital rules for the service trade. Article 14 (b) states that if measures are necessary to protect the health or life of humans, animals, and plants, the members of the World Trade Organization shall be authorized to adopt such measures, even if they are contrary to the Gates. Contrary to Article 20 of the GATT, the regulation does not contain any restrictive restrictions, which is one of the significant benefits of this document. Gates, for example, facilitated the relocation of individuals to provide and use services to protect the environment. The Gates Appendix sets out environmental commitments such as wastewater management, waste management and recycling, and sanitation. The critique of this article is the narrow interpretation it has received. Because Article 14 of the agreement is similar to Article 20 of the GATT, the WTO will likely follow the same approach and ignore new developments. In addition, the Gates Agreement only mentions one of the restrictions set out in Article 20 of the GATT, and there is no other restriction, the exception relating to the protection of limited and non-renewable natural resources.

Agreement on Trade-Related Aspects of Intellectual Property Rights (TRIPS)

The Agreement on Trade-Related Aspects of Intellectual Property Rights, discussed at the "Uruguay Round" negotiations, addresses issues related to intellectual property rights. The second and third paragraphs of Article 23 of this Agreement relating to the environment. The second paragraph of this article deals with patentable inventions. Governments are committed to issuing patents in all areas related to technology unless the invention results in severe damage to the environment. It is necessary to protect the life or health of humans, animals, and plants from Avoid patents. In the third paragraph, the cases in which governments should not grant patents, states "diagnostic, therapeutic and surgical methods for the treatment of humans and animals, methods related to the preparation and production of plants and animals except for microorganisms "Biological processes necessary for the production of plants and animals, except for non-biological and microbial processes."

Positive effects such as access to new and environmentally friendly technologies are the results of this agreement, but on the contrary, spending a lot of money by countries for these technologies and the lack of guarantees to control technologies that are harmful to the environment can be considered as one of the technical points to be considered "thrips." Part "B" of the third paragraph of Article 27," TRIPS," has been criticized by experts. Critics say that biotechnology for the production of breeding organisms leads to ecological and environmental problems, the results of which include the extinction of traditional plant species. (Alexander, 1993)

Other criticisms of the TRIPS agreement include the failure to grant oversight to environmental institutions. For example, the TRIPS Council's negative response to the United Nations Environment Program (UNEP) request for oversight of the council. (Helfer, 2004) The non-implementation of the second paragraph of Article 66 of this document also raises concerns about the effectiveness of this agreement for the environment. According to the article, "developing countries are required to establish institutions and companies to promote and encourage the transfer of technology to less developed countries." Due to many implementation problems, especially in the transfer of environmentally friendly technologies, this article has not had a scientific function.

A REVIEW OF WTO PERFORMANCE

Trade and Environment Committee

The Moroccan Ministerial Conference established the Trade and Environment Committee in 1994 to look at the relationship between the environment and trade in WTO-related matters, replacing the GATT Trade and Environment Group. The establishment of this committee can be considered the beginning of severe and transparent discussions on trade and the environment in the World Trade Organization.

It has tasks: identifying the relationship between business and environmental measures to promote sustainable development and providing appropriate advice on the reforms that should be made to the rules of the business system. These reforms include the development of rules that enhance the interrelationships between trade and environmental measures, avoiding protectionist trade practices, and implementing effective multilateral regulations to ensure that the collective trade system adheres to the environmental objectives set out in agenda 21 of the Rio Declaration, in particular Article 12 of that Declaration, referred to the monitoring of commercial activities used for environmental purposes, the review of the commercial aspects of environmental measures, and the effective implementation of the multilateral regulations governing those measures.

A clear and unequivocal exchange of views on trade and environmental issues could have enhanced the committee's position, but so far no important and effective decisions have been made and its analytical results have been poor. One of the criticisms of the committee is that no ministers or environmental officials have been invited to participate in the committee's talks. There is also no opportunity for representatives of international environmental organizations and organizations to comment. This is why some believe that a clear and appropriate meeting agenda needs to be adopted to clarify the relationship between trade and the environment(Shaffer, 2001).

General Council of the World Trade Organization

The General Council of the World Trade Organization shall establish appropriate arrangements for practical cooperation with the competent organizations associated with the World Trade Organization. The council's cooperation with other organizations includes signing cooperation agreements with the International Monetary Fund, the World Intellectual Property Organization, the World Bank, and the United Nations Conference on Trade and Development. One of the criticisms of the council is the failure to agree with the competent organizations in the field of environment, which has been a sign of ignorance and disregard for environmental issues.

CONCLUSION

One of the questions that have always been asked is to explain the relationship between international trade law and international environmental law. Given that business activities have led to environmental crises, some belief in the conflict between these two branches of jurisprudence. International trade companies are also pessimistic about environmental approaches, such as strict regulation, because they see the commitment to meet environmental commitments and standards as costly. Thus, according to the pollution shelter theory, these companies seek to evade hazardous waste or start companies in developing and less developed countries to evade strict rules and regulations, mainly enforced in developing countries. They are going. These countries are also seeking to attract these companies due to the urgent need for foreign capital to develop the economy and resolve related crises without strict regulations, such as the unemployment crisis. In other words, these countries will become a haven for industries for polluting industries in the world. Proponents of the conflict between international trade law and international environmental law believe that implementing international trade law objectives in developed countries has been hampered by government interference through strict laws and regulations to transfer contaminants from developed countries to other countries. It seems that developed countries specialize in sanitary goods and development. Less developed countries specialize in producing polluted goods and become a haven for polluting industries in the world. In addition, the view that international trade law and international environmental law are in conflict points to the high costs of enforcing environmental commitments and standards, which lead to a desire for these companies to protect themselves. Do not have the environment. Therefore, it can be said that according to this view, due to government interference in the enactment of strict environmental laws and regulations, as well as the high costs of their implementation, these companies have sought to relocate hazardous waste and their companies. In practice, it leads to infecting developing and less developed countries.

In contrast, there is a view that international trade law and international environmental law are interactions. The presence of the state as a powerful player in the world today can prevent the destruction of the environment and the goals of international trade law, which is in line with the principles of sustainable development to be achieved. The government and organizations among the new players in international relations can play a valuable and constructive role. An example is the International Criminal Court, which criminalizes destructive environmental activities and imposes penalties on perpetrators.

Attention to the environment emerged as a new achievement of the global community facing environmental crises. Attention to the environment can be seen in documents such as the Convention on the Reduction of Customs Duties and Customs Duties, the Convention on Trade Prohibitions, the essential document of the International Union for Conservation of Nature and Natural Resources 1948, the 1946 Fishing Tours Convention and the 1946 International Whales Convention. After World War II, at the founding of the International Trade Organization (ITO), an environmental approach was adopted in the ITO Charter. Although the organization was not established, environmental concerns were addressed in another GATT document. According to the document, the "Environmental Action and International Trade Group" group was set up to study the relationship between the environment and trade. However, it did not have a significant impact.

After establishing the World Trade Organization, in the preamble to its founding document, it was explicitly stated that "the parties to the treaty recognize that a view to raising living standards must accompany their trade relations and economic activities... while making optimal use of resources. The world is in line with sustainable development goals and seeks to protect the environment and increase the tools and means of this protection at various levels of sustainable development." Environmental regulations in documents such as the GATT, the Agreement on Technical Barriers to Trade, the Agreement on Plant Health and Hygiene, the Agreement on Agriculture, the General Agreement on Trade in Services or the GATS, and the Agreement on Trade-Related Aspects of Intellectual Property Rights "Thrips" is considered to be one of the binding documents for the World Trade Organization. In addition to the above documents, the environmental approach can be seen in the following bodies of the World Trade Organization, such as the Committee on Trade and Environment and the General Council of the World Trade Organization. Thus, attention to developments in international trade law shows that the approach to environmental protection in business activities has gradually become a concern of the international community and competent organizations such as the World Trade Organization. However, there have been criticisms of how it works. There is a WTO, but its constructive actions cannot be ignored.

REFERENCES

Akinyemi, O. E., Osabuohien, E. S., Alege, P. O., & Ogundipe, A. A. (2017). Energy security, trade and transition to green economy in Africa. *International Journal of Energy Economics and Policy*, *7*(3), 127–136.

Alexander, D. (1993). Some Themes in Intellectual Property the Environment and the Environment. *Rev. Eur. Comp. & Int'l Envtl. L.*, *2*(2), 113–120. doi:10.1111/j.1467-9388.1993.tb00100.x

Alvarez, J. E. (2002). The WTO as linkage machine. *The American Journal of International Law*, *96*(1), 146–158. doi:10.2307/2686131

Bown, C. P. (2017). Mega-regional Trade Agreements and the Future of the WTO. *Global Policy*, *8*(1), 107–112. doi:10.1111/1758-5899.12391

Bucher, H., Drake-Brockman, J., Kasterine, A., & Sugathan, M. (2014). *Trade in environmental goods and services: Opportunities and challenges*. ITC.

Helfer, L. R. (2004). Mediating Interactions in an Expanding International Intellectual Property Regime. *Case W. Res. J. Int'l L.*, *36*, 123.

Hilf, M. (2001). Power, rules and principles-which orientation for WTO/GATT law? *Journal of International Economic Law*, *4*(1), 111–130. doi:10.1093/jiel/4.1.111

Movahedian, H., Norouzi, N., & Ataei, E. (2021). *Energy law and environmental law in oil and gas contracts with an emphasis on the MENA region*.

Norouzi, N. (2021a). Post-COVID-19 and globalization of oil and natural gas trade: Challenges, opportunities, lessons, regulations, and strategies. *International Journal of Energy Research*, *45*(10), 14338–14356. doi:10.1002/er.6762 PMID:34219899

Norouzi, N. (2022a). Regulating Sustainable Economics: A Legal and Policy Analysis in the Light of the United Nations Sustainable Development Goals. In Handbook of Research on Changing Dynamics in Responsible and Sustainable Business in the Post-COVID-19 Era (pp. 266-287). IGI Global. doi:10.4018/978-1-6684-2523-7.ch013

Norouzi, N. (2022b). A Practical and Analytic View on Legal Framework of Circular Economics as One of the Recent Economic Law Insights: A Comparative Legal Study. *Circular Economy and Sustainability*, 1-26.

Norouzi, N., & Fani, M. (2021). Monopoly and competition in the energy market: A legal analysis. *Global Journal of Business Management, 15*(2), 001-007.

Norouzi, N., & Fani, M. (2022). Globalization and the oil market: An overview on considering petroleum as a trade commodity. *Journal of Energy Management and Technology*, 6(1), 54–62.

Norouzi, N., & Sheikhi, M. (2021). Achieving Sustainable Development from the Perspective of International Environmental Law. *Eurasian Journal of Environmental Research*, 5(1), 1–13.

Posner, E. A., & Sykes, A. O. (2013). *Economic foundations of international law*. Harvard University Press. doi:10.2307/j.ctt2jbtsp

Salzman, J., & Ruhl, J. B. (2000). Currencies and the commodification of environmental law. *Stanford Law Review*, *53*(3), 607. doi:10.2307/1229470

Shaffer, G. C. (2001). The world trade organization under challenge: Democracy and the law and politics of the WTO's treatment of trade and environment matters. *Harv. Envt'l L. Rev.*, *25*, 1.

Staiger, R. W. (2004). *Report on the international trade regime for the International Task Force on Global Public Goods*.

ADDITIONAL READING

Helfer, L. R. (2004). Mediating Interactions in an Expanding International Intellectual Property Regime. *Case W. Res. J. Int'l L.*, *36*, 123.

Norouzi, N. (2021a). Post-COVID-19 and globalization of oil and natural gas trade: Challenges, opportunities, lessons, regulations, and strategies. *International Journal of Energy Research*, *45*(10), 14338–14356. doi:10.1002/er.6762 PMID:34219899

Norouzi, N. (2022a). Regulating Sustainable Economics: A Legal and Policy Analysis in the Light of the United Nations Sustainable Development Goals. In Handbook of Research on Changing Dynamics in Responsible and Sustainable Business in the Post-COVID-19 Era (pp. 266-287). IGI Global. doi:10.4018/978-1-6684-2523-7.ch013

Norouzi, N. (2022b). A Practical and Analytic View on Legal Framework of Circular Economics as One of the Recent Economic Law Insights: A Comparative Legal Study. *Circular Economy and Sustainability*, 1-26.

Norouzi, N., & Fani, M. (2021). Monopoly and competition in the energy market: A legal analysis. *Global Journal of Business Management, 15*(2), 001-007.

Norouzi, N., & Fani, M. (2022). Globalization and the oil market: An overview on considering petroleum as a trade commodity. *Journal of Energy Management and Technology*, 6(1), 54–62.

Norouzi, N., & Sheikhi, M. (2021). Achieving Sustainable Development from the Perspective of International Environmental Law. *Eurasian Journal of Environmental Research*, *5*(1), 1–13.

Posner, E. A., & Sykes, A. O. (2013). *Economic foundations of international law*. Harvard University Press. doi:10.2307/j.ctt2jbtsp

Shaffer, G. C. (2001). The world trade organization under challenge: Democracy and the law and politics of the WTO's treatment of trade and environment matters. *Harv. Envt'l L. Rev.*, *25*, 1.

KEY TERMS AND DEFINITIONS

Equity: Defined by UNEP to include intergenerational equity - "the right of future generations to enjoy a fair level of the common patrimony" - and intragenerational equity - "the right of all people within the current generation to fair access to the current generation's entitlement to the Earth's natural resources" - environmental equity considers the present generation under an obligation to account for long-term impacts of activities and to act to sustain the global environment and resource base for future generations. Pollution control and resource management laws may be assessed against this principle.

Polluter pays principle: The polluter pays principle stands for the idea that "the environmental costs of economic activities, including the cost of preventing potential harm, should be internalized rather than imposed upon society at large." All issues related to responsibility for environmental remediation costs and compliance with pollution control regulations involve this principle.

Precautionary principle: One of the most commonly encountered and controversial principles of environmental law, the Rio Declaration formulated the precautionary principle: To protect the environment, the precautionary approach shall be widely applied by States according to their capabilities. Where there are threats of serious or irreversible damage, lack of complete scientific certainty shall not be used as a reason for postponing cost-effective measures to prevent environmental degradation. The principle may play a role in any debate over the need for environmental regulation.

Prevention: The concept of prevention can perhaps better be considered an overarching aim that gives rise to a multitude of legal mechanisms, including prior assessment of environmental harm, licensing or authorization that set out the conditions for operation and the consequences for violation of the conditions, as well as the adoption of strategies and policies. Emission limits and other product or process standards, the use of best available techniques, and similar techniques can all be seen as applications of the concept of prevention.

Public participation and transparency: identified as necessary conditions for "accountable governments,... industrial concerns," and organizations generally, public participation and transparency are presented by UNEP as requiring "effective protection of the human right to hold and express opinions and to seek, receive and impart ideas,... a right of access to appropriate, comprehensible and timely information held by governments and industrial concerns on economic and social policies regarding the sustainable use of natural resources and the protection of the environment, without imposing undue financial burdens upon the applicants and with adequate protection of privacy and business confidentiality," and "effective judicial and administrative proceedings." These principles are present in environmental impact assessment, laws requiring publication and access to relevant environmental data, and administrative procedures.

Sustainable development: Defined by the United Nations Environment Programme as "development that meets the needs of the present without compromising the ability of future generations to meet their own needs," sustainable development may be considered together with the concepts of "integration" (development cannot be considered in isolation from sustainability) and "interdependence" (social and economic development, and environmental protection, are interdependent). Laws mandating environmental impact assessment and requiring or encouraging development to minimize environmental impacts may be assessed against this principle.

Transboundary responsibility: Defined in the international law context as an obligation to protect one's environment and prevent damage to neighboring environments, UNEP considers transboundary responsibility at the international level as a potential limitation on the sovereign state's rights. Laws that limit externalities imposed upon human health and the environment may be assessed against this principle.

Chapter 7
Environmental Public Finance Law

ABSTRACT

With the increasing recognition of external costs caused by environmental pollution and its calculation in the total price of goods and services, there is a need to review as soon as possible the expansion of green tax bases as well as long-term policies to protect the environment. Although the country's environmental protection laws have a long history, new tax instruments are needed to achieve sustainable development. To achieve this goal, studying the experiences of other countries in the use of these new tools such as awareness-raising and political coordination in the field of green tax legislation and the authority to buy and sell pollution can be helpful. This chapter describes the status of green tax and its history in Iran and discusses new tools in environmental policy implementation.

INTRODUCTION

In all countries, the bulk of government revenue comes from taxes. The share of taxes in total public revenues varies from country to country, depending on their level of development and economic structure. Taxes are sometimes used as a tool of economic policy [from the point of view of Keynesian theorists] to achieve the goals of economic policymakers. Since Iran's primary source of government revenue is mainly through exporting petroleum products, this sector is constantly fluctuating. It is essential to carefully review tax revenues to stabilize the government budget and reduce fluctuations. In addition to the importance of taxes in economic growth, one of the most important types of taxes economists have considered to achieve sustainable development is called green tax(Barry, 2007).

Sustainable development is the development that meets current needs without compromising the ability of future generations to meet their own needs. Therefore, economies that rely on non-renewable resources are always exposed to economic instability. Green taxes provide a good tool to compensate for the external costs of pollution damage caused by institutions and factories to solve this problem. Reduce pollution emissions to a stable level(Bovenberg & De Mooij, 1994).

DOI: 10.4018/978-1-6684-4158-9.ch007

Many economists believe that if the green tax is appropriately designed and implemented for each country based on its characteristics, it will impose the least cost on society (Baumol & Owat, 1988; Pierce & Turner, 1990).

However, the costs of this tax are borne by the polluters rather than the legislators, so they will strongly oppose such a tax. The most critical goal in overcoming political controversy is designing and imposing a green tax to resolve conflicts between producers and legislators and implement it successfully. Three of the most basic conditions for imposing a green tax are how to repay the tax Receipts by the government, taxes on production institutions or final products, and political controls over tax revenues.

This article first deals with the status and history of taxes in Iran and then the importance of green taxes in achieving sustainable growth and existing laws in Iran and other countries; It then refers to the optimal policy in the field of environmental protection and finally provides policy conclusions and recommendations (Bovenberg & van der Ploeg, 1994).

ENVIRONMENTAL LAW IN IRAN

In a coherent and modern context, environmental protection in Iran does not have a long history. The first step in establishing a new and organized organization to protect the environment in Iran dates back to 1956. The Hunting Center of Iran was established, and the first steps were taken to protect wildlife and monitor the implementation of related regulations. After that, until the victory of the Islamic Revolution, the essential measures in the field of environment are as follows(Cropper & Oates, 1992):

1967: Adoption of the Hunting and Fishing Law of the Hunting and Fishing Supervision Organization.

1971: Renaming of the Hunting and Fishing Supervision Organization to the Environmental Protection Organization 1974: Adoption of the Law on Environmental Protection and Improvement and increase of the powers and duties of this organization with the approval of the Constitution of the Islamic Republic of Iran and emphasis of Article 50 of the Constitution on protection The environment, which at the time was one of the leading laws in the world, raised the marginal view of environmental protection. Article 50 addresses the issue of the environment as follows:

In the Islamic Republic of Iran, protecting the environment, in which the present and future generations have a prosperous social life, is considered a public duty. Hence, economic activities are prohibited, except in the case of environmental pollution or irreparable damage.

In the first development plan, the environmental sector was presented with qualitative goals and did not have any quantitative goals. The quality objectives of the environment in the first program were as follows:

The overall goal of the environment is to provide factors to improve and enhance the quality of human life, prevent irreparable damage to the environment, and repair the harmful effects of the past on the environment. According to Article 50 of the Constitution; Economic and cultural programs and plans with goals, strategies, and policies of environmental protection, following Article 50 of the Constitution, programs, economic and cultural plans with goals, strategies, and policies of protection The environment is aligned and harmonized (Daugbjerg & Svendsen, 2001).

The emergence and intensification of environmental problems on the one hand and the issues and views raised at the Rio de Janeiro Conference, in which representatives of the Islamic Republic of Iran also actively participated, on the other hand, caused environmental attitudes in the second development plan more than the first Pay attention.

The third development plan started while good experiences had been gained from previous environmental programs, and because of this, the program was more comprehensive. This program is considered as a turning point in the process of environmental changes in the country, so that from the beginning of the program, one of the sub-sectoral committees, called the Environmental Policy Committee, was responsible for formulating environmental policies and strategies and presented the policies of the Third Plan in the form of legal materials (104 and 105) and executive solutions.

One of the essential functions of the government in the third development plan is to formulate environmental rules and standards for the establishment of industrial and service units and continuous monitoring of some significant polluting industries.

Paragraph (c) of Article (104) of the Third Development Plan Law also states: To reduce environmental pollutants, especially in the country's natural resources and water resources, production units are required to comply with their technical specifications with environmental standards. And take action to reduce pollution. Expenses paid in this regard are considered acceptable costs of the units.

The fourth development plan, approved from 2005 to 2009, contains 161 articles, 34 notes, and nine appendices. The law of the fourth plan has six main parts that express the axes of the primary plan, and the following 15 chapters are.

The second part or axis of the program law, entitled Environmental Protection, Land Management, and Regional Balance, shows the importance of the environment and land management in the fourth program. The fifth chapter is dedicated to environmental protection. This chapter contains 14 articles and a note (from Articles 58 to 71), and since Article (71) is only the implementation of two articles of the Third Plan Law (Article 105 and paragraph "c" of Article 104). The number of main items of the Fourth Plan Law in environmental protection reaches 15 articles. In contrast, other articles in this law are related to the environment, which in some cases are no less important than the articles of the fifth chapter.

As a result, the fourth development plan of the country in the field of environmental protection, both in terms of quality and in terms of explaining the macro-environmental policies governing it, has grown significantly compared to the third development plan and, consequently,, the first two plans, for example. Prevention of excessive use of pesticides and chemical fertilizers, plan to prepare criteria for entry, manufacture, formulation, and use of chemical fertilizers and pesticides for environmental impact by the Ministries of Jihad, Agriculture, Health and Medical Education, Environmental Protection Organization Biology, and the Institute of Standards and Industrial Research of Iran are included in paragraph "b" of Article (61). Also notable for green resource management is the economic value of environmental resources in national accounts and the calculation of the values and costs of priority items such as forest, water, soil, energy, biodiversity, and environmental pollution. Sensitive points and consideration of environmental costs calculated in the feasibility of capital asset acquisition projects were considered the most critical environmental achievements in the Fifth Development Plan(Eskeland, 1993).

In the Fifth Development Plan, as materials (187 to 193) are devoted to environmental protection, although most of the provisions of this plan are the same as in the Fourth Development Plan; Among the reforms made in this program, we can mention the formulation and implementation of an integrated ecosystem management plan and an operational plan for the protection and sustainable exploitation of the biodiversity of sensitive and fragile ecosystems and the amendment of the statute of the National Environment Fund.

Law of Aggregation of Tolls

From the beginning of 2004, the collection and receipt of any funds, including taxes and duties, both national and local, from producers of goods, service providers, and imported goods, was done only following the Law on Consolidation Tolls. Receipt by the State Tax Affairs Organization, and according to the existing agreement, a percentage of it was transferred to the municipal account. The rest was given to the Tax Affairs Organization. In 2009, the VAT law replaced the toll consolidation scheme. Note 1 of Article (38) of the VAT Law regarding the production units of environmental pollutants that do not comply with the standards and criteria of environmental protection; In addition to paying the duties related to the law, one percent of the sale price of these factories is considered as pollution duties. Despite the measures taken, factories and other environmental polluters prefer to pay one percent VAT instead of optimizing the production cycle and using environmentally friendly technologies to eliminate pollution. Given the current state of environmental law deficiencies, the legislature has to plan so that both the amount of pollution is minimized and the incentive to produce is not taken away from economic units(Gervais, 2005).

Environmental Tax Laws in Iran

No case in tax law can be linked to a green tax; Only in tax exemptions have arrangements been made for institutions and companies that move their industries from large cities to less developed areas. Also, for companies and institutions that spend on research and development to improve production methods and reduce pollution, these costs will be considered tax-eligible. The rules and regulations related to this article in the Direct Taxes Law or Budget Laws of previous years are as follows:

Article 132: Income taxable income due to production and mining activities, etc., in less developed areas, are 100% exempt from tax subject to Article (105) of this law for ten years. Note 2; Exemptions subject to this article include production and mineral revenues located within a radius of 120 km from the center of Tehran and 50 km from the center of Isfahan, and 30 km from provinces with a population of more than 300,000 based on the latest census except for industrial towns located within a radius of 30 km. The provincial capitals of the mentioned cities will not be(Gomez, 2001).

Article 138: The part of the profit expressed by cooperatives and private companies that are used for development, renovation, and reconstruction or completion of their existing industrial and mining units or creation of new industrial and mining units in that year will be exempted from 50% tax belonging to 105 of this law. Note 3; Factories in the catchment area of Tehran with several employees not less than 50, if they move their facilities outside the radius of 120 km from Tehran, according to the plan established by the Ministry of Economy and Finance and the relevant Ministry as appropriate. Be exempt from tax for up to ten years from the date of operation in the new location.

The implementation of macro-plans in the field of taxation shows its importance in the country's economy, if achieving a level of coverage of 85% of current government expenditures is one of the goals of the 20-year comprehensive vision document of the government, taking into account other policies of the government and tax organization. A country that is to the people's satisfaction is one of the logical strategies to expand the tax base instead of increasing the level of taxes.

At first glance, new tax bases may be justified to earn revenue for the government. Still, attention to the impact on orientations suggests a purpose beyond acquisition beyond the attention of legislators. For example, the tax on environmental pollutants is one of the new tax bases that, in addition to the income effect, has significant allocation effects and is now one of the axes of sustainable development in countries, including effective policies to control the factors of the environment using economic tools. Therefore, it should be possible to move forward with the introduction of new foundations, extensive studies, the establishment of appropriate laws, and cultural context with the requirements of the time and to achieve a proper tax system to achieve the desired result.

Green Tax in Agriculture

However, according to Article (81) of the Law on Direct Taxes, income from all agricultural activities, animal husbandry, fish and beekeeping and poultry farming, fishing, and fishery, nomadic, pasture and forest rehabilitation, orchards, trees such as palm trees It is exempt from paying taxes and supporting this sector is one of the significant development policies of the country. Still, it is possible to tax the type of activities that cause environmental degradation.

Tax on fertilizer and pesticide consumption in the land unit, improper use of agricultural institutions and fertilizer and pesticide, causes water, soil and air pollution, affects the growth and life of other useful plants and animals, and the destruction of rare plant and animal species.

Taxation on water consumption per unit of land, excessive use of water and unscientific and out-of-substitution harvest, especially from aquifers, has caused the water level in most of the country's aquifers to decrease that in many The plains have been declared dangerous.

Taxes on changing the region's ecosystem, turning forest areas and pastures into gardens and fields, changing the type of plants and introducing non-native plants destroy rare animal and plant species, increase soil erosion, landslides, etc. unevenness of the region(Jaffe & Stavins, 1995).

Taxes on agricultural waste agricultural activities during the production and storage stages contain waste materials that can cause environmental pollution. Spoiled fruits and products, remnants of harvested crops, twigs, and wood left in nature are among these pollutants.

However, according to Article (81) of the Law on Direct Taxes, income from all agricultural activities, animal husbandry, fish farming, beekeeping and poultry farming, fishing and fishery, livestock, pasture and forest rehabilitation, orchards, trees such as palm trees It is exempt from paying taxes and supporting this sector is one of the significant development policies of the country. Still, it is possible to tax the type of activities that cause environmental degradation.

Tax on fertilizer and pesticide consumption per unit of land, improper use of agricultural institutions, fertilizer, and pesticide cause water, soil, and air pollution and affect the growth and green life of valuable plants and animals and the destruction of rare plants and animal species. Taxes on water consumption per unit of land, excessive use of water and unscientific harvesting, and out of alternative capacity, especially from the country's aquifers, have become so dangerous that many country plains. Taxes on changing the region's ecology, converting forest areas and pastures with gardens and fields, changing the type of plants, and introducing non-native plants cause the destruction of rare animal and plant species, increase soil erosion and landslides and change the unevenness of the region. Agricultural waste tax is an agricultural activity during the production and storage stages of waste that can cause environmental pollution. Spoiled fruits and products, leftover spoiled products, harvested products, and wood left in nature are among the mentioned pollutants(Kasa, 2000).

Impact of Green Tax on Pollutants

Because the green tax creates a financial burden for businesses and individuals and is also a new issue, its application will face resistance. On the other hand, all sectors of the economy that somehow benefit from government exemptions and subsidies and whose lives depend on government support will react to this new tax base. Of course, many resistances can have non-economic reasons related to socio-cultural and managerial issues. Also, it is not possible to accurately estimate the cost of pollution, and it is not well known.

Concept of Sustainable Development

In general, any factor that reduces the quality of the environment is called pollution. Economic pollution occurs when this reduction in the quality of the environment harms the health, human health, and efficiency of production and society or reduces human well-being due to the damage done in a way that outweighs the benefits of production. Thus, pollution from an economic point of view depends on the physical impact of waste on the environment and human response to it. Hence, the physical presence of pollution does not mean economic pollution. Of course, there is the fact that most environmental pollution is caused by excessive consumption of energy carriers. One of the most important reasons for this is the lack of appropriate economic policies for energy pricing. The issue is not what economic growth rate should be achieved but what type of growth should be chosen(Krass et al., 2013).

Underdevelopment can be decisive in environmental degradation and conditions, just as rapid economic growth can have decisive effects on the environment and social status. There is a broad perspective of development in front of the world today that seeks solutions to overcome critical environmental conditions. In line with goals such as sustainable growth, creating a precise balance between the needs of today and tomorrow, etc., thinkers seek to find new socio-economic development models. These new development models must be cost-effective based on clean technology. Therefore, we need to reflect the actual value of the environment in all decision-making processes with solid motivations. The only effective way to determine the value of the environment is the economical pricing of natural resources, especially how to use energy at the national level should be reconsidered to use mobile energy much more efficiently. In this way, it is possible to save consumption significantly(Kopczuk et al., 2013).

The literature and experiences of world tax developments can be a good guide and basis for helping tax reform in Iran. Studying and analyzing this literature and experiences can significantly help improve and increase essential tax revenues or the environmental tax system known as green tax. This tax system becomes essential when public sector theories show that this type of tax increases efficiency in the economy.

Sustainable development is a development that meets the needs of the present without compromising the ability of future generations to meet their own needs, which has two key concepts at its core: First; The concept of needs - especially the basic needs of the world's poor - must be prioritized, and the other is related to the constraints that governments must impose on technology and social organization for the environment to meet present and future needs.

A Review of Social Theories and Environmental Management

Environmental considerations have a special place when it comes to sustainable development. Various measures have been taken in various fields to protect environmental resources, and various tools such as advice, order, training, law of encouragement, and punishment have been used. For this reason, in recent decades, the emphasis of environmental economists on the issue of green taxes, which is also referred to as pollution tax, has emerged. Theoretically, pollution tax is an excellent tool to compensate for external costs of pollution damage caused by institutions and factories. It reduces the rate of pollution to an optimal and stable level (Lilliestam et al., 2021).

Concept of Green Productivity

In a simple sense, productivity is the output ratio to the input used in all products and services areas. In recent years, attention to environmental issues in the field of productivity has become particularly important because the meaning of the input of the same renewable resources, the undesirable use of which respectively leads to the destruction of resources and depletion of resources and lack of access to the future to the mentioned sources. Output refers to the production of goods and services and environmental pollutants, waste, and goods and services. Because of this, a new category of productivity emerged called green productivity. The definition of productivity organization and environmental protection guarantees economic and social development (Nannerup, 2001).

IMPACTS OF GREEN TAXATION

Green taxes can be imposed on various environmental pollution processes in Iran, the most important of which, given the level of pollution and the economic importance of each sector, are as follows:

Industry and Mining Section

Iran is known as a mining country whose economy depends on the revenues of the mining sector, especially oil. Many industries have been formed to discover, extract, process, store, and transport minerals. The most critical cases of green tax can be applied as follows:

Tax on the type of solids and particles emitted from factory chimneys, particles are taxed based on pollution and the percentage of stability—the tax on the number of gases and vapors leaving the chimneys of factories. Refineries, the amount and volume of smoke and fumes are the basis for assessing the amount of tax. In this regard, the relevant organizations consider a certain amount as permissible and, in addition, subject it to tax. Taxes on discharged materials on running water, groundwater, and soil, the scope of this tax is extensive and includes a large group of industrial and mining activities taxes on changes in the temperature of running water and seas, industries that use water as cooling for their devices and products, and water that plays a role as part of the production cycle, such as combined cycle, nuclear power plants, Metal smelting industries, chemical industries, refineries and so on(Norouzi et al., 2022a).

Tax on regional degradation and change resulting from mining activities, many mines are located in ecologically significant areas. Mining activities have irreversible effects on the surrounding environment without environmental considerations. Will have. Taxes change the landscape. Factories, access roads, power lines, water, and communications change the region's landscape. Construction of dams and flooding of a large area behind the dam, valleys, and windy plains where wind turbines are installed are some of the things that can be mentioned(Zhang & Wen, 2008):

Noise tax on industries and mines; This type of tax will be applied to the amount of noise produced outside the normal range continuously or at short intervals in the environment.

Tax on industrial and consumer materials with a long cultivation period to the environment; For example, plastics, aluminum coatings, waste from mining activities, smelting and condensation overheads, and production line waste.

Tax on gas produced with oil; Significant amounts of methane gas are extracted simultaneously as oil production. If no action is taken to control it, in addition to destroying this energy carrier, pollution caused by this gas will also be created (greenhouse phenomenon).

Another solution is to create a market for the sale of pollution production licenses (clean air incentive market), implemented in many developed countries(Norouzi & Sheikhi, 2021).

Accordingly, any industry that can reduce its pollution level (especially in carbon monoxide, nitrogen oxide, and sulfur oxide) below the allowable level will sell the difference between its pollution level and the standard level in the market where the supplier will be allowed. The buyers of this quota are industries and enterprises that could not reduce their pollution levels to the permissible level and will be the buyers of these licenses.

GREEN ENVIRONMENTAL TAX IMPACTS ON POLLUTANTS

As green taxes create a financial burden on businesses and individuals and are a new issue, their application will face resistance. On the other hand, all sectors of the economy that somehow benefit from government aid, exemptions, and subsidies and whose lives depend on government support will react to this new tax base(Shohani et al., 2021). Of course, many resistances can have non-economic reasons related to social, cultural, and managerial issues. It is also impossible to accurately estimate the cost of pollution, and it is not well known. Types of environmental taxes are used in different countries of the world, which are: taxes on waste and waste, taxes on energy (electricity and gas), taxes on carbon dioxide emissions, taxes on fossil fuels, taxes on mineral oils and oil. Crude, tax on nitrous oxide and sulfur dioxide, tax on the table of environmental policies and programs in the second development plan (1996-2000) groundwater, tax on sand and climate change effects of industry and application Energy in jobs is pristine cliffs(Norouzi, 2021a).

Green Tax Legal Infrastructure

- Green tax as a tax base
- Law on Environmental Protection and Improvement approved in 1974, consisting of 21 articles
- Law on how to prevent air pollution approved in 1996, consisting of 36 articles
- Environmental policies and programs in the first development plan (1990-1994)
- Environmental policies and programs in the second development plan (1996-2000)
- Environmental policies and programs in the second development plan (2000-2004)
- Environmental policies and programs in the second development plan (2005-2009)
- Twenty-year vision document of economic and social development, Islamic Republic of Iran

- Waste Management Law: Approved in 2004
- Environmental Considerations in Budget Laws (2004 and 2007)
- Law on how to calculate the collection of customs duties, commercial profits, and taxes on all types of vehicles
- In the tax laws, no case can be related to green tax.

In tax, exemptions are arrangements for institutions and companies to relocate their industries from large cities to less developed areas(Norouzi, 2022a).

- Law on how to calculate and collect customs duties, commercial profits, and taxes on all types of vehicles
- civil law
- Islamic Penal Code, Punishments

Accordingly, there are some problems in the process of drafting laws for the environment, the most important of which are(Norouzi, 2022b):

- Some legal gaps in environmental issues still exist.
- Overlapping and parallel laws in the field of environment
- How to formulate laws, regulations, rules, and environmental standards

EXPERIENCE OF OTHER COUNTRIES IN IMPLEMENTING GREEN TAX

Pigovian tax, mainly used in European countries, is levied on Austria, Belgium, Denmark, France, and Norway to dispose of polluting waste. The audio created depends. Indirect environmental taxes, which are widely used only in industrialized countries and rarely used in developing countries, include specific energy taxes implemented by some European countries; Such as Denmark, Finland, the Netherlands, Norway, and Sweden. Fertilizers have also been levied in Austria, Finland, the Netherlands, Sweden, and beverage taxes have been levied in Canada, Denmark, Finland, Norway, and Sweden.

Tax incentives for the use of renewable energy in Portugal and Switzerland, increasing depreciation regulations to invest in pollution reduction equipment in Japan, and car taxes in Japan, Russia, Italy, and Portugal Examples of environmental regulations in other taxes In industrialized countries.

Developing countries are also encouraging safe investments for the environment. Kenya and Tanzania have introduced increasing depreciation regulations to prevent soil erosion and plant sustainable crops. In Latin America, Brazil, Chile, and Colombia have provided incentives for afforestation.

Many industrialized and transition countries and a small number of developing countries use indirect taxes on motor fuels to reduce environmental damage. For example, in many countries, indirect tax rates are high for leaded fuels and low for lead-free fuels(Norouzi & Fani, 2021).

OPTIMAL POLICY IN IMPLEMENTING GREEN TAX POLICY

Most environmental economists believe that if the green tax is appropriately designed and implemented, it can reduce environmental pollution at the lowest cost to society (Baumol & Quatt, 1988; Pierce & Turner, 1990). From this point of view, from the macroeconomic point of view, green taxes are an efficient method until the profits from implementing this policy are returned to the polluters. Otherwise, the costs of implementing this policy will be more on the polluters. Due to the lack of Acceptance of pollution-producing groups, policy implementation will be complex, and it will be impossible to achieve the goals of green tax policy implementation.

The second point for the optimal implementation of this policy is its transparency, simplicity, and fairness (Norouzi & Fani, 2022). For example, research in Denmark has shown that due to differences in rainfall, soil type, and type of crop, a tax on the use of chemical fertilizers is not a fair and equitable solution for taxation. Instead, pesticides and Pesticides are viable alternatives to green taxation (Oates, 1995).

To solve the contradictions created by implementing the Green Tax Law, economists have introduced a solution to allow the purchase and sale of pollution, which has many advantages compared to the enactment of green laws. The successful experience of the United States in implementing this policy in controlling sulfur dioxide and carbon monoxide, which have been the leading cause of acid rain in the country, so that the amount of sulfur dioxide production between the years (1990-2000) Has decreased by 50% (Oates & Baumol, 1975). Therefore, paying attention to the issue of allowing the sale and purchase of pollution and using the experience of other countries can be of great help in enacting laws and implementing environmental policies in the country.

CONCLUSION

Given the importance of environmental protection and the high external costs of environmental degradation, and its importance in sustainable development, green taxes provide an excellent tool to offset external costs of pollution damage to reduce pollution emissions to a stable level. Environmental taxes are one of the economic tools to reduce the external costs of economic activities. In this type of tax, there is a big gap between theory and application. There is a significant gap in the efficiency dimension between environmental materials in development programs and implemented cases. Economic development programs are vibrant and complete in terms of environmental rights. It has met most of the country's environmental needs. Still, in practice, due to the lack of good tools and a lack of a national will, it has not provided the expected results, and the need for legislation and oversight in this area has become more apparent. Is. Accordingly, it is necessary to reflect the absolute value of the environment in all decision-making processes with strong motivations to make consumption significantly economical. Also, considering the positive effects of implementing a pollution control policy in developed countries, their experiences can be used to adopt an appropriate policy to control pollution in our country.

REFERENCES

Barry, J. (2007). *Environment and social theory*. Routledge. doi:10.4324/9780203946923

Bovenberg, A. L., & De Mooij, R. A. (1994). Environmental levies and distortionary taxation. *The American Economic Review, 84*(4), 1085–1089.

Bovenberg, A. L., & van der Ploeg, F. (1994). Environmental policy, public finance and the labour market in a second-best world. *Journal of Public Economics, 55*(3), 349–390. doi:10.1016/0047-2727(93)01398-T

Cropper, M. L., & Oates, W. E. (1992). Environmental economics: A survey. *Journal of Economic Literature, 30*(2), 675–740.

Daugbjerg, C., & Svendsen, G. T. (2001). Designing Green Taxation. In Green Taxation in Question (pp. 117-135). Palgrave Macmillan, London. doi:10.1057/9780230595538_5

Eskeland, G. S. (1993). *A presumptive pigovian tax on gasoline: analysis of an air pollution control program for Mexico City (No. 1076)*. The World Bank.

Gervais, D. J. (2005). Intellectual Property, Trade & (and) Development: The State of Play. *Fordham Law Review, 74*, 505.

Gomez, C. M. G. (2001). On optimal environmental taxation and enforcement: Information, monitoring and efficiency. *Natural Resource Modeling, 14*(1), 5–30. doi:10.1111/j.1939-7445.2001.tb00048.x

Jaffe, A. B., & Stavins, R. N. (1995). Dynamic incentives of environmental regulations: The effects of alternative policy instruments on technology diffusion. *Journal of Environmental Economics and Management, 29*(3), S43–S63. doi:10.1006/jeem.1995.1060

Kasa, S. (2000). Policy networks as barriers to green tax reform: The case of CO2-taxes in Norway. *Environmental Politics, 9*(4), 104–122. doi:10.1080/09644010008414553

Kopczuk, W., Marion, J., Muehlegger, E., & Slemrod, J. (2013). *Do the laws of tax incidence hold? point of collection and the pass-through of state diesel taxes (No. w19410)*. National Bureau of Economic Research. doi:10.3386/w19410

Krass, D., Nedorezov, T., & Ovchinnikov, A. (2013). Environmental taxes and the choice of green technology. *Production and Operations Management, 22*(5), 1035–1055. doi:10.1111/poms.12023

Lilliestam, J., Patt, A., & Bersalli, G. (2021). The effect of carbon pricing on technological change for full energy decarbonization: A review of empirical ex-post evidence. *Wiley Interdisciplinary Reviews: Climate Change, 12*(1), e681. doi:10.1002/wcc.681

Nannerup, N. (2001). Equilibrium pollution taxes in a two industry open economy. *European Economic Review, 45*(3), 519–532. doi:10.1016/S0014-2921(99)00028-8

Norouzi, N. (2021a). Post-COVID-19 and globalization of oil and natural gas trade: Challenges, opportunities, lessons, regulations, and strategies. *International Journal of Energy Research, 45*(10), 14338–14356. doi:10.1002/er.6762 PMID:34219899

Norouzi, N. (2022a). Regulating Sustainable Economics: A Legal and Policy Analysis in the Light of the United Nations Sustainable Development Goals. In Handbook of Research on Changing Dynamics in Responsible and Sustainable Business in the Post-COVID-19 Era (pp. 266-287). IGI Global. doi:10.4018/978-1-6684-2523-7.ch013

Norouzi, N. (2022b). A Practical and Analytic View on Legal Framework of Circular Economics as One of the Recent Economic Law Insights: A Comparative Legal Study. *Circular Economy and Sustainability*, 1-26.

Norouzi, N., & Fani, M. (2021). Monopoly and competition in the energy market: A legal analysis. *Global Journal of Business Management, 15*(2), 001-007.

Norouzi, N., & Fani, M. (2022). Globalization and the oil market: An overview on considering petroleum as a trade commodity. *Journal of Energy Management and Technology*, 6(1), 54–62.

Norouzi, N., Fani, M., & Talebi, S. (2022). Green tax as a path to greener economy: A game theory approach on energy and final goods in Iran. *Renewable & Sustainable Energy Reviews*, *156*, 111968. doi:10.1016/j.rser.2021.111968

Norouzi, N., & Sheikhi, M. (2021). Achieving Sustainable Development from the Perspective of International Environmental Law. *Eurasian Journal of Environmental Research*, *5*(1), 1–13.

Oates, W., & Baumol, W. (1975). The instruments for environmental policy. In *Economic analysis of environmental problems* (pp. 95–132). NBER.

Oates, W. E. (1995). Green taxes: Can we protect the environment and improve the tax system at the same time? *Southern Economic Journal*, *61*(4), 915–922. doi:10.2307/1060731

Shohani, A., Ataei, E., & Norouzi, N. (2021). Prevention and Suppression of Environmental Crimes in the Light of the Actions of Non-Governmental Organizations in the Iranian Legal System. *Research Journal of Ecology and Environmental Sciences*, *1*(1), 57–70.

Zhang, K. M., & Wen, Z. G. (2008). Review and challenges of policies of environmental protection and sustainable development in China. *Journal of Environmental Management*, *88*(4), 1249–1261. doi:10.1016/j.jenvman.2007.06.019 PMID:17767999

ADDITIONAL READING

Helfer, L. R. (2004). Mediating Interactions in an Expanding International Intellectual Property Regime. *Case W. Res. J. Int'l L.*, *36*, 123.

Movahedian, H., Norouzi, N., & Ataei, E. (2021). *Energy law and environmental law in oil and gas contracts with an emphasis on the MENA region.*

Norouzi, N. (2021a). Post-COVID-19 and globalization of oil and natural gas trade: Challenges, opportunities, lessons, regulations, and strategies. *International Journal of Energy Research*, *45*(10), 14338–14356. doi:10.1002/er.6762 PMID:34219899

Norouzi, N. (2022a). Regulating Sustainable Economics: A Legal and Policy Analysis in the Light of the United Nations Sustainable Development Goals. In Handbook of Research on Changing Dynamics in Responsible and Sustainable Business in the Post-COVID-19 Era (pp. 266-287). IGI Global. doi:10.4018/978-1-6684-2523-7.ch013

Norouzi, N. (2022b). A Practical and Analytic View on Legal Framework of Circular Economics as One of the Recent Economic Law Insights: A Comparative Legal Study. *Circular Economy and Sustainability*, 1-26.

Norouzi, N., & Fani, M. (2021). Monopoly and competition in the energy market: A legal analysis. *Global Journal of Business Management, 15*(2), 001-007.

Norouzi, N., & Fani, M. (2022). Globalization and the oil market: An overview on considering petroleum as a trade commodity. *Journal of Energy Management and Technology*, 6(1), 54–62.

Norouzi, N., & Sheikhi, M. (2021). Achieving Sustainable Development from the Perspective of International Environmental Law. *Eurasian Journal of Environmental Research*, 5(1), 1–13.

Posner, E. A., & Sykes, A. O. (2013). *Economic foundations of international law*. Harvard University Press. doi:10.2307/j.ctt2jbtsp

Shaffer, G. C. (2001). The world trade organization under challenge: Democracy and the law and politics of the WTO's treatment of trade and environment matters. *Harv. Envt'l L. Rev., 25*, 1.

KEY TERMS AND DEFINITIONS

Equity: Defined by UNEP to include intergenerational equity - "the right of future generations to enjoy a fair level of the common patrimony" - and intragenerational equity - "the right of all people within the current generation to fair access to the current generation's entitlement to the Earth's natural resources" - environmental equity considers the present generation under an obligation to account for long-term impacts of activities and to act to sustain the global environment and resource base for future generations. Pollution control and resource management laws may be assessed against this principle.

Polluter pays principle: The polluter pays principle stands for the idea that "the environmental costs of economic activities, including the cost of preventing potential harm, should be internalized rather than imposed upon society at large." All issues related to responsibility for environmental remediation costs and compliance with pollution control regulations involve this principle.

Precautionary principle: One of the most commonly encountered and controversial principles of environmental law, the Rio Declaration formulated the precautionary principle: To protect the environment, the precautionary approach shall be widely applied by States according to their capabilities. Where there are threats of serious or irreversible damage, lack of complete scientific certainty shall not be used as a reason for postponing cost-effective measures to prevent environmental degradation. The principle may play a role in any debate over the need for environmental regulation.

Prevention: The concept of prevention can perhaps better be considered an overarching aim that gives rise to a multitude of legal mechanisms, including prior assessment of environmental harm, licensing or authorization that set out the conditions for operation and the consequences for violation of the conditions, as well as the adoption of strategies and policies. Emission limits and other product or process standards, the use of best available techniques, and similar techniques can all be seen as applications of the concept of prevention.

Public participation and transparency: identified as necessary conditions for "accountable governments,... industrial concerns," and organizations generally, public participation and transparency are presented by UNEP as requiring "effective protection of the human right to hold and express opinions and to seek, receive and impart ideas,... a right of access to appropriate, comprehensible and timely information held by governments and industrial concerns on economic and social policies regarding the sustainable use of natural resources and the protection of the environment, without imposing undue financial burdens upon the applicants and with adequate protection of privacy and business confidentiality," and "effective judicial and administrative proceedings." These principles are present in environmental impact assessment, laws requiring publication and access to relevant environmental data, and administrative procedures.

Sustainable development: Defined by the United Nations Environment Programme as "development that meets the needs of the present without compromising the ability of future generations to meet their own needs," sustainable development may be considered together with the concepts of "integration" (development cannot be considered in isolation from sustainability) and "interdependence" (social and economic development, and environmental protection, are interdependent). Laws mandating environmental impact assessment and requiring or encouraging development to minimize environmental impacts may be assessed against this principle.

Transboundary responsibility: Defined in the international law context as an obligation to protect one's environment and prevent damage to neighboring environments, UNEP considers transboundary responsibility at the international level as a potential limitation on the sovereign state's rights. Laws that limit externalities imposed upon human health and the environment may be assessed against this principle.

Chapter 8
Environmental Intellectual Property Law

ABSTRACT

Climate change is a phenomenon that affects humanity as a whole. One step in its solution is the development of technologies that allow us to adapt and mitigate its effects. However, most of these environmentally friendly technologies are protected by intellectual property rights, which, far from allowing their access, sometimes constitute an obstacle, fundamentally, for developing countries. In this sense, it is up to the States to dictate public policies that promote the research, development, dissemination, and transfer of these technologies through the use of the tools of the intellectual property system.

INTRODUCTION

There is no doubt that one of the most pressing issues of recent times is the complex challenges posed by environmental protection, climate change, and the search for solutions to mitigate its effects in the short and medium term. Ban Ki-Moon, Secretary-General of the United Nations, asserts that climate change is one of the most complex, multifaceted, and serious threats facing the planet. The response to this threat is closely related to the pressing issues of sustainable development and global justice, the economy, poverty reduction, society, and the world we want to leave our children.

In 1992, the United Nations Framework Convention on Climate Change was approved in Rio de Janeiro; It established the general objectives and the institutional foundations for international efforts to combat climate change. Likewise, it was found that the development of technological and ecologically sound innovation is, to a large extent, one of the mechanisms to solve the problems we face. In the words of Francis Gurry, Director-General of the World Intellectual Property Organization:

The power of ingenuity is our best hope for restoring the delicate balance between ourselves and our environment. It is the only weapon we have to face this global challenge and the one that will allow us to move from the gray technologies of the past, based on carbon, to the green innovation of the future, without carbon dioxide emissions(Nagel, 2017).

DOI: 10.4018/978-1-6684-4158-9.ch008

In Chapter 34 of the United Nations Program of Action Rio, 1992, also known as Program 21, the definition of environmentally sound technologies is established, these "... protect the environment, are less polluting, sustainably use all resources, recycle more of their waste and products and treat residual waste more acceptably than the technologies it has come to replace. Section 34.3 of the same document states that "... environmentally sound technologies are not merely isolated technologies, but total systems that include technical knowledge, procedures, goods and services, and equipment, as well as organization and management procedures(Momtaz, 1996)."

Technological innovations that respect the environment are protected, in most cases, by utilizing Intellectual Property Rights. However, while on the one hand, the Intellectual Property system contributes to promoting the creation and diffusion of new technical advances, it can also become an obstacle to their acquisition, fundamentally for developing countries.

There is no doubt about the need for these technologies to be accessible to all. For this reason, it is essential not only to accelerate its transfer and dissemination but also to have financial resources, obtained through the promotion of investment, destined for its research, development, application, and dissemination. This position was taken into account in 2007, at the Bali Conference, when the Bali Action Plan was approved. In said plan, the countries were encouraged to intensify the work related to the development and transfer of technology to support measures to mitigate and adapt to climate change. In turn, the following measures were adopted:

- Remove barriers and create incentives to encourage the use of clean and affordable technologies;
- Accelerate the deployment, diffusion, and transfer of environmentally responsible technologies;
- Cooperate in research to obtain innovative technologies and examine the effectiveness of the mechanisms established at the level of technical cooperation.

Consequently, Intellectual Property Rights have been, in recent times, a controversial issue in the debate on the transfer and development of environmentally responsible technology. Fundamentally, regarding the most appropriate way to take advantage of this system or carry out the necessary reforms to develop and spread. However, "ensuring access to climate-friendly technologies, at affordable prices, is a crucial issue in international public policy, which affects economic, legal, geopolitical and security issues(Cullet, 2005)."

This chapter aims to analyze the existing relationship between Industrial Property and access to environmentally friendly technologies; without proposing formulas, outline some elements to be taken into account by developing countries for their access. In the first part of the present work, a recount of the means of protection of technology that respects the environment will be made; in the second part, it will be explained how the use of the Intellectual Property system contributes to access to technology that respects the environment; in the third part, the different ways of accessing environmentally friendly technologies will be analyzed, and, finally, in the fourth part, the negotiations under debate in the framework of the United Nations Framework Convention on Climate Change will be presented. Climate and Intellectual Property.

PROTECTION OF ENVIRONMENTALLY FRIENDLY TECHNOLOGIES

Legal Instruments for the Protection of Environmentally Friendly Technologies

There are various legal instruments for protecting technologies that can be used to protect environmentally responsible technologies. Through these, "... it allows the creators of new technologies to appropriate the results of their inventive efforts, and protects them from the competition of potential users of the same technologies...". However, there are two opposing positions regarding the role played by the Intellectual Property System in the protection of environmentally friendly technologies. On the one hand, one proposes stricter Intellectual Property rights regimes that stimulate innovation in climate technologies. On the other hand, he considers that the Intellectual Property system should be more flexible to expand access to this technology, especially in developing countries.

Despite being one of the most used, Patents are not the only legal instrument for protecting technologies. In addition to these, others have the same purpose. Among them, it is worth highlighting the protection through industrial secrecy, utility models, industrial designs and models, and plant varieties. However, it is essential to specify that these protection instruments cannot be seen or analyzed isolation. Different legal instruments for their protection can converge in the same technology(Cullet, 2005).

Inventions and Patents

The view that patents are intended to protect technologies and promote their development and dissemination is widespread. That is why, when the applicant for a patent meets the formal requirements and the technical solution meets the patentability requirements established by law, the public power, that is, the state, makes an administrative concession called a patent. In this way, it grants its owner an exclusive right to exploit the invention, temporarily and territorially, and "... is limited to establishing the conditions of exclusive exploitation that will favor the appropriation by the inventor of the benefits derived from the patented technology." That is, the patent owner acquires a negative right through which he can prevent third parties, without his consent, from using, manufacturing, selling, offering for sale, or importing the patented product.

In exchange, the owner makes a sufficiently clear and complete disclosure of the invention, which allows its subsequent reproduction and contributes to humanity's technological development. In Mathély's opinion:

The inventor publishes his invention and thus releases it into the public domain; but, in exchange, he receives for a limited duration the exclusive right to exploit that invention for his benefit or to authorize the exploitation also for his benefit. The patent is the title, which describes and defines the protected invention and confers the owner the exclusive right to exploit that invention(List & Qui, 2004).

On the other hand, the right to prohibit, held by the patent owner, has its limitations, regulated in each of the laws of the different countries. These exceptions to the scope of patent rights allow third parties to use the invention without the need for the owner's prior consent. Generally, those uses are justified for research and teaching purposes and even for non-commercial purposes.

The disclosure of the invention is essential for developing new environmentally responsible technologies. Through the disclosure made by the applicant for the registration of the patent, it is allowed that people trained in said technical area reproduce the technology. This way, access to information is achieved and can carry out new investigations. In addition, it is guaranteed that the technology can be used effectively once the patent expires and passes into the public domain.

Ilian Iliev believes that a registered patent can attract venture capital, stimulate strategic alliances, offer protection against litigation, and create opportunities for mergers and acquisitions. Despite the patent system being an incentive for the development of environmentally friendly technologies, some consider that in some cases:

… may be an obstacle in transferring technologies to developing countries at affordable prices. Especially when it is the developed countries that possess and own the vast majority of said technologies. There are examples of developing countries that have been hampered in introducing climate-friendly technologies or products because (…) they were protected by patents and because patent holders made unreasonable demands.

In recent years, applications for and grants of invention patents to protect environmentally friendly technologies have increased significantly. For example, the Derwent patent database reveals that between 2003 and 2008, inventions to reduce energy consumption, protected by patents, reached 1,200, compared to only 481 in the previous five years. Patents are of different ranges of technologies, for example, new methods of energy generation products to increase the efficiency of processes or energy conservation.

Alan MacDougall, a partner at the Intellectual Property law firm Mathys&Squire, believes that "What we are seeing is typical of any technology: the deployment of different types of technologies is growing very much in line with the registration of patents. There is a correlation of the whole evident between the number of patent applications and the number of products coming onto the market".

However, the fundamental problem in applying environmentally responsible technologies is how to bring them to market. Although many companies develop new environmentally friendly technologies, it is estimated that it takes years to commercialize and use them. It is essential to point out that environmentally responsible technologies take time to be marketed or used goes against the speed with which it is necessary to protect the environment.

On the other hand, most patent applications to protect environmentally responsible technologies come mainly from developed countries. In contrast, developing countries have very little representation in the statistics on the generators of patent applications and patents that protect environmentally responsible technologies. Still, it does not exempt them from being claimants of said technologies. For example, in 2005, the percentage of patents on renewable energy had been conceived fundamentally in countries such as the United States, Japan, countries of the European Union, among others.

Lastly, with the aim of more agile protection of inventions related to environmentally friendly technologies, in the Intellectual Property Offices of countries such as Australia, the Republic of Korea, and the United Kingdom, a "fast window" system was established " to grant the registration of the patent. For example, in the United Kingdom since 2009, in its Intellectual Property Office, a prompt window system was established, where applicants can request that part or all of the patent application process be accelerated. This makes it possible to significantly reduce the period between the application and granting the patent, from three to five years to only eight or nine months of processing the application(Anderson, 2009).

Trade Secrecy

The technical knowledge that gives its owner a value or a competitive advantage with respect to third parties in the market is protected through the industrial secret. The industrial secret "… implies that whoever develops certain technology can keep it in his possession and factually and legally prevent third parties from having access to it. However, it cannot prevent third parties from independently developing the same technology." The owner of the industrial secret must establish the necessary mechanisms to preserve the secrecy of the technology. In case of losing said condition, it will cease to have economic value.

However, if environmentally responsible technology is protected by industrial secrecy, dissemination and circulation are prevented. Consequently, while it is kept as an industrial secret, it can only be used by its owner or transferred, with his consent, through voluntary licenses to third parties, and its disclosure and access will be difficult.

Utility Models

Utility models are legal figures intended to protect the functional aspects of technical creations, with an inventive level lower than that required for patentable inventions. They are always associated with an object, making it impossible to protect chemical substances or procedures in this way.

Not all legislations regulate this legal figure, or they only do so through the patent system, which makes their protection difficult as they must comply with the same patentability criteria as for patents. This lack of protection means that researchers do not feel motivated in these "minor inventions," and trade, disclosure, and access are indirectly affected.

Guillermo Cabanellas de las Cuevas, citing S. Matsui, states that "Utility models have been recommended as a means to protect the type of innovative activity typical of developing countries." The author himself explains that the preceding is doubtful and that in certain limits of the figure in patent laws, he increases doubts about its potential benefits and costs.

Industrial Designs and Models

This legal figure protects the ornamental and aesthetic elements of industrial designs, drawings, and models, separating the protection from its functional aspects. According to the doctrine, "... greatly limits its potential as an instrument for the protection of technological innovations". The Paris Convention requires the protection of this figure and the TRIPS Agreement reaffirms it, an aspect that guarantees its knowledge and dissemination worldwide(Dreyfuss & Strandburg, 2009).

Biodiversity Protection

This legal figure protects new plant varieties different from known, homogeneous, stable, and innovative commercialization and denomination. The TRIPS Agreement forced the member countries of the WTO to protect this figure in three ways: through patents, through an effective sui generis system, or a combination of both.

The effects that these TRIPS regulations have had on sustainable development are undeniable. At the point where we find ourselves, it is evident how free varieties have been used for protected varieties. The owner has the right to receive royalties for their use, limiting their access.

Industrial Property systems have also encouraged the development of transgenic plants that, when released irrationally, could cause harmful and irreversible effects on the environment, such as the emergence of new pests. Hence the need for States to dictate strong regulations on biosafety.

But the TRIPS Agreement is not the only international mechanism that protects plant varieties: the International Treaty for the Protection of New Varieties of Plants, administered by UPOV, is another tool that seeks this end. This international instrument has been revised three times: 1972, 1978, and 1991. The 1991 Act strengthens the Intellectual Property rights of plant variety breeders much more. In this sense, and to cite one example, farmers are prevented from using the product of their harvests for subsequent plantings, forcing them to repurchase the seeds, which is a brake on access to the variety(Dutfield, 2000).

INTELLECTUAL PROPERTY AND ACHIEVING ENVIRONMENTAL TECHNOLOGIES

There are two dominant positions in the controversy regarding access to environmentally friendly technologies and the Industrial Property system. On the one hand, those who consider that the system constitutes an obstacle to their access and diffusion "... because the countries that most need them simply cannot afford to pay market prices" or because they do not have the technological capabilities to develop them. Khor even asserts that … the fact that companies from developed countries own most of the intellectual property rights of environmentally sound technologies is an obstacle for developing countries to have meaningful and affordable access to these technologies. Among the obstacles identified (...) it is worth mentioning: a) high royalties; b) the refusal of the patent holder to grant voluntary licenses; the perpetuation of patents (evergreening); d) increase in patent litigation; d) impediments to innovation.

An example of the actual existence of such obstacles is the inability of Indian and Korean companies to obtain production rights for substitutes for chlorofluorocarbons, which damage the ozone layer of the atmosphere. This is because the patent holders have established in the negotiation to grant licenses conditions that the companies cannot comply with.

On the other hand, those who consider that Intellectual Property rights and the incentives they offer are essential for the development of innovations, since "... they are essential to stimulate investment in 'green' innovations and contribute to the rapid diffusion around the world of new technologies and new knowledge". Industrial Property attracts investments, facilitates entry into new markets, and allows effective collaborations. Misconceptions or lack of knowledge of how technology transfer works and the role of Industrial Property in the process hinder effective collaboration and promote knowledge exchange. However, we share the criteria of Pedro Roffe, who states that Intellectual Property is important, but in its fair measure, since too many Intellectual Property rights endanger innovation. Too much can kill your spread. So he demands a double challenge (...) an accelerated and large-scale implementation of clean technologies and policies that support the diffusion and rapid transfer of these technologies and offer incentives for innovation and investment in new ones(Cullet & Raja, 2004).

TRIPS Agreement for Access to Environmentally Friendly Technologies

Although it does not establish a uniform regime for the protection of Industrial Property, the entry into force of the TRIPS Agreement did harmonize this right by instituting minimum standards to be taken into account in national legislation. However, with the use of flexibilities, countries have a margin of maneuver to interpret and put into practice the various standards and, therefore, gain access to environmentally responsible technologies.

Professor Carlos Correa believes that the laws should only reflect the minimum provisions required by the TRIPS and have a legal figure for compulsory licenses without limiting the grounds on which it is applied27, thus achieving the creation of mechanisms to be able to exploit environmentally responsible technologies. To have mechanisms to access these technologies, they can use the principle of exhaustion of rights and parallel imports, in turn, of the patentability exceptions. The author himself asserts that countries can: … adopt a series of measures within the existing norms, for the application of other measures of protection of the environment and the promotion of sustainable development, as well as, ensure changes in the existing normative regime, with the purpose of:

- Exclude from patentability all substances existing in nature;
- Strengthen the system of compulsory licenses, even when patents are not being worked on and also for reasons of environmental protection;
- The sui generis protection of the knowledge of local and indigenous communities;
- The revision of the TRIPS agreement, for example, concerning duration, to promote the effective transfer of environmentally sound technologies(Cullet, 2001).

Patentability Exclusions in Environmentally Friendly Technologies

The issue of patentability exclusions has been discussed internationally. For example, at the Summit of the Group of Eight plus Five, held in Gleneagles (Scotland). In this framework, the delegation of India proposed two options for the exclusion of patents: the general exclusion of patentability for environmentally sound technologies; or applying only the exclusion in developing countries.

The regulation of patentability exclusions in national laws will depend on their legislative policies. In the TRIPS Agreement, countries are given a margin of autonomy, which allows them to decide which exclusions to include or not. Article 27, paragraph 2, establishes the possibility that Members may exclude from patentability inventions whose commercial exploitation in their territory must necessarily be prevented to protect public order or morality, including the health or life of persons. Or of animals or to preserve plants or avoid serious damage to the environment, provided that this exclusion is not made merely because the exploitation is prohibited by its legislation.

This article undoubtedly offers the possibility that the technologies included within the exclusions of patentability can be used, to the extent that they have access, without the obligation to grant protection employing a patent. Khor affirms that it is necessary "… to make the corresponding amendment to the TRIPS Agreement so that its members (…) can exclude these environmentally responsible technologies from the patent system." However, we do not share this opinion.

No amendment to Article 27 paragraph 2 is necessary for members to exclude environmentally responsible technologies from patentability. The agreement allows States to freely Remember that it establishes minimum requirements to be taken into account but that it will be the internal policies of the countries that will define what will be regulated in their national legislation.

The situation of environmental protection and climate change is a phenomenon that affects humanity and endangers people's health and life. Therefore, granting exclusive rights that prevent the dissemination and use of environmentally responsible technologies in a territory can endanger the health and life of people. Therefore, its inclusion as a reason for patentability exclusion may promote access to and use of environmentally responsible technologies(Matthews, 2003).

Exhaustion of Duty and Parallel Imports in Environmentally Friendly Technologies

The exhaustion of the right is a limitation to the exclusive right held by the owner of the Industrial Property. Through this: … the owner of the technology protected by an Industrial Property right may not prevent further marketing (offer and placing on the market as well as import and possession for these purposes) or use (which in principle are acts of exploitation that are reserved in exclusive) to those who have acquired products protected by the patent (whether patented products or products directly obtained through the use of the patented procedure) marketed by him or by a third party with his consent.

There are three types of exhaustion of rights, which will be defined in the national provisions of each country. These types of exhaustion are:

- National exhaustion: it is the most rigorous version of this figure. The right to the technology is exhausted only in the country where the owner or an authorized third party is placed on the market. That is to say, and the owner cannot prevent that once placed on the market, it is marketed within the national geography by any distributor in the conditions and places it deems. However, the right comes back to life or is not understood to be exhausted when the technology is put on the market in a second country with the owner's consent, and a third person tries to import it to the country where the registration is located. In this case, the owner can prevent the entry of that merchandise into his territory since his right has not been exhausted in the second country.
- Regional depletion: This is similar to national depletion. Through it, the owner of the technology cannot prevent its circulation in the region in question, once it has been marketed by itself or through an authorized third party; but if he has the power to prevent their entry into the region when the placing in the trade has been made outside the region in question. For example, this is what happens in the European Union.
- International exhaustion: the owner, once the technology is marketed for the first time, either by himself or through an authorized third party, loses any possibility of control or intervention over his destiny. That is to say, the place of commercialization of the legitimate good is irrelevant since the owner of the Intellectual Property right lacks the legal tools to intervene in the subsequent stages of the trade. In this way, the bases are created for the full freedom of parallel imports.

The exhaustion of rights is another possibility that developing countries have to access technologies related to the environment. Note that, with the international exhaustion of rights, any technology lawfully offered could be accessed by the patent owner or an authorized third party, regardless of where the technology was offered, which can lower the costs of obtaining. Also, because with the international exhaustion of rights, the legitimacy of parallel imports is recognized, developing countries can use it and access environmentally sound technologies(Khor, 2002).

Compulsory Licenses in Environmentally Friendly Technologies

The granting, by the States, of compulsory licenses can be an effective mechanism for the acquisition, by developing countries, of environmentally friendly technologies. The TRIPS agreement does not expressly mention the compulsory licensing regime nor the causes for its granting; only in its Article 31 does it establish a set of regulations related to "Other uses without authorization of the owner of the rights." The Paris Convention, with a very limited mention of the grounds, makes literal mention of compulsory licenses.

Within the provisions established in the article mentioned above 31 of the TRIPS Agreement, it is detailed that in case of national emergency or other circumstances of extreme urgency, to grant a compulsory license, it will not be necessary for the applicant to have previously requested, from the holder of the patent, the granting of a voluntary license. The truth is that, in the understanding of the authors, the state doesn't need to declare itself in a situation of this type to grant compulsory licenses, more so if the cause refers to obtaining technologies to prevent or mitigate the harmful effects to the environment and adaptation to climate change. More than an emergency that concerns a country or region, this is an aspect that affects the world. Especially when in countries like the United States or the United Kingdom, governments have granted compulsory licenses to promote cheaper technologies and products in the industrial sector. In the United States, specific provisions have been enacted, such as the Atomic Energy Act (42 USC Sec. 2183), the amendment to the Patent and Trademark Act of 1980, or the Bayh-Dole Act (42 USC Sec. 7608). and the Air Quality Protection Law. The latter establishes the granting of compulsory licenses when patented technologies are necessary to comply with environmental standards. This reaffirms the idea that developed countries even use this legal figure(Khor, 2002).

Undoubtedly, this is an option to be adopted by developing countries for access to patented environmentally friendly technologies, when they are expensive, it is not possible to negotiate a reasonable price with the patent holder, they are not in exploitation in the country or for the license of the desired generic product. Therefore, governments have to exercise their rights to grant compulsory licenses. It is up to the political will of each state to enact laws that simplify obtaining compulsory licenses for this type of technology.

However, despite the possibility of using compulsory licenses, developing countries often do not have the skills and experience to produce and use such technologies. This prevents them from being materially unable to incorporate them, despite having the legal mechanisms to access said technologies.

A solution to this problem could be to establish a system similar to the one established in paragraph 6 of the Doha Declaration on the TRIPS Agreement and Public Health, but in this case, for environmentally responsible technologies. It establishes special compulsory licenses for exporting medicines to countries whose manufacturing capacities in the pharmaceutical sector are insufficient or non-existent to use compulsory licenses under the TRIPS Agreement effectively(Khor, 2002).

Exceptions to Patent Rights in Environmentally Friendly Technologies

Article 30 of the TRIPS Agreement allows States to have exceptions to the exclusive rights conferred by a patent, provided that they do not unjustifiably interfere with its normal exploitation or cause unjustified damage to the legitimate interests of the owner. Among the exceptions regulated by governments, it is worth highlighting the exception for teaching and scientific or technological research. Professor Massaguer asserts that the research exception is based on: … the need to prevent the Industrial Property system from becoming - contrary to its postulates - a mechanism that stops technical progress and, in particular, the use of patented technical rules for the development of innovations and improvements; it only covers experimentation (rectius: acts of exploitation carried out on the occasion of experimentation) carried out precisely with the products or procedures protected by the patent and precisely to know, test or developing the patented invention itself.

Therefore, we consider that this is another means for States to obtain or improve products or technologies that are friendly to the environment. The regulation of this exception in the national legislation allows that, during the term of the patent, the research institutes can carry out studies of the patented inventions, prepare them for their exploitation once the term of validity expires, and even obtain improvements. At the same time, this contributes to technological progress, the diffusion or dissemination of technology, and the knowledge and awareness of the population of the problems that afflict the environment and their possible solutions.

However, in this case, the same thing happens with compulsory licenses; if the capacities created for research in the sector are insufficient or deficient in the countries, the exceptions to patent rights cannot be used. Bear in mind, furthermore, that generally, developing countries are not only not generators of technology but also do not have trained personnel to carry out research, which makes it difficult to make effective use of the said exception to patent rights(Cullet, 2001).

WAYS TO ACCESS ENVIRONMENTALLY RESPONSIBLE TECHNOLOGIES

Collaboration Mechanisms in Environmentally Friendly Technologies

At present, international organizations and countries are immersed in creating more appropriate collaboration mechanisms than the existing ones to share environmentally responsible technologies. In this scenario, dissimilar ideas have been born, which have the diffusion and transfer of ecologically clean technologies as a common objective. These include the creation of: … the patent pool, the free patent pool, open-source innovation, open license agreements, and, finally, (…) commitments to waive legal action for infringement of patent rights. However, these (…) are generally voluntary and have been devised by the technology owners themselves, who understand that the benefit of pooling technologies from various sources far outweighs the immediate benefit that would accrue to restrict the use of those inventions(Cullet, 2001).

Patent Pool in Environmentally Friendly Technologies

A patent pool is "…an exchange of patent rights by several companies. One or more of the patent holders, or a separate entity, have the right to grant licenses to third parties under such pooled patents". However, according to the criteria of Krattiger and Kowalski, the patent "... simplifies the articulation of Intellectual Property, but does not necessarily lead to technology transfer or access and distribution within the market".

There are a variety of types of a patent pool, but "... the common denominator among them is that the holders agree to reciprocally assign the respective licenses, for which in some cases the figure is also called 'reciprocal license agreement'". Among the types of consortium can be found those of a closed and open nature. Those of a closed nature can constitute an obstacle for the diffusion of environmentally responsible technology since only the consortium members will have access; however, those of an open nature guarantee that anyone can make use of the technologies shared by the patent holders.

For example, in 2010, large companies dedicated to the information and communication technology (ICT) sector created Green TouchTM. The global consortium aimed to create new technologies to make communication networks 1,000 times more efficient. In their energy consumption than those currently used. Its members include AT&T, China Mobile, Freescale Semiconductor, Huawei, Samsung Advanced Institute of Technology, Swisscom, University of Melbourne's Institute for a Broadband-Enabled Society (IBES). Dr. Steven Chu, Secretary of Energy of the United States, believes that "the best way to tackle truly global challenges has always been to bring together the best minds in a creative environment and without any pressure (…) Green TouchTM is an example of this type of action, bringing together scientists and technologists from all over the world, and from different fields, in an environment of open innovation to approach the problem from many different perspectives".

In Krattiger and Kowalky's view, patent pools are "…competitively beneficial, as they can help resolve patent conflicts, make pooled patents available to others, or resolve disputes over patents they block. On the other hand, (...) being a horizontal agreement between competitors (...) it has the potential for abuse and as a screen for anti-competitive monopoly". Which, without a doubt, will imply an obstacle for the diffusion and transfer of ecologically sound technology.

Despite this, there is no doubt that alliances are created through patent pools, thus ensuring that innovators work together. This is how collective knowledge and creativity seek solutions to face environmental problems(Cullet, 2001).

Public Domain in Environmentally Friendly Technologies

One of the ways to achieve the transmission and diffusion of ecologically friendly technologies is to place them directly in the public domain. The territorial nature of Industrial Property rights means that not all technologies are protected in all countries. The choice of the territories in which they are protected will depend largely on the marketing interests of their owner.

Therefore, in countries where the technology has not been protected, it will be considered in the public domain and may be used freely. However, there is a paradox, the countries that can make use of public domain technologies often do not have the financing, with trained personnel to be able to produce it, or have null or insufficient production capacities and situations that will hinder access to environmentally sound technologies, fundamentally for developing countries(Cullet, 2001).

Open Innovations in Environmentally Friendly Technologies

Open source has its origins in the method used to develop new computer programs. Its users were allowed to use it and modify it according to their interests and needs. Likewise, its distribution is permitted in its original or modified form.

According to the doctrine, "By analogy, the term 'open source is also currently applied to other fields of innovation in which a technological platform is constituted that allows others to use and adapt..." technologies. The same author states that "... 'open innovation' has also emerged (...) a mechanism with characteristics similar to 'open source,' but broader, by which numerous companies come together to take advantage of synergies and collaborate with other parties that work in related technologies.

In his book Open Innovation, Henry Chesbrough created open innovation: The new imperative for creating and profiting from technology. Said author points out that it is the opening of the investigation process of a company to third parties. He, in turn, understands that, in a world characterized by distributed knowledge, companies can create more value and better exploit their research processes if they integrate knowledge, Intellectual Property objects, and external products in their innovation work. He adds that the products, technologies, knowledge, and objects of Intellectual Property that are not used in the company can be made available to third parties through figures such as licenses, joint ventures, or spin-offs. In this way, it is possible to make them profitable. According to its creator, "The concept of open innovation consists of working in a world full of knowledge in which not all the talents work for the company, so it is advisable to look for them, get in touch with them and take advantage of their abilities. "

An example of the first open innovation business model was the one carried out by Procter & Gamble in 2000 when it created the Connect+Develop platform. Three connections were created to link R&D activities and internal resources with the outside world: technology sector entrepreneurs, Internet platforms, and retirees(Cullet, 2001).

INITIATIVES TO SHARE ENVIRONMENTALLY RESPONSIBLE TECHNOLOGIES

Eco-Patent Commons in Environmentally Friendly Technologies

An initiative called Eco-Patent Commons was launched in January 2008. The main objective is to create an ecologically responsible patent pool. It arose within the World Business Council for Sustainable Development (WBCSD) framework. "The idea is to contribute to the consortium patents that have environmental advantages for other manufacturers and that are freely available to other contributors - as well as to companies and people who do not belong to the consortium." At its inception, it had a total of 31 patents provided, free of charge, by the companies; and already in 2009, they had a total of 100 patents, maintaining their growth today(Cullet, 2001).

The donated patents are published through its website, which will be freely available to the public. "The companies participating in the project are expressly obliged to desist from legal actions for patent infringement, provided that the inventions are used for beneficial purposes for the conservation of the environment. Among these purposes, the reduction or elimination of the consumption of natural resources and (...) the production of waste and pollution". In this way, not only can these patents be used and disseminated, but they are also used as a tool to stimulate innovation.

In this case, the people or companies that access these technologies use them and do not have to register. Its fundamental objective is to be "... easy to administer and that its use should not present difficulties for potential users. It is very important that Eco-Patent Commons be easy to use by others". In the words of Donal O'Connell, Nokia's representative on the pool, ... the fact of participating in the plan provides benefits that are not financial, in addition to those that can be obtained from the technological point of view by being able to take advantage of the technology of others. Conversations with people and companies with similar ideas have made it possible to create a valuable network of experience that is making itself known collectively and committed to environmental initiatives. For example, Nokia telecommunications engineers have been studying the environmental benefits of Bosch patents concerning automotive technology and the possibility of wider application.

Although this initiative stimulates the dissemination and dissemination of environmentally responsible technologies and collaboration among its members for the development and continuation of new technologies and the improvement of existing ones, the volunteerism of people and companies is required to participate in them. Although it has its merits, it is still not fundamentally the solution to our problems. It does not seek to transfer technology, mainly to developing countries that do not have funding for research and trained personnel(Cullet, 2001).

WipoGreen in Environmentally Friendly Technologies

WIPO Green is a free platform to disseminate and access environmentally responsible technologies. It "...seeks to accelerate the adaptation, adoption, and distribution of climate-sound technologies, especially in developing and emerging countries". Through it, the worldwide diffusion of ecologically responsible technologies is facilitated.

The platform not only facilitates access to patent information. In addition, it provides information on integrated technology solutions, such as technical expertise and specialized knowledge. Likewise, "... it is designed to improve knowledge of existing environmentally sound technologies and increase access to them, as well as to contribute to the search for solutions to specific problems in the field of technologies related to climate change, and offers complimentary opportunities of commercialization and creation of alliances". WIPO Green also provides other services to stimulate the diffusion and transfer of environmentally sound technology. Additional services provided include:

- Access to complementary information on the patenting of environmentally sound technologies and the technology transfer process;
- case study materials illustrating the different types of agreements that arise from the many and varied circumstances in which technology transfer occurs;
- training, for example, in technology licensing;
- individualized dispute resolution procedures;
- information on potential funding sources (e.g., national governments, international organizations, foundations or philanthropic institutions, private sector entities);
- Licensing tools, such as model clauses, to support license negotiations.

However, like the Eco-Patent Commons, Wipo Green is not a sufficient mechanism to achieve the diffusion and transfer of technology as required by Agenda 21 because it depends largely on the active participation of stakeholders. However, "...it is not only a practical way forward in the fight against climate change, but it will also help strengthen and expand technology networks." What is certain is that both one and the other simplify the search for the essential technologies to combat climate change. Negotiations under discussion in the framework of the united nations framework convention on climate change and intellectual property since December 2007, at the thirteenth session of the conference of the Parties to the United Nations Framework Convention on Climate Change (UNFCCC), developing countries and groups made up of them have submitted proposals on development and technology transfer. The most important was submitted in September 2008 by the Group of 77 and China, calling for a new technology mechanism under the UNFCCC to accelerate the development and transfer of technology and support the effective implementation of the UNFCCC relating to technology and financing. In addition, they set out the rationale, criteria, and institutional arrangements for putting in place a new technology mechanism, including a new technology executive body, a multilateral climate technology fund, a technology action plan, and activities eligible for funding.

Concerning Intellectual Property rights, other countries highlighted the need to address technology transfer problems. They were the delegations from China, Cuba, India, Indonesia, and the United Republic of Tanzania(Cullet, 2001).

In June 2008, at the UNFCCC Conference on Climate Change, held in Bonn, concerning patented technologies, Brazil proposed establishing a multilateral public fund to acquire licenses. Likewise, the possibility of resorting to compulsory licenses and elaborating a declaration similar to the Doha Declaration on the TRIPS Agreement and public health was explored. For its part, India stated that an adequate regime must accompany new technologies, the transfer of technology and technical knowledge in terms of Intellectual Property rights, and proposed a system of joint development with the exchange of these rights; He added that global funding initiatives require the public acquisition of intellectual property rights globally to ensure that products and services are available at affordable prices.

Pakistan proposed the establishment of an international agreement or system on compulsory licensing of climate-friendly technologies similar to that for health and the establishment of technology pools or patents that allow technology transfer to developing countries at affordable prices. In addition, he suggested that the period of validity of patents be reduced, in the case of climate-friendly technologies; and incentives were given to the owners of the technologies so that they could put into practice a system of differentiated prices so that the developing countries could pay less for access to the technologies.

At the Poznam Conference in December 2008, the debate on technology transfer and the role of Intellectual Property continued. Reforms were proposed to prevent it from being an obstacle to the transfer of environmentally friendly technology, recognizing that the current Intellectual Property system is essential for the development and effective dissemination of new technologies necessary to face climate change. In this framework, although the issues discussed continue without an effective solution, it is worth highlighting the approval of the Strategic Program for Technology Transfer at the initiative of the Global Environment Fund(Dutfield, 2000).

In June 2009, at the conference held in Bonn, the Group of 77 and China, the Philippines, and Bolivia independently submitted proposals on Intellectual Property rights. In general, it is proposed to initiate discussion forums to exclude climate-friendly technologies from patentability compulsorily; create a global technology fund for climate change; and exclude from the patent system biological resources, such as the different varieties and species of animals, plants, and microorganisms and their parts that are used to mitigate or adapt to climate change. In addition, it was suggested that specific measures be taken and mechanisms developed to eliminate obstacles to the development and transfer of technology to developing countries, derived from the protection of Intellectual Property rights; likewise, immediately guarantee that the transfer of technology to developing countries is carried out adequately and that it allows for the effective use of the technologies.

Similar proposals to the previous ones were made by Bolivia, Bangladesh, and India, in November 2009, during the conference on climate change held in Barcelona. On the other hand, at the fifteenth session of the Conference of the Parties, which took place in December 2009 in Copenhagen, it was urged to eliminate them on several occasions and include them in the draft decision on technology.

During these sessions, although the developing countries and the groups united with them have advocated addressing and adopting measures to solve the existing problems between technology transfer and Intellectual Property rights, the developed countries have remained the idea of adhering to a strong rights regime to make way for innovation and technology transfer to developing countries. The United States and the European Union have not accepted the proposals that have been made and deny the obstacle that Intellectual Property rights represent for the transfer of technology.

At the Cancún Conference in December 2010, no mention was made of the term Intellectual Property, which undoubtedly meant a setback for developing countries as the subject was completely excluded from the text. At the sixteenth session of the Conference of the Parties, some countries led by India tried to reintroduce the issue of Intellectual Property rights in the negotiating agenda of the Framework Convention on Climate Change. At the sessions held in June 2011, India proposed that Intellectual Property rights be included in the provisional agenda of the seventeenth session of the Conference of the Parties to be held in Durban, South Africa, considering, contrary to the United States, that these issues had not been adequately addressed at the Cancun Conference. India's proposal reaffirms that a global regime that efficiently and effectively manages the Intellectual Property rights of climate-friendly technologies is urgently needed to promote their development, use, diffusion, and transfer.

Most developing countries supported the proposal. However, many developed countries and others, such as Mexico, refused to include India's proposal in the Durban program. The United States argued that these issues had already been discussed without any agreement being reached or expected to be reached. Contrary to what is believed, a strict regime in terms of Intellectual Property rights does not hinder the transfer of technology but rather cements it.

In 2011, the adoption of the decision known as the Durbam Platform favored the developed countries to eliminate the terms equity and common but differentiated responsibilities, clearly enunciated in the United Nations Framework Convention on Climate Change and its main decisions. This issue brought about strong contradictions and debates at the Climate Change Conference held in Lima in 2014 until these terms were reintroduced.

Likewise, the developing countries managed to include in the new text that the contributions to be made by the countries will maintain a balance between mitigation, adaptation, and the transfer of financial and technological resources; the text should not "focus on mitigation"; the issue of "loss and damage" caused by climate change is duly mentioned in the new draft text. In turn, there will be no formally approved process for assessing each country's intended contributions before the 21st session of the Conference of the Parties in Paris; and the conditions and information related to the "contributions" that each country will make will not be as demanding for developing countries as those foreseen in the previous draft.

Finally, at the 2015 Paris Conference, the agreement known as the Paris Agreement was signed. It is established that the Parties' developed countries will provide financial resources to developing countries for assistance, both in mitigation and adaptation. However, this support remains voluntary.

On the other hand, the importance of effectively developing and transferring technology to improve resilience to climate change and reduce greenhouse gas emissions is recognized. It is urged to strengthen cooperation actions for the development and transfer of technology, with the aim of its dissemination and deployment, also taking into account the importance of technology for the implementation of mitigation and adaptation measures. A Technology Mechanism is established to promote and facilitate technology development and transfer to support the agreement's implementation.

Likewise, it is recognized that it is essential to enable, encourage and accelerate innovation, which will be supported, among others, by the Technology Mechanism and, with the financial means, by the Financial Mechanism of the Convention(Cullet & Raja, 2004).

All this to promote the collaborative approach in research and development work, and facilitate access to developing countries, Parties to the Convention. It also provides financial support to these countries to strengthen cooperation activities in the development and transfer of technology.

Furthermore, it is recognized that capacity-building, under the agreement, should enhance the capacity and competence, in particular, of small island developing States, which are particularly vulnerable to the adverse effects of climate change, to develop an effective action against it and apply adaptation and mitigation measures, which facilitate the development, diffusion, and deployment of technologies, access to climate finance, aspects pertinent to education, training and public awareness and the communication of information in a transparent, timely and accurate manner. Finally, the express recognition in the agreement that developed countries must provide information on support in financing, technology transfer, and capacity-building provided to developing countries is significant.

CONCLUSION

Although it is true that Intellectual Property rights encourage innovation and transfer of environmentally friendly technologies, they also constitute an obstacle to their access, fundamentally for developing countries. However, we agree with the criteria of Ilian Illiev, who states that "It is very important to turn the way Intellectual Property is used, and by this we do not mean that the patent system has to be changed, but the example of other sectors in which the role of Intellectual Property has been fundamental could be followed"66. In this case, it refers to the development achieved in the mobile phone technology sector and its wide expansion, which has been thanks to the standardization of technology through the use of the Intellectual Property system and specifically of patents. Something similar could be done with environmentally responsible technologies, through the use of patent pools and patent pools. Although for the development of environmentally responsible technologies the use of these figures is transcendental, they are usually generated by companies in developed countries, because, unfortunately, in developing and underdeveloped countries they do not have the capacities and the experience to research, develop, and even incorporate and use the technologies.

Given this reality, it is necessary to continue promoting the inclusion of access to access to environmentally responsible technologies and Intellectual Property in the negotiation process of the United Nations Framework Convention on Climate Change. At the same time, there must be an international consensus aimed at seeking a global solution to this problem, encouraging not only policies to promote the financing of research and development of technologies related to climate change, but also the transfer of technology, eliminating the obstacles to its access and promoting its use on a global scale.

On the other hand, even though, under the TRIPS agreement, countries must make use of the flexibilities and establish national policies that favor the granting of compulsory licenses and favor parallel imports at more accessible prices, not always the use of these mechanisms is viable for some countries, fundamentally, the underdeveloped and the developing ones. As we have already said, the development, manufacturing, use, and knowledge capacities are insufficient or non-existent.

In this situation, one solution would be to establish a system similar to the one established in paragraph 6 of the Doha Declaration on the TRIPS Agreement and Public Health, but specifically for environmentally friendly technologies. In this way, developing and underdeveloped countries could access them through imports at more affordable prices.

Another solution would be for the registration authorities to emphasize the fulfillment of the sufficiency requirement in the description. In this way, it is guaranteed that the reproduction of the invention is easier for those trained in the field of art once the patent registration has expired or to work on the investigation of future improvements.

In turn, States must establish public policies aimed at favoring the rapid granting of patent rights for the protection of environmentally friendly technologies and their rapid exploitation.

Likewise, the countries must establish policies aimed at financing the research and development of technologies related to climate change, since, in the end, they will result in social benefits that exceed the costs of development and implementation; reduce energy costs; they improve efficiency in manufacturing processes, and create new jobs, thus reducing the impact of climate change.

Finally, we believe that Intellectual Property should not be understood as something sacred that must be preserved at all costs. Climate change is a global problem that requires the commitment and ingenuity of humanity. Therefore, the priority must be to transfer technologies to developing countries to combat it.

REFERENCES

Anderson, J. E. (2009). *Law, knowledge, culture: The production of indigenous knowledge in intellectual property law*. Edward Elgar Publishing. doi:10.4337/9781848447196

Cullet, P. (2001). *Intellectual property and environment: impacts of the TRIPS agreement on environmental law making in India*. In Global environmental change and the nation state. Proceedings of the 2001 Berlin Human Dimensions of Global Environmental Change Conference, Postdam Institute for Climate Impacts Research, Postdam.

Cullet, P. (2005). *Intellectual property protection and sustainable development. LexisNexis*. Butterworths.

Cullet, P., & Raja, J. (2004). Intellectual property rights and biodiversity management: The case of India. *Global Environmental Politics*, *4*(1), 97–114. doi:10.1162/152638004773730239

Dreyfuss, R. C., & Strandburg, K. J. (Eds.). (2011). *The law and theory of trade secrecy: a handbook of contemporary research*. Edward Elgar Publishing. doi:10.4337/9780857933072

Dutfield, G. (2000). *Intellectual property rights trade and biodiversity*. Routledge. doi:10.4324/9781849776233

Khor, M. (2002). Rethinking intellectual property rights and TRIPS. In *Global Intellectual Property Rights* (pp. 201–213). Palgrave Macmillan. doi:10.1057/9780230522923_12

List, J. A., & Qui, L. D. (2004). Intellectual property rights, environmental regulations, and foreign direct investment. *Land Economics*, *80*(2), 153–173. doi:10.2307/3654736

Matthews, D. (2003). *Globalising intellectual property rights: the TRIPS Agreement*. Routledge. doi:10.4324/9780203165683

Momtaz, D. (1996). The United Nations and the protection of the environment: From Stockholm to Rio de Janeiro. *Political Geography*, *15*(3-4), 261–271. doi:10.1016/0962-6298(95)00109-3

Nagel, T. (2017). The problem of global justice. In *Global Justice* (pp. 173–207). Routledge. doi:10.4324/9781315254210-9

ADDITIONAL READING

Anderson, J. E. (2009). *Law, knowledge, culture: The production of indigenous knowledge in intellectual property law*. Edward Elgar Publishing. doi:10.4337/9781848447196

Cullet, P. (2001). *Intellectual property and environment: impacts of the TRIPS agreement on environmental law making in India.* In Global environmental change and the nation state. Proceedings of the 2001 Berlin Human Dimensions of Global Environmental Change Conference, Postdam Institute for Climate Impacts Research, Postdam.

Cullet, P. (2005). *Intellectual property protection and sustainable development. LexisNexis.* Butterworths.

Cullet, P., & Raja, J. (2004). Intellectual property rights and biodiversity management: The case of India. *Global Environmental Politics*, *4*(1), 97–114. doi:10.1162/152638004773730239

Dreyfuss, R. C., & Strandburg, K. J. (Eds.). (2011). *The law and theory of trade secrecy: a handbook of contemporary research*. Edward Elgar Publishing. doi:10.4337/9780857933072

Dutfield, G. (2000). *Intellectual property rights trade and biodiversity*. Routledge. doi:10.4324/9781849776233

Khor, M. (2002). Rethinking intellectual property rights and TRIPS. In *Global Intellectual Property Rights* (pp. 201–213). Palgrave Macmillan. doi:10.1057/9780230522923_12

List, J. A., & Qui, L. D. (2004). Intellectual property rights, environmental regulations, and foreign direct investment. *Land Economics*, *80*(2), 153–173. doi:10.2307/3654736

Matthews, D. (2003). *Globalising intellectual property rights: the TRIPS Agreement*. Routledge. doi:10.4324/9780203165683

Momtaz, D. (1996). The United Nations and the protection of the environment: From Stockholm to Rio de Janeiro. *Political Geography*, *15*(3-4), 261–271. doi:10.1016/0962-6298(95)00109-3

Nagel, T. (2017). The problem of global justice. In *Global Justice* (pp. 173–207). Routledge. doi:10.4324/9781315254210-9

KEY TERMS AND DEFINITIONS

Equity: Defined by UNEP to include intergenerational equity - "the right of future generations to enjoy a fair level of the common patrimony" - and intragenerational equity - "the right of all people within the current generation to fair access to the current generation's entitlement to the Earth's natural resources" - environmental equity considers the present generation under an obligation to account for long-term impacts of activities and to act to sustain the global environment and resource base for future generations. Pollution control and resource management laws may be assessed against this principle.

Polluter pays principle: The polluter pays principle stands for the idea that "the environmental costs of economic activities, including the cost of preventing potential harm, should be internalized rather than imposed upon society at large." All issues related to responsibility for environmental remediation costs and compliance with pollution control regulations involve this principle.

Precautionary principle: One of the most commonly encountered and controversial principles of environmental law, the Rio Declaration formulated the precautionary principle: To protect the environment, the precautionary approach shall be widely applied by States according to their capabilities. Where there are threats of serious or irreversible damage, lack of complete scientific certainty shall not be used as a reason for postponing cost-effective measures to prevent environmental degradation. The principle may play a role in any debate over the need for environmental regulation.

Prevention: The concept of prevention can perhaps better be considered an overarching aim that gives rise to a multitude of legal mechanisms, including prior assessment of environmental harm, licensing or authorization that set out the conditions for operation and the consequences for violation of the conditions, as well as the adoption of strategies and policies. Emission limits and other product or process standards, the use of best available techniques, and similar techniques can all be seen as applications of the concept of prevention.

Public participation and transparency: identified as necessary conditions for "accountable governments,... industrial concerns," and organizations generally, public participation and transparency are presented by UNEP as requiring "effective protection of the human right to hold and express opinions and to seek, receive and impart ideas,... a right of access to appropriate, comprehensible and timely information held by governments and industrial concerns on economic and social policies regarding the sustainable use of natural resources and the protection of the environment, without imposing undue financial burdens upon the applicants and with adequate protection of privacy and business confidentiality," and "effective judicial and administrative proceedings." These principles are present in environmental impact assessment, laws requiring publication and access to relevant environmental data, and administrative procedures.

Transboundary responsibility: Defined in the international law context as an obligation to protect one's environment and prevent damage to neighboring environments, UNEP considers transboundary responsibility at the international level as a potential limitation on the sovereign state's rights. Laws that limit externalities imposed upon human health and the environment may be assessed against this principle.

Chapter 9
Environmental and Competition Law

ABSTRACT

Across every continent, climate change is the crisis of our time. The European Green Deal makes 'sustainability and competition law' one of the most discussed topics in the EU right now. This chapter explores the interaction between competition law and sustainability in the EU in a historical context that includes 'competition law as a shield' and 'competition law as a sword' perspectives. While the European Commission may have indicated its intention to pursue more sustainable competitive practices, it has not yet explicitly expressed a position (except in its policy brief). This chapter charts trailblazing national initiatives being proposed by the EU.

INTRODUCTION

A crisis of our time, climate change affects every country on every continent(Vedder, 2000). In 2015, the United Nations General Assembly adopted the 2030 Agenda for Sustainable Development (including its 17 Sustainable Development Goals, SDGs), which provides a shared plan of action for people, the planet, and prosperity that will guide all countries policies towards sustainable development until 2030. Through appropriate financial flows, a revised technology framework, and a strengthened capacity-building framework, the Paris Agreement also seeks to strengthen the global response to climate change. In response to these international commitments, the EU launched its European Green Deal, an integral part of the Commission's strategy for implementing the United Nations' 2030 Agenda, particularly the SDGs(Monti & Mulder, 2017). Competition law discussions about sustainability have not received much attention until recently, particularly in light of the European Green Deal. European Green Deal, launched in December 2019, seeks to make Europe the first carbon-neutral continent by 2050 by setting a plan for sustainable economic growth based on environmental and social priorities, decarbonizing not only electricity but buildings, transportation, agriculture, and industry. Since the European Green Deal, the European Commission has also instigated debates on greening competition law and policy. This is

DOI: 10.4018/978-1-6684-4158-9.ch009

how competition policy can support the EU's focus on sustainability and progression towards climate neutrality by 2050.

Vestager noted in her recent Keynote that 'green policies like regulations, taxes, and investments are the key to the Green Deal. With so much to do in such a short time, everyone - including competition enforcers - needs to ensure they are doing their part. This message suggests an 'all-hands-on-deck' approach, implying that competition law also plays a role. Even though the message here is clear that competition law is not the main tool for tackling climate change issues, it can nevertheless play a role, perhaps more than just supporting, in achieving green policy objectives. The European Commission has conducted several public consultations. As a matter of fact, in the Better Regulation consultation, there is an assessment of competition rules regarding horizontal agreements (Commission Regulations (EU) No 1217/2010 (Research & Development Block Exemption Regulation) and 1218/2010 (Specialisation Block Exemption Regulation), commonly known as the horizontal block exemption regulations (HBERs) (Monti & Mulder, 2017).

Furthermore, according to the Commission's Guidelines on the Applicability of Article 101 TFEU to Horizontal Cooperation Agreements (Horizontal Guidelines) held from November 2019-February 2020, the most important development was 'climate change and the corresponding challenging environmental goals.' This results in increased demand from consumers and businesses for sustainable, ethical, and environmentally friendly business practices. To embrace a green competition policy, the most recent consultation, October–November 2020, explored how competition law and policy can contribute to the European Green Deal. Over 200 contributions came from companies, social partners, governments, public administrations, competition authorities, and civil society across the EU and beyond, indicating significant interest in this field. Participants of this consultation, as well as those attending the EU competition law and sustainability conference held virtually on 4 February 2021, agreed that competition law and policy play a key role in delivering the Green Deal objectives, especially by "driving green innovation and bringing about the technological revolution required to achieve sustainable jobs and growth, consistent with EU rules and values.". The HBERs, which are due to expire in December 2022, along with the Horizontal Guidelines (covering a wider range of provisions related to abuse of a dominant position and objective justification), provide important legal certainty for businesses. Therefore, this chapter, among other things, will shed light on the policy options as proposed in the recently issued documents, such as the Inception Impact Assessment, Staff Working Document (SWD), and Competition Policy Brief(Monti, 2020).

The sword and shield paradigm is considered a traditional approach to competition law. Like a sword, competition law can aid in achieving sustainability (i.e., to prevent the degradation of the environment), where the provisions are interpreted so that harmful measures from a sustainability standpoint are prevented/prohibited. Sustainability cannot be used to hide cartels (known as greenwashing). Implementing an environmental initiative concerning laundry detergents in the Consumer Detergents case resulted in a cartel coordinating price increases. Using the competition law as a supportive tool, sustainable measures can counterbalance anticompetitive effects or shield measures from competition law prohibitions as long as they achieve sustainability. Businesses are part of the 'green economy and can contribute to sustainability. In competitive markets, businesses are encouraged to produce at the lowest cost, use scarce resources efficiently, innovate and adopt more energy-efficient technologies, reduce CO_2 emissions, and contribute to environmental and climate policies. To achieve low-carbon economies, circularity-inspired solutions must be adopted. However, a circular economy, in which recycling and recovering materials in production, distribution, and consumption processes are brought back to the market by their definition, is characterized by "first-mover disadvantages" and high investment costs. Thus, cooperation between economic agents, holding a long-term perspective on economic relations, and relying on an understanding of corporate responsibility beyond economic profit is essential. Circularity can, however, cause tensions with competition law. Several studies have shown that businesses fear unnecessarily restrictive or unpredictable competition law enforcement(Gehring, 2006).

Competition law is traditionally focused on economic goals, assuming that other areas of law should deal with non-economic public interests. Building on the European Green Deal, this approach should change since competition law and policy contribute to sustainability debates. Specifically, competition law should not be a barrier to industry initiatives to achieve sustainability objectives. It is unclear what is meant by the notion of 'sustainability' or 'sustainable development' (used interchangeably). The literature is defined by a three-pillar model that addresses social, economic, and environmental issues. The Brundtland Report, Agenda 21, and the 2002 World Summit on Sustainable Development also fall under these pillars. The Brundtland Report defines sustainable development as "development that meets present needs without compromising the ability of future generations to meet their own needs." Kuhlman and Farrington define sustainability as the welfare of generations and the fair use of limited natural resources. The proportionality test can be used to balance different pillars (i.e., environmental and economic), which implies that preserving ecological functions is proportional, despite economic losses. This chapter does not define because sustainability remains an open concept with numerous interpretations and context-specific understandings. The study notes differences in the notions of sustainability in different jurisdictions in the EU, with some focusing exclusively on environmental or climate-related issues. In contrast, others are willing to incorporate a broader meaning, encompassing both an environmental and a social aspect (e.g., improving working conditions) (Kloosterhuis & Mulder, 2015).

Sustainability and competition law are two topics of increasing academic interest. Added to the existing debates about the need for green competition policies, this chapter explores the European Commission's experience in applying sustainability-related issues to competition law from a historical perspective that embraces both a 'shield' and a 'sword' approach to competition law. This chapter maps out national initiatives to integrate different tools to navigate the sustainable development and competition law debates, in contrast to previous studies.

CHALLENGES IN THE EUROPEAN REGION

Through various strategies and action plans (i.e., Circular Economy Action Plan), taxation and investment (including discretionary enforcement of State aid rules, e.g., to promote renewable energy projects), the European Commission and the EU Member States have promoted sustainability-related goals in the past (including measures to transition to a green economy). Neither the European Commission nor the NCAs (National Competition Authorities) has used competition rules to advance these aims, except in isolated cases. Taking a European legal perspective, scholars reflect on the constitutional context of the EU to give directions towards sustainability, 'at least a normative space'. Sustainability and environmental protection are undoubtedly among the primary objectives of EU law embodied in the Treaties (i.e., TEU and TFEU) and the Charter of Fundamental Rights of the EU (the Charter), also known as the 'constitution.' In particular, Article 37 of the Charter states: "environmental protection and improvement of the quality of the environment must be integrated into the policies of the Union.". TEU Article 3(3) clearly states that the Union strives for sustainable development. Article 7 TFEU states that "the Union shall ensure consistency between its policies and activities, taking all its objectives into consideration," with a priority placed on environmental protection as "environmental protection requirements must be integrated into the definition and implementation of the Union's policies and activities(Bennett, 2000)."

Recently, environmental regulations have shifted from command-and-control to market-based measures, such as environmental taxes, green subsidies, emissions trading, and other voluntary initiatives. As emphasized in a 2002 communication, environmental agreements are defined as "those arrangements in which stakeholders undertake to achieve pollution abatement, as defined by environmental law, or environmental objectives outlined in Article 191 TFEU.". The competition law provisions do not explicitly address sustainability issues, such as abuse of dominant positions, restrictive agreements, and merger control. In addition to the former Horizontal Cooperation Guidelines, which discussed different types of cooperation and their potential to boost efficiency, the former guidelines also included agreements on R&D, production, purchasing, commercialization, standardization, and, most importantly, environmental agreements. There would be a separate section on environmental agreements, such as standards on the environmental performance of products (inputs or outputs) or production processes, or horizontal agreements for the common achievement of an environmental target, such as recycling certain materials and reducing emissions or improving energy efficiency. Different scenarios were highlighted where environmental agreements might be covered by Article 101 TFEU and when relevant. The panel noted that environmental agreements fall under Article 101(1) simply by their nature if the cooperation does not pursue genuine environmental objectives but serves as a tool to engage in a disguised cartel that is otherwise prohibited price-fixing, output limitation, or market allocation, or if the cooperation forms part of a broader restrictive agreement that excludes actual or potential competitors(Monti & Mulder, 2017). The 2011 Guidelines for Horizontal Cooperation omitted this chapter. Still, the assessment of these agreements has not been downgraded. Furthermore, it has been noted that the Guidelines' generic provisions or standardization agreements, as far as they pertain to environmental standards, are sufficient to cover these kinds of agreements(Odudu, 2010).

Competition law is traditionally used as a sword or shield from two different perspectives. Using competition law as a sword in the context of sustainability, such as prohibiting certain measures from a sustainability perspective, can be beneficial. As part of this preventive interpretation, there is also a balancing approach to determine whether the harm to competition and sustainability outweighs the measure's benefits. From a supporting perspective, competition law can provide the means to counterbalance any anticompetitive effects using measures aimed at achieving sustainability. Sustainability measures may not even be subject to competition law prohibition, falling outside the scope of Article 101 TFEU (also known as the Albany route, where the CJEU ruled that Article 101 TFEU does not apply to collective bargaining) (Monti & Mulder, 2017). Several associations of automobile manufacturers, such as ACEA, JAMA, and KAMA, have committed to reducing CO_2 emissions from automobiles; the targets are set collectively on behalf of all members, rather than individually. Independently and in competition with each other, car manufacturers developed and introduced new CO_2-efficient technologies. Accordingly, the Commission concluded that they did not restrict competition in violation of Article 101(1) TFEU. The ancillary restraints/objective necessity doctrine exempts sustainability agreements from Article 101 TFEU. Under Article 101(3) TFEU, even agreements under Article 101(1) can be exempted. Under Article 101(3) TFEU, the Exxon/Shell case confirmed, for instance, that environmental cooperation to reduce pollution often leads to both technical and economic progress. As the agreement in the CECED (European Council of Manufacturers of Domestic Appliances) case restricted producing and importing less energy-efficient washing machines, the agreement was restrictive by essence. Despite this, it was upheld under Article 101(3) TFEU because new machines would reduce energy consumption, decreasing electricity and water bills for consumers. In addition to creating new technically efficient machines, future R&D on energy efficiency would be stimulated. Similar to the environmental objectives notified by the CECED, the CEMEP (the European Committee of Manufacturers of Electrical Machines and Power Electronics) agreement aimed at driving the EU market towards higher efficiency motors and saving energy. Unlike CECED, where the definition of and labeling of energy-efficient washing machines had already been established, there was no established definition of the energy efficiency of standardized low voltage motors on the market, so there was not much competition on this product characteristic. This resulted in negative clearance to the agreement without examining whether Article 101(3) TFEU was met(Monti & Mulder, 2017).

Lastly, competition law can also be used as a sword to prevent cartels from hiding behind green initiatives. Consumer Detergents, for instance, resulted in a prohibition under Article 101 TFEU against three major detergent manufacturers: Henkel, Unilever, and Procter & Gamble, which engaged in the infringing behavior to ensure that none of them would gain a competitive advantage over others through the environmental initiative. Additionally, they coordinated prices on washing powder. During the different phases of the environmental initiative, for instance, the parties agreed to keep the price the same (namely, when products were 'compacted' (i.e., reduced in weight) when the product quantity was reduced (i.e., decreased in volume), or when the number of scoops (wash loads) per package was reduced). Therefore, their behavior was determined to have the purpose of restricting competition.

Past cases illustrate how environmental agreements have been considered using existing competition tools without further in-depth assessments, often adhering to environmental benefits as complementary to economic benefits without quantifying them.

STRATEGIES OF EUROPEAN PLANS TOWARD SUSTAINABLE FUTURE

Thanks to the European Green Deal, businesses have been given a new impetus to pursue sustainability initiatives. Following the public consultations, the Commission published the Inception Impact Assessment and Staff Working Document (SWD) as part of the fit-for-purpose HBERs regulation review. As indicated in the SWD, many respondents believed businesses must cooperate for sustainability initiatives to succeed. However, it is equally essential to clarify when such cooperation is compatible with EU competition rules. According to the accompanying evaluation study, the NCAs have examined several cases involving sustainability and joint bidding agreements, which are not explicitly covered in the Horizontal Guidelines, leading to divergent approaches between EU member states. It seems the Commission also agrees that the provisions in the HBERs and Horizontal Agreements Guidelines are outdated and not sufficiently adapted to recent market developments, especially concerning sustainability goals. Horizontal agreements related to sustainability do not provide sufficient legal certainty. However, the policy options in this context remain unclear. However, the Commission promises specific guidelines on horizontal cooperation to pursue sustainability goals(Monti & Mulder, 2017).

In light of the recent public consultation on the European Green Deal, the Competition Policy Brief (Brief) has been released. The constructive feedback gathered during the public consultation has been used to frame three main ongoing reform work-streams, incorporating examples from the three competition instruments. There are three main items to be addressed: (i) state aid aimed at funding non-fossil fuels; clarifying and simplifying the rulebook; and enhancing opportunities to support innovation; ((ii) antitrust, where further clarification is needed on how and whether to assess sustainability benefits; improved guidance and an open-door policy; (iii) mergers with enhanced enforcement regarding possible harm to innovation (green 'killer acquisitions'); reflecting sustainability aspects/features prevailing in the market and consumer preferences for sustainability(De Stefano, 2020). In this Brief, the Commission commits to providing concrete examples of how sustainability objectives can be pursued through different types of cooperation agreements (e.g., joint production/purchasing agreements, specifications, etc.) without restricting competition. The results are promising and can reassure businesses and help unlock further investments. To assess the Article 101(3) TFEU, the Brief notes that benefits from sustainability can be assessed as qualitative efficiency that results in an increased quality or longevity (e.g., replacing plastic with wood in toys or using recycled materials for clothing), as well as potential cost efficiencies passed on to consumers (e.g., reducing plastic packaging may result in reduced materials, transportation, and storage costs). There seems to be a change of approach from the Commission regarding sustainability benefits not having to be immediate or noticeable improvements in product quality or cost savings, provided users appreciate the sustainability benefits and are willing to pay a higher price solely for this reason(Jacobs, 1993).

Regarding 'out-of-market' efficiencies (e.g., the societal benefits accrued by carbon emission reduction), the Brief proposes that the assessment of the anticompetitive effects and benefits of practice should be confined to the same relevant market. Based on the current anchored consumer welfare theory, restricting competition for a product can only be justified if the product users are not, on balance, worse off. The national positions of several Member States contradict this interpretation. In brief, it is acknowledged that benefits generated on separate markets can be taken into account, provided that the group of consumers affected by the restriction and the group of consumers who benefit from the restriction is substantially the same. The benefits are sufficient to compensate the consumers for the harm (the former is part of society). We will see how the Commission addresses these efficiencies shortly(De Stefano, 2020).

There have not been any 'pure' sustainability cases in recent years. As a result, a good example of how competition can be used as a sword to limit innovation is the recent German carmaker's infringement case, in which the European Commission fined Daimler, BMW, and Volkswagen group (i.e., Volkswagen, Audi, and Porsche) EUR 875,189,000 for violating EU antitrust rules (notably, Article 101(1)(b) TFEU/ Article 53(1)(b) of EEA-Agreement) by colluding on technical development in the

In particular, these car manufacturers over five years (2009-2014) held technical meetings, where they agreed on AdBlue tank sizes and ranges and agreed on a common understanding of the average estimated AdBlue consumption, eliminating the uncertainty about their future market conduct regarding NOx-emissions cleaning beyond and above the legal requirements and AdBlue-refill ranges. In other words, they conspired to avoid competition on cleaning better than what is required by law, despite having access to the relevant technology. By their very nature, the parties in question limited competition on product characteristics and in terms of technical developments in the field of NOx-cleaning for new diesel passenger cars, thus constituting an infringement 'by object' prohibited under Article 101(1)(b) TFEU. Although neither undertaking had introduced SCR-systems with AdBlue-Tanks of uniform size or range or used the same cleaning strategies, the Commission concluded that the effects of the agreement (and/or concerted practice) were not relevant. The European Commission has determined that conspiracy on technical development constitutes a cartel for the first time. In this case, innovation was referred to as the key to achieving ambitious Green Deal objectives, with competition being of paramount importance to fostering such innovation. According to Vestager, a car's ability to emit less pollution has become an important characteristic in today's world. The cartel sought to limit competition in this important area. [..] competition and innovation are also essential for Europe to meet its ambitious Green Deal goals(Nowag, 2016).

The preventive interpretation of Article 101 TFEU also questions whether the harm to competition and sustainability outweighs the benefits. The parties, in this case, argued that their contacts relating to the development of SCR-technology (selective catalytic reduction) served to enhance the marketability of the environment-friendly SCR technology and build a customer-friendly AdBlue infrastructure. According to the Commission, the agreements (or concerted practices) of the parties concerning the product characteristics of diesel passenger cars with SCR-systems did not comply with the requirements of Article 101(3) TFEU because it was doubtful in how far they agreed to certain AdBlue tank sizes and refill ranges. Likewise, insufficiently anonymized or aggregated exchanges of data on the assumed average AdBlue consumption of their new diesel passenger car models with SCR-Systems were capable of achieving the claimed advantages of "customer-friendly AdBlue infrastructure" and "marketability of SCR technology." Regardless, the Commission did not deem this conduct essential to achieving these objectives.

One must take note of the 'shield' aspects in this case. Since DAIMLER, VW, and BMW also discussed other issues related to the development of SCR-systems, the European Commission accompanied its prohibition decision with a letter to the relevant parties, in which it provided much-needed guidance on aspects that do not raise competition concerns, such as the joint development of an AdBlue dosing software platform; Collaboration on the development of liquid SCR-systems; the standardization of the AdBlue filler neck; the preparation of charge sheets for parts of SCR-systems; and the discussion of quality standards for AdBlue, monitoring strategies to ensure the timely refill of AdBlue, and the establishment of an AdBlue supply infrastructure(Rizzuto, 2010).

PROTECTION OF ENVIRONMENTALLY FRIENDLY TECHNOLOGIES

Hellenic Competition Commission

As with the ACM, the Hellenic Competition Commission (HCC) can also be seen as a leading authority in this debate, noting that the HCC has a role to play and should support innovation within a Green economy while taking into account the possibility of externalities from generation to generation through the use of new tools and approaches to understand consumer behavior. Article 24 of the Greek Constitution states that the state is obliged to protect the environment. The HCC published a discussion chapter (discussion chapter) to start an open dialogue about business practices and their impacts on the environment. It discusses convergence areas and conflicts between sustainable development and competition law.

According to the discussion chapter, competition law should become more sensitive to broader constitutional values and programmatic aims regarding sustainability, and NCAs should take an active role by reevaluating their aims and objectives with a broader perspective, including externalities and intergenerational effects, as well as monetarily assessing them. By constructing a competition law sustainability sandbox, the chapter further explores how the theory of harm may be redressed with sustainability considerations in the long run. This sandbox is designed to provide businesses with a safe environment (free from regulatory penalties) to experiment with new formats to achieve sustainable objectives faster and more efficiently, even if it may entangle cooperation between competing companies or even more permanent changes in the market structure. These experiments would be time-constrained and supervised by the HCC, which would balance the possible anticompetitive effects with the need to provide incentives for the sustainability investment.

As opposed to the ACM, the HCC noted that sustainability consideration should have a broader scope (beyond Article 101(3) TFEU) and should cover all its aspects, for example, under Article 102 TFEU (domestic equivalent) or merger control. Furthermore, the chapter suggests that additional consideration should be given to whether abuse of dominance violations may also constitute environmental law violations or restrict sustainable development. Sustainability considerations, for example, could justify otherwise illegal foreclosure based on the efficiency or objective necessity, or more broadly in line with Article 11 TFEU's requirement that environmental protection is integrated into all EU policies. A broader interpretation of the prohibition of 'unfair purchase or selling prices or other unfair trading conditions in Article 102 TFEU could cover practices with negative sustainability impacts, such as dominant buyers paying excessively low purchase prices for inputs. In merger control, it questions the extent to which sustainability issues can be considered when assessing mergers. Consequently, it addresses sustainability in several ways: (i) under substantive assessment according to Article 2 of the European Union Merger Regulation (EUMR); (ii) under 'efficiencies'; (iii) under the provision of relevant 'remedies'; (iv) by applying Article 21(4) of the EUMR; or (v) through the review of mergers under national competition law. While the HCC has already engaged in sustainability-related arguments in past merger cases, sustainability has never played a significant role in the final decision.

Due to upcoming legislation changes and the inclusion of no-action letters in its Competition Law, the HCC plans to adopt sustainability guidelines and design the competition law sustainability sandbox. Moreover, the HCC proposed establishing a 'Common Advice Unit' composed of experts from different regulatory institutions to offer informal consultation on proposed sustainability initiatives. To achieve the goals of the Green Deal, the HCC emphasized the importance of collaborations between competition authorities at national and supranational levels. ACM and HCC commissioned a joint technical report that illustrates this.

In conclusion, the HCC's exploratory discussion chapter proposes novel approaches, such as (i) the creation of a competition law sustainability 'sandbox' within which market participants could collaborate on sustainable business projects while still being protected from competition regulation; and (ii) the creation of an 'Advice Unit' comprising experts from different regulatory authorities providing informal advice on sustainability-related initiatives. Further guidelines will define the contours of legitimate cooperation between rivals on sustainability projects(Maisin & Meagher, 2020).

Dutch Competition Authority

This debate is dominated by the Dutch competition authority/authority for consumers and markets (Autoriteit Consument & Markt, ACM). ACM published the 'Vision Document Competition and Sustainability' in 2014, proposing a radical change in the approach to price-centric competition law assessment by stating that lower prices don't always result in increased consumer welfare. According to the Energy Agreement for Sustainable Growth (Energieakkoord vor duurzame groei), four electricity manufacturers consulted the ACM about their intention to close up their five old coal-fired power plants to reduce environmental damage. As a result of this coordinated approach, the total energy supply was expected to decrease by 10% and electricity prices. ACM found that this initiative would violate Article 6(1) of the Dutch Competition Act because no convincing evidence of consumer benefits was provided (no significant effect on health or carbon emissions was evident). It was in 2015 that chicken producers requested to become referred to as the "Tomorrow's Chicken (Kip van Morgen)" case, where chicken producers (covering approximately 95% of the relevant market) agreed to raise chickens more organically, resulting in higher costs. In the absence of an exemption, the ACM focused on consumer preferences and their sensitivity to price and reduced choice, which resulted in lower consumer welfare. Based on the ex-post analysis, this largely criticized decision turned out to be a good one, as it noted that the current market for sustainable chicken is healthier and more competitive than it would have been if the alliance went forward.

The cases have led to further discussion about whether competition law represents a barrier to sustainability. As a result, the ACM issued the Sustainability Guidelines Draft in 2020 with a further revision in 2021. In this proposal, the benefits for society as a whole are taken into account in the competition law analysis rather than just the users of the products in question, thus encouraging businesses to enter into sustainable agreements which contribute to achieving climate objectives (e.g., reducing carbon emissions). The connotation of 'sustainability' has been influenced largely by previous cases-serving the purpose of identifying, preventing, restricting, or mitigating the adverse effects of economic activities on people, animals, the environment, or nature. According to the draft guidelines, the business can meet the sustainability objective in three ways.

The first point is that under Article 101(1) TFEU (the domestic equivalent of Section 6(1) MW), all agreements apart from certain types of agreements (i.e., price-fixing, customer-sharing, distribution, collective distribution, production restraints, and collective refusal to buy or supply) are exempt from the cartel prohibition. As per the draft guidelines, sustainability agreements are unlikely to be anticompetitive "if they do not or do not substantially affect key competition parameters such as price, quality, diversity, service, and distribution method.".

Second, the guidelines propose the revised conditions under Article 101(3) TFEU (section 6(3) MW). As part of the 'efficiency gains' notion, the first condition explicitly refers to sustainability benefits. Even though efficiencies should be 'objective,' quantitative data (i.e., the reduction rate of carbon footprints) and qualitative data (i.e., related to animal welfare) could be considered. As part of the second condition, 'a fair share' of the benefits of the product/service must be shared with society at large. Deviation from the traditional interpretation is only applicable to environmental-damage agreements and if the agreement helps meet an international or national standard or helps achieve a concrete policy goal (i.e., to prevent such damage). There is no need always to quantify the pros and cons of sustainability agreements. When the undertaking in question has a limited, combined market share (up to 30%), or when the harm to competition is smaller than the agreement's benefits. The last two conditions of indispensability and preservation of competition are based mainly on current practices.

Furthermore, undertakings can conduct a self-assessment of their agreement or request an ACM assessment. Undertakings could submit sustainability initiatives to the legislature if they are incompatible with the MW. It has also been announced that undertakings will have the opportunity to contact the Minister of Economic Affairs and Climate Policy shortly. There is no doubt that the ACM sends a positive message to industry and businesses regarding its commitment to sustainability.

This NCA is the first to develop a different perspective on Article 101(3) TFEU by distinguishing sustainable agreements from cartel agreements using novel tools, such as a willingness-to-pay test and an environmental cost calculation. Essentially, the ACM opened the door to assessing sustainability agreements under Article 101(3) TFEU.

ACM also has responsibility for consumer protection, so it published draft guidelines on sustainability claims from a consumer protection perspective. The guidelines advise businesses and consumers on different types of claims that could be misleading (or incorrect) and how to avoid 'greenwashing' (i.e., claiming a product is more sustainable than it is)(Kloosterhuis & Mulder, 2015).

Austrian Competition Authority

In contrast to the previous two soft approaches initiated by the ACM and HCC, Austria has taken a more radical approach. It is the first EU Member State to incorporate sustainability-related matters through a legislative route proving sustainability's increasing importance. The Austrian legislator explored sustainability-related considerations in competition law enforcement practice and questioned whether agreements between undertakings that may (partially) restrict competition could be exempted due to broader sustainability benefits.

Accordingly, the legislator proposed to expand the scope of the exemption from the cartel prohibition under section 2(1) of the Cartel and Competition Law Amendment Act 2021 (Kartell und Wettbewerbsrechts Änderungsgesetz 2021, the domestic equivalent of Article 101(3) TFEU), which implements the ECN+ Directive. According to Austrian antitrust law, section 2(1) explicitly allows out-of-market efficiencies if the environmental benefits benefit society rather than consumers on the relevant market. The provision states that consumers are entitled to a fair share of benefits if advances in production or distribution of goods or the advancement of technical or economic progress lead to an ecologically sustainable or climate-neutral economy. Unlike a wide range of SDGs, the focus seems to be on environmental issues. During the consultation process, the Austrian federal competition authority (BWB) (which is an independent and autonomous authority at the Austrian Federal Ministry of Science, Research and Economy to conduct investigations into the possible violations of antitrust law and European competition law) indicated the necessity of a broad understanding of consumer welfare by taking more account of impacts on quality, variety, and innovation, rather than adhering to the short-term effects of low prices in the enforcement of competition law. The BWB noted that the new concepts proposed and the meaning of the phrase 'positive effects on the environment or climate' need to be clarified to avoid legal uncertainties. Despite the lack of regulatory guidance, the legislative materials provide an overview of the types of environmental benefits that may be considered sufficient to 'escape' the cartel prohibition. A 'future generation' can benefit from the ecological advantages of cooperation since immediate benefits can be considered and those that will materialize in the short term. There are several specific examples of advantages that contribute to an 'ecologically sustainable and climate-neutral economy,' such as renewable energy and emission reductions, the sustainable use of natural resources, measures contributing to the transition to a circular economy, or measures to save or restore ecosystems and biodiversity. The calculation of the costs of environmental impacts on society is expected as proof of a "significant contribution."

In contrast to the ACM, however, an exact calculation of environmental benefits will not be required if the disadvantages resulting from the competition agreement are minimal, while the environmental contribution is substantial. To the same extent as the ACM's draft guidelines, the Austrian materials exclude hardcore restrictions captured by the cartel prohibition from an individual exemption under sustainability principles. As most cooperation agreements have cross-border effects within the EU, it will be important to see how this new national provision will be implemented in practice and its impact on Article 101(3) TFEU.

Based on practice, it appears that the BWB will need to provide further guidance to the Federal Ministry of Climate Protection on the new sustainability exemption from the cartel prohibition(Robertson, 2022).

Hungarian Competition Authority

Hungarian Competition Authority, Gazdasági Versenyhivatal (GVH), provides three pillars of activity: (i) competition supervision (under both local competition law and EU law); (ii) competitive advocacy; and (iii) development of competition culture, which encompasses consumer decision-making as well. The GVH aims to become a green authority by 2020, with its green strategy expected to be approved in 2021. This sets an example for other public authorities and businesses alike, but the GVH has also taken steps related to the interplay between competition law and sustainability. Specifically, it has recently amended its notice regarding fines (the so-called New Notice), which introduces a proactive remedy with an undertaking to provide partial or full compensation for the negative impact of its competition law violation to obtain a reduction in the fine (or the abolition of the fine in its entirety). Under the New Notice, the GVH may also consider the positive impact of pro-active remedies on sustainability and environmental protection, even if businesses make commitments unrelated to the violation, if such commitments serve sustainability and environmental objectives. Even though there is an out-of-market approach here to support sustainability initiatives, there is no further clarification as to what is meant by 'sustainability objectives,' how these benefits will be calculated, or how these commitments will be assessed or monitored. It will be interesting to see how effective this new development is short. To combat greenwashing, as part of the GVH's other task to ensure conscious consumer decision-making and consumer awareness, the GVH issued a 'green marketing notice.' This notice provides 'do' s and 'donts' and sets a clear environmentally-friendly advertising practice based on 'green claims (greenwashing)' in response to increased numbers of environmentally-conscious customers(Tóth, 2018).

UK Case

In the wake of the UK leaving the EU, the Competition and Markets Authority (CMA), an independent non-ministerial department in charge of competition and consumer protection, has been faced with new challenges due to an expected increase in merger control and competition law enforcement cases, which are addressed in the Competition and Markets Authority's 2020/2021 annual plan. The UK follows the pattern of the EU and the other Member States when it comes to sustainability. The CMA report notes that one of its key priorities is to support the transition to a low-carbon economy due to the UK's legally binding commitment to net zero emissions by 2050. The CMA launched a consultation similar to the European Commission on how competition and consumer regimes (the CMA's other role) can better support the UK's Net Zero and sustainability goals. Interestingly, the CMA excludes State aid, an aspect included in the European Commission's consultation(Bennett, 2000).

Although sustainability is a broad concept that encompasses several objectives beyond addressing climate change, the CMA rightly prioritizes the 'environmental aspect of sustainability agreements.' It is reflected in its recent guidance on sustainability agreements and competition law, covering industry-wide initiatives and decisions of trade associations related to sustainability, including climate change initiatives. According to the CMA, it has drawn inspiration from the ACM guidelines. It is keen to support businesses in adapting to climate change while ensuring that markets remain open to disruptive innovation that can help meet climate change and sustainability goals. According to CMA's new guidance, businesses may have to cooperate to achieve sustainability goals. Businesses may, for example, combine expertise to make their products more energy-efficient or choose packaging materials that meet certain standards to facilitate recycling and reduce waste. The guidance remarks that businesses should use a fair standard-setting process when determining industry-based standards to achieve sustainability goals, ensuring access to 'the standard is on fair, reasonable and non-discriminatory terms.' Similar to the other jurisdictions discussed above, the CMA also warns that sustainability agreements must not be used to cover a cartel; no commercially sensitive information beyond what is necessary to set the standard is allowed. Also, the guidance reminds that some sustainability initiatives may fall under one of the general categories of agreements (such as research and development agreements or specialization agreements) which are exempt from the prohibition on anticompetitive agreements. In the case of an individual exemption, sustainability agreements may still be permitted if they generate benefits that exceed the disadvantages of restricted competition under four traditional conditions: (i) the agreement should generate efficiencies (i.e., better products); (ii) these efficiencies cannot be achieved through less restrictive methods; (iii) they should benefit consumers; and (iv) the agreement should not eliminate competition. The UK guidance, in contrast to the detailed Dutch guidelines, is somewhat vague with many unanswered questions raised by other NCAs, such as whether out-of-market efficiencies may be taken into account or if benefits as a whole may be included rather than just a group of consumers affected by the agreement, and if so, how these efficiencies should be evaluated and measured. In 2020, Pinsent Masons conducted a survey that found that 72% of the respondents wanted more explicit guidance on what is and is not permissible in competition law. This guidance is unlikely to provide legal certainty to businesses.

Last but not least, given the CMA's role in consumer protection, it recently published a Green Claims Code that aims to protect consumers from misleading environmental claims amid concerns over 'green-washing,' similar to notices previously published by the Dutch and Hungarian authorities.

Although green claims and environmental information disclosure are intended to prevent misleading claims that undermine consumers' trust, the UK code also signals that it intends to protect businesses against unfair competition and ensure a level playing field. In addition, the CMA is also co-leading a project under the auspices of the International Consumer Protection Enforcement Network (ICPEN) looking at misleading green claims made online(Johnston, 2012).

CONCLUSION

Sustainability debates are inevitable in light of the European Green Deal and international commitments (i.e., the Paris Agreement) and should be placed in a competition law context. Although sector-specific regulations, taxes, and investments (including State aid due to the first-mover disadvantage associated with high investment costs) are the main tools to facilitate the transition to a green economy, an 'all hands on deck approach is needed to tackle the climate emergency, and isolated sustainability exceptions are no longer acceptable. Sustainability-related matters should be conceptualized within competition law, providing legal certainty to the industry by defining clear rules.

The chapter examines the previous and current approaches to sustainability-related matters, especially the European Commission's recent call to address the interplay between competition law and sustainability. Despite the European Commission's indications of an intention to promote more sustainability-friendly competition practices, no official position has been stated, apart from hints in its recent Policy Brief. Accordingly, as discussed in this article, the NCAs are trailblazing their agendas. A chapter identifies five assertive national proposals that demonstrate a willingness to incorporate sustainability into competition law enforcement. However, the NCAs in the EU and the UK have vastly different approaches, ranging from explicit national draft guidelines (the Netherlands) to less detailed ones (the UK); a draft staff discussion chapter with further suggestions for experimental sustainability sandbox tools (Greece); Sustainability and environmental protection can be addressed via proactive solutions (recently proposed in Hungary), or via a completely different route-legislation, as proposed in Austria (that is, the inclusion of an explicit condition of an ecologically sustainable or climate-neutral aspect to the national equivalent of Article 101(3) TFEU). Although the interface debate is still in its infancy, the near future will show which approach the EU and UK will favor. Different national approaches create legal uncertainty for businesses, so a more uniform stance is most definitely needed.

REFERENCES

Bennett, P. (2000). Anti-Trust? European Competition Law and Mutual Environmental Insurance. *Economic Geography*, *76*(1), 50–67. doi:10.2307/144540

De StefanoG. (2020). Measurable Environmental Protection As A Necessity For Competition Law. Available at SSRN 3533499. doi:10.2139/ssrn.3533499

Gehring, M. W. (2006). Competition for sustainability: Sustainable development concerns in national and EC competition law. *Review of European Community & International Environmental Law*, *15*(2), 172–184. doi:10.1111/j.1467-9388.2006.00519.x

Jacobs, R. (1993). EEC Competition Law and the Protection of the Environment. Legal Issues of Eur. *Integration (Tokyo, Japan)*, *20*, 37.

Johnston, A. (2012). The Interface between EU Energy, Environmental and Competition Law in the UK. *Oil, Gas & Energy Law, 10*(4).

Kloosterhuis, E., & Mulder, M. (2015). Competition law and environmental protection: The Dutch agreement on coal-fired power plants. *Journal of Competition Law & Economics*, *11*(4), 855–880. doi:10.1093/joclec/nhv017

Maisin, J. B., & Meagher, M. (2020). Sustainable development and competition law: Towards a Green Growth regulatory osmosis. In Sustainable development and competition law.

Monti, G. (2020). Four options for a greener competition law. *Journal of European Competition Law & Practice, 11*(3-4), 124–132. doi:10.1093/jeclap/lpaa007

Monti, G., & Mulder, J. (2017). Escaping the clutches of EU competition law. *European Law Review*, *42*(5), 635–656.

Nowag, J. (2016). *Environmental integration in competition and free-movement laws*. Oxford University Press. doi:10.1093/acprof:oso/9780198753803.001.0001

Odudu, O. (2010). The wider concerns of competition law. *Oxford Journal of Legal Studies*, *30*(3), 599–613. doi:10.1093/ojls/gqq020

Rizzuto, F. (2010). The Private Enforcement of European Competition Law: What Next? *Global Competition Litigation Review*, *3*(2), 57–68.

Robertson, V. H. (2022). Sustainability: A World-First Green Exemption in Austrian Competition Law. *Journal of European Competition Law & Practice*.

Tóth, T. (2018). Life after Menarini: The conformity of the Hungarian Competition Law enforcement system with human rights principles. [YARS]. *Yearbook of Antitrust and Regulatory Studies*, *11*(18), 35–60. doi:10.7172/1689-9024.YARS.2018.11.18.2

Vedder, H. (2003). *Competition law and environmental protection in Europe: towards sustainability?* (Vol. 3). Europa Law Publishing.

VedderH. H. (2000). Voluntary Agreements and Competition Law: What are, and What Should Be the Boundaries to Va's Imposed by Competition Law? SSRN 253333. doi:10.2139/ssrn.253333

ADDITIONAL READING

Monti, G., & Mulder, J. (2017). Escaping the clutches of EU competition law. *European Law Review*, *42*(5), 635–656.

Nowag, J. (2016). *Environmental integration in competition and free-movement laws*. Oxford University Press. doi:10.1093/acprof:oso/9780198753803.001.0001

Odudu, O. (2010). The wider concerns of competition law. *Oxford Journal of Legal Studies*, *30*(3), 599–613. doi:10.1093/ojls/gqq020

Rizzuto, F. (2010). The Private Enforcement of European Competition Law: What Next? *Global Competition Litigation Review*, *3*(2), 57–68.

Robertson, V. H. (2022). Sustainability: A World-First Green Exemption in Austrian Competition Law. *Journal of European Competition Law & Practice*.

Tóth, T. (2018). Life after Menarini: The conformity of the Hungarian Competition Law enforcement system with human rights principles. [YARS]. *Yearbook of Antitrust and Regulatory Studies*, *11*(18), 35–60. doi:10.7172/1689-9024.YARS.2018.11.18.2

Vedder, H. (2003). *Competition law and environmental protection in Europe: towards sustainability?* (Vol. 3). Europa Law Publishing.

VedderH. H. (2000). Voluntary Agreements and Competition Law: What are, and What Should Be the Boundaries to Va's Imposed by Competition Law? SSRN 253333. doi:10.2139/ssrn.253333

KEY TERMS AND DEFINITIONS

Equity: Defined by UNEP to include intergenerational equity - "the right of future generations to enjoy a fair level of the common patrimony" - and intragenerational equity - "the right of all people within the current generation to fair access to the current generation's entitlement to the Earth's natural resources" - environmental equity considers the present generation under an obligation to account for long-term impacts of activities and to act to sustain the global environment and resource base for future generations. Pollution control and resource management laws may be assessed against this principle.

Polluter pays principle: The polluter pays principle stands for the idea that "the environmental costs of economic activities, including the cost of preventing potential harm, should be internalized rather than imposed upon society at large." All issues related to responsibility for environmental remediation costs and compliance with pollution control regulations involve this principle.

Precautionary principle: One of the most commonly encountered and controversial principles of environmental law, the Rio Declaration formulated the precautionary principle: To protect the environment, the precautionary approach shall be widely applied by States according to their capabilities. Where there are threats of serious or irreversible damage, lack of complete scientific certainty shall not be used as a reason for postponing cost-effective measures to prevent environmental degradation. The principle may play a role in any debate over the need for environmental regulation.

Prevention: The concept of prevention can perhaps better be considered an overarching aim that gives rise to a multitude of legal mechanisms, including prior assessment of environmental harm, licensing or authorization that set out the conditions for operation and the consequences for violation of the conditions, as well as the adoption of strategies and policies. Emission limits and other product or process standards, the use of best available techniques, and similar techniques can all be seen as applications of the concept of prevention.

Public participation and transparency: identified as necessary conditions for "accountable governments,... industrial concerns," and organizations generally, public participation and transparency are presented by UNEP as requiring "effective protection of the human right to hold and express opinions and to seek, receive and impart ideas,... a right of access to appropriate, comprehensible and timely information held by governments and industrial concerns on economic and social policies regarding the sustainable use of natural resources and the protection of the environment, without imposing undue financial burdens upon the applicants and with adequate protection of privacy and business confidentiality," and "effective judicial and administrative proceedings." These principles are present in environmental impact assessment, laws requiring publication and access to relevant environmental data, and administrative procedures.

Transboundary responsibility: Defined in the international law context as an obligation to protect one's environment and prevent damage to neighboring environments, UNEP considers transboundary responsibility at the international level as a potential limitation on the sovereign state's rights. Laws that limit externalities imposed upon human health and the environment may be assessed against this principle.

Chapter 10
Environmental and Contract Law

ABSTRACT

Most oil reserves are located in developing countries, which often do not have the regimes required to comply with international standards. In developed countries, new technologies such as framing explain the need to develop the necessary laws and regulations and their reflection in the drilling and development contracts. In the present chapter, with a comparative study of oil contracts, including the new generation of oil contracts of the Islamic Republic of Iran (IPC), the environmental conditions contained in these contracts have been studied and analyzed. The main purpose of this study is to review the solutions provided in international agreements, analyze the current status of laws and regulations of the Islamic Republic of Iran, and provide the necessary solutions according to the current situation of oil operations areas in Iran. The results of this chapter, besides its ethical value, can be useful for policymakers and lawyers practicing in the field of oil and gas contracts.

INTRODUCTION

Oil exploration was a golden opportunity for developing countries to generate prosperity for their people with huge revenues from oil and gas sales. Despite the positive impact of oil activities in countries with oil resources, its negative effects on the environment cannot be ignored. Oil operations have always put many parts of the environment at risk. The implementation of oil activities has caused inevitable harmful effects at sea or on land. The effects of oil operations have increased greenhouse gas emissions, air pollution, infiltration of soil, sea, and groundwater, and respiratory diseases(Amani, 2010). The harmful effects of oil activities on the environment resulted from the mere attention of host countries and international oil companies to economic benefits. This can be illustrated briefly by Concession Contracts to extract oil in the early years of oil exploration, which do not essentially include environmental commitments(Dabiri & Poorhashemi, 2009).

After a while, the environment suffered a lot due to the detrimental effects of the oil companies' operations and their destructive activities(Berkhout & Smith, 2003). Hence, it was time for Oil market players to pay attention to the environment and their duty to preserve and maintain it in their industrial activities. As a result, the desire to protect the environment as a concern of countries Despite interna-

DOI: 10.4018/978-1-6684-4158-9.ch010

tional environmental treaties and the adhesion of oil-owning countries to them, the ideal results in terms of environmental protection have not been achieved, and new problems arose. Environmental rules and conditions that were no longer enforced in oil contracts always required high costs for oil companies. Due to Increased costs, oil companies were reluctant to continue their activities in the developing countries with oil resources. Having an oil-based economy and the absence of oil companies lead to new difficulties for these countries. The crucial challenge was the imbalance between oil production and compliance with environmental regulations stems from a conflict of interest and a desire for maximum profit(DiMentom, 2021; Fuentes, 2002).

Moreover, domestic laws had serious weaknesses in covering all environmental principles. This drawback in legislation was mainly in developing countries. Despite these problems, some countries, such as Norway, one of the countries with oil resources, have taken positive steps to protect the environment by enacting comprehensive laws and adopting appropriate policies. Norway illustrates this point clearly that environmental protection is one of the requirements of oil activities, and it is not acceptable to overlook them under the pretext of economic interests(ADNOC, 2020).

On this basis, this paper examines the environmental considerations in oil contracts in three main sections; In the first part, the environmental effects of oil and gas activities are mentioned; In the second part, the method of forecasting environmental considerations in international oil contracts are examined, and then in the third part, the environmental criteria of Iranian oil contracts are examined, and finally, according to the problems in the adopted methods, a solution is presented.

ENVIRONMENTAL PROTECTION AND OIL CONTRACTS

Although the discovery of oil has positively influenced human life and caused a lot of income to enter the countries with oil resources, its negative effects on the environment cannot be ignored. Drilling wells and extracting oil have had devastating traces in the environment. Examples include pollution of land, sea, and groundwater. Oil drilling requires a lot of equipment, which makes a big hole in the ground. The infiltration of oil into the water and soil around the oil pipelines causes people living near the wells to contract diseases. Transporting oil by sea and oceans may risk that crude oil will leak out of its reservoir and pollute the environment. Oil spills were the result of oil infiltration into the waters of the seas or oceans. After a while, seabirds and mammals are trapped in the oil and die. "In 1989, the Exxon Valdez oil spill in Prince William of Alaska caused the worst oil spill in North America several years ago. This caused 38,800 tons of reservoir oil to expand to 1,200 miles. About 1,000 otters and 300,000 to 400,000 seabirds were killed off the coast"(Sabour, 2015). The problems do not end there. The transfer of oil to industrial sectors of society and its consumption in cars or factories causes air pollution, acid rain, and climate change.

Early oil contracts, such as Concession Contracts in the first half of the twentieth century, did not mention the environment. For example, in the Darcy Concession Contracts, there are no conditions regarding environmental issues. "Other similar contracts do not require environmental considerations(Gao & Kao, 1998)." "The 1933 Concession Contracts, which replaced the Darcy contract, also does not contain an environmental clause"(Shiravi, 2016). The length of the Concession Contracts and the size of the contract area made it impossible for the host states to monitor the implementation of the international oil companies' obligations related to environmental protection. Kuwaiti Concession Contract is a good illustration to clarify this point. "In the Kuwaiti Concession Contracts in 1934, all the islands and coastal waters belonging to Kuwait was part of the contract area. In the Iran Concession Contracts, which was concluded in 1933, the contract area was about one hundred thousand square miles for sixty years" (Amani, 2010). In Production Sharing Contracts, by accepting the sovereignty of governments over natural resources, including oil resources, it became possible for the host government to assess the obligations of the other party to the contract to protect the environment. Environmental obligations were also considered in the Service Contract. But the features of these contracts were the mere focus on economic matters rather than environmental duties.

The deleterious effects of some oil operations in various parts of the world on the environment have raised concerns. International treaties and domestic laws on environmental protection were an answer to this concern. Thus, a positive step was taken towards protecting the environment, and principles such as the principles of international environmental law, which were accepted by many countries, affected oil contracts. Considering the guarantee of legal sanction in contracts causes the parties to adhere to their obligations to protect the environment. Using environmental terms means adhering to the principles of international environmental law, such as the principle of prevention, the principle of precaution, the Polluter Pays Principle and Sustainable Development.

Precaution Principle

The precautionary principle is one of the most important principles of international environmental law. According to this principle, governments must take many precautionary measures in proportion to their capabilities to protect the environment. The purpose of the principle of environmental precaution is to protect the environment and prevent its degradation and pollution. "In general, the environment, by its nature, is closely related to science and knowledge. Among these, the principle of precaution is more related to science and knowledge than other principles of the environment, and from this perspective, this principle is considered a very good incentive for the development of scientific research in unknown fields"(Ramezani, 2013).

This principle has been mentioned in various treaties such as the Declaration of November 25th, 1987, Ministerial Declaration of the Second International Conference on the Protection of the North Sea: London, November 25th, 1987, International Conference on the Protection of the North Sea 1987, the Convention for the Protection of the Marine Environment of the North-East Atlantic (the 'OSPAR Convention') September 22nd, 1992, the Maastricht Treaty of European Union, and1992, and Rio Declaration. Following Principle 15 of the Rio Declaration to protect the environment, the precautionary approach shall be widely applied by States according to their capabilities. Where there are threats of serious or irreversible damage, lack of full scientific certainty shall not be used as a reason for postponing cost-effective measures to prevent environmental degradation.

Prevention Principle

This principle demonstrates that instead of compensating for environmental damage, the appropriate solution is to prevent such damage from occurring. Given that it is difficult to compensate for the environmental damage caused by oil activities, rules on the need to prevent such damage were established in international environmental law. For instance, "The cost of cleaning up oil spills was at least $ 3 billion, much of which was paid for by the US government, but due to the high cost of this cost, many cannot afford it" (Sabour, 2015).

Both experience and the opinion of scientific experts prove that the principle of prevention for the environment, both ecologically and economically, is a golden rule(Habibi, 2005). Article 206 of the United Nations Convention on the Law of the Sea (1982) stated: " When States have reasonable grounds for believing that planned activities under their jurisdiction or control may cause substantial pollution of or significant and harmful changes to the marine environment, they shall, as far as practicable, assess the potential effects of such activities on the marine environment and shall communicate reports of the results of such assessments in the manner provided in article 205".

Polluter Pays Principle

It is necessary to interpret who is a polluter. "Polluters are responsible for the pollution they have caused. Therefore, polluters should bear the cost of measures aimed at preventing and reducing pollution"(Lindhout & Van den Broek, 2014). Article 16 of the Rio Declaration notes: "National authorities should endeavor to promote the internalization of environmental costs and the use of economic instruments, taking into account the approach that the polluter should, in principle, bear the cost of pollution, with due regard to the public interest and without distorting international trade and investment."

This principle has been introduced since the introduction of civil liability in international treaties on accidents caused by nuclear and oil energy for persons whose activities have led to the destruction of the environment. "In implementing this principle, the UN Security Council held Iraq responsible for the destruction of the environment during the invasion of Kuwait. Therefore, the extent of the damage to the region's environment was assessed by the affected countries and reported to the Security Council's Compensation Committee" (Dabiri & Poorhashemi, 2009).

Sustainable Development Principle

Sustainable Development means that in addition to meeting the needs of current generations, development should not impair the ability of future generations to meet their own needs. World Commission on Environment and Development (the Brundtland Commission1987), which played an important role in popularizing the concept of sustainable development in the world, in its report called Our Common Future, defined sustainable development as follows: "Humanity can make development sustainable to ensure that it meets the needs of the present without compromising the ability of future generations to meet their own needs." "The Rio and Stockholm Declarations state that achieving sustainable development requires the protection of the environment and that the development agenda has made the two concepts inseparable(Poorhashemi, 2013)."

Stockholm Declaration on the Human Environment of the United Nations Conference on the Human Environment (1972) was the first step to connect environmental protection and sustainable development. For example, Article 8 of the Declaration considers economic and social development desirable for the environment and necessary for life and work. Article 11 emphasizes: "The Declaration on the Environmental Policy of Governments The environmental policies of all States should enhance and not adversely affect the present or future development potential of developing countries, nor should they hamper the attainment."

Developing countries were able to place the right to development in the third principle of the Rio Declaration. This principle states that: "The right to development must be fulfilled to equitably meet developmental and environmental needs of present and future generations." The fourth principle, which stipulated the Rio Declaration, states that: "To achieve sustainable development, environmental protection shall constitute an integral part of the development process and cannot be considered in isolation from it." These two principles are the most important principles of the Rio Declaration and must be considered together. Principle 3 is very similar to the definition of sustainable development in the Brundtland Commission. Principle 4 also considers the environment to be an integral part of development. The only difference between the two principles is that the former emphasizes the development and the latter on the environment. Principle 25 also highlights that "Peace, development and environmental protection are interdependent and indivisible. States shall resolve all their environmental disputes peacefully and by appropriate means following the Charter of the United Nations" (Dabiri & Poorhashemi, 2009).

Paragraph 6 of the Copenhagen Declaration states, "Development, social development, and environmental protection are interdependent and mutually reinforcing components of sustainable development, which is the framework for our efforts to achieve a higher quality of life for all people.... ". The fifth paragraph of the Johannesburg Declaration states "Accordingly, we assume a collective responsibility to advance and strengthen the interdependent and mutually reinforcing pillars of sustainable development—economic development, social development, and environmental protection— at the local, national, regional and global levels." It is also stated in Principle 16, "We are determined to ensure that our rich diversity, which is our collective strength, will be used for the constructive partnership for change and the achievement of the common goal of sustainable development(Pūraitė, 2012)."

ENVIRONMENTAL CONDITIONS IN OIL CONTRACTS

The first common form of contract between oil-rich countries and oil companies was Concession Contracts. These contracts did not consider environmental principles. The Darcy Contract and the Concession Contracts in Saudi Arabia did not address the environment. Since the second half of the twentieth century, countries have paid attention to the environment and the need to preserve it. Therefore, environmental principles were gradually considered in the oil contracts of that time. In recent years, modern oil contracts such as Norwegian contracts have paid special attention to the environment regarding energy policy. Environment and climate change are an important part of the Norwegian government's policy towards the oil industry. Therefore, the standards and necessary criteria for environmental protection have always been applied in these industries. The law governing oil activities in Norway is the 1996 Oil Law(Gray et al., 1999). This law gives powers to regulate oil activities. For example, the Norwegian government in Activities Regulation plan and prevent environmental damage during oil operations. Norway, as a pioneer country, introduces a carbon tax. The Norwegian oil industry is committed to paying a carbon tax. Burning oil by the oil industry emits CO_2 (carbon dioxide), CH_4 (methane), NOx (nitrogen oxides), and NMVOCs (non-methane volatile organic compounds) to the air. Therefore, the tax rate in this area is higher than in other industries.

ENVIRONMENTAL PROVISIONS IN INTERNATIONAL OIL CONTRACTS

Oil contracts include terms, each of which refers to the principles of international environmental law to some extent. For instance, Standard requirements, the Stabilization Clause, Clause on Gas flaring,

Clause on Access to Water & Others Natural Resources, Liability, Compensation, and Insurance, and Environmental Impact Assessments are the provisions that apply to oil contracts(ADNOC, 2020).

Requirements in International Standards

Kyla Tienhaara divided environmental standards in oil and gas contracts divided into five items of reference to domestic environmental law only; Reference to international industry standards only; Reference to both domestic law and international industry standards; Reference to domestic law and/or industry standards and international environmental agreements; and Development of project-specific environmental standards(Tienhaara, 2011).

It has been customary in oil contracts to refer to the domestic law of the host country. But because in some developing countries, environmental laws are not up to date, and oil standards are often used as a benchmark in oil contracts rather than domestic law. These standards are modern and cover topics that are not normally referred to as environmental law. International oil industry standards are provisions used by some institutions and are set out in guidelines. A case in point is the American Petroleum Institute (API), which determines that its members are dedicated to continuous efforts to improve the compatibility of their operations with the environment while economically developing energy resources and supplying high-quality products and services to consumers. For instance, in Blowout Prevention Equipment (BOPE), API stated that All BOPE should be selected, installed, and properly maintained to prevent uncontrolled releases to the environment.

Another example is ISO which published some standards related to the environment. These standards include the ISO's environmental standards (ISO 14000 family), ISO 14004:2016, and ISO 14012:1996. "In several contracts in the sample, the parties instead included a reference to international industry standards." For example, international standards such as "Good Oilfield Practices" or "Good Production Practices" are used in oil contracts. What is important about these standards is that they should be clearly defined to avoid any ambiguity. A 2002 Cambodian contract provides a rare example of a definition: "Good Petroleum Industry Practices means the standards and practices, and exercise of that degree of skill, prudence, and foresight that would reasonably be expected of persons carrying out international petroleum operations, and adherence to generally accepted standards of the international petroleum industry, including sound environmental provisions(Tienhaara, 2011)." In the Concession Contract of the Republic of Mozambique, the International Oil Company undertook to take the necessary measures to prevent environmental damage following the accepted standards in the oil industry, modern techniques, and methods.

References to international standards cause problems with interpretation and implementation due to their ambiguity. Although reference to standards benefits the parties to the oil contracts, the problem is that no specific authority is willing to provide a comprehensive definition of these standards. Therefore, given the many institutions that set these standards, there is a big difference between defining these standards from one contract to another(Wawryk, 2002). As a solution, contracts refer to domestic law and international standards at the same time. The advantage of this approach is that the action can reduce the existing ambiguities and cause the contract to be executed without any problems. Under the 2007 Indian Production Sharing Contracts, the International Petroleum Corporation must comply with international oil industry standards and domestic law to carry out its environmental protection role. The majority of contracts reviewed for this article contained a reference to domestic environmental law and international industry standards. Article 21.1 of Brazil's 2001 Model Concession Contract indicates that industry standards are only intended to act as a supplement to domestic legislation: "The Concession-aire shall adopt, at its own cost and risk, all the necessary measures for the conservation of reservoirs and other natural resources and the protection of the air, soil and water in the surface or the subsurface, subject to Brazilian legislation and rules about the environment and, in their absence or lack, adopting Oil Industry Best Practice in this regard(Tienhaara, 2011)."

The last type of standard is the development of a project-specific environment. A contract between Azerbaijan and a Consortium stipulated that the contractor, The State Oil Company of the Azerbaijan Republic (SOCAR), and the State Committee for the Exploitation of Natural Resources would jointly agree on environmental standards. The use of this type of standard can speed up the implementation of oil projects because to determine this type of standard, the parties to the contract need to agree on ways to protect the environment. Obviously, due to the specific characteristics of each project and oil field, this method will be useful(Young, 2017).

Stabilization Clause

Abdullah Faruque defines the stabilization clause, "a form of government guarantee in a negotiated petroleum contract, which usually provides that the terms negotiated under the contract between a state and an International oil Company will not be altered unilaterally or terminated by the state through the promulgation of legislation or regulation. In essence, stabilization clauses aim at the prevention of legislative intervention in the negotiated contract regime"(Faruque, 2006). Today, the condition for the Stabilization Clause is that if the host government makes conflicting laws that affect the economic condition of the contract, the government will compensate the other party's losses equally. By way of illustration, if, under the new law, the number of taxes increases from the time the contract is concluded, the government must pay the additional amount to the other party to maintain and stabilize the economic condition of the contract.

Gas Flaring Clause

Flaring is one of the problems caused by oil operations. Due to the increasing demand for crude oil, many materials are burned. Burning associated gases that are harmful to the environment is inevitable for safety reasons. "The World Bank has estimated that the annual volume of associated gas being flared and vented is about 110 billion cubic meters (bcm), which is enough fuel to provide the combined annual natural gas consumption of Germany and France. Flaring in Africa (37 bcm in 2000) could produce 200 Terawatt hours (TWh) of electricity, which is about 50 percent of the current power consumption of the African continent and more than twice the level of power consumption in Sub-Saharan Africa, except for the Republic of South Africa"(Ismail & Umukoro, 2012). "In the Middle East where the largest volumes of gas flaring occur in Iran and Iraq. However, Saudi Arabia, Kuwait, Qatar, and other parts of the region have flaring issues. Both satellite and reported data sources indicated 7 to 10 bcm per year of gas flaring. Flaring during oil production operations emits CO_2 (carbon dioxide), CH_4 (methane), and other forms of gases which contribute to global warming causing climate change, and this affects the environmental quality and health of the vicinity of the flares" (Ismail & Umukoro, 2016).

"Despite the global campaign against the flaring of Associated Petroleum Gas (APG) during crude oil production and the resulting environmental degradation, gas flaring remains a major disposal option for unwanted APG. Flaring of associated petroleum gas (natural gas) is an age-long environmental concern that remains unabated. One reason is that flaring can sometimes be more economical than using this gas in some oil fields or regions. Another reason is that some flare facilities believe that flaring is a highly efficient combustion process, especially steam-assisted or air-assisted flare. However, this has serious implications on emissions and ultimately on the environment" (Ismail & Umukoro, 2016). Although environmental laws in Iran as a Middle Eastern country refer to gas flaring, a significant part of associated gas is burned due to oil operations. Hence, Applying strict and precise environmental standards is necessary to reduce these actions.

Access to Water & Others Natural Resources

Natural materials such as water are needed to carry out oil operations. One of the concerns about oil contracts is that the host governments focus only on oil and gas and neglect other natural resources such as water. The nationalization of the oil industry in oil-rich countries, mainly developing countries, led to the acceptance of permanent sovereignty over natural resources as a principle of international law. This principle is reflected in international documents such as UN resolutions. On December 20th, 1952, the General Assembly adopted Resolution 626 (7), which explicitly stated: "Remembering that the right of people freely to use and exploit their natural wealth and resources is inherent in the sovereignty and is following the Purpose and Principles of the Charter of the UN(ADNOC, 2020)." Subsequently, on December 14th, 1954, the General Assembly, at the requests of the Commission on Human Rights, stated in Resolution 837(9) that: "The permanent sovereignty over natural wealth and resources has been part of the right of peoples and nations to self-determination, which is also part of human rights. Resolution 3171 also protects the sovereignty of developing countries over natural resources". This principle has been emphasized in international treaties. Article 193 of the United Nations Convention on the Law of Sea, December 10th, 1982 notes: "States have the sovereign right to exploit their natural resources according to their environmental policies and following their duty to protect and preserve the marine environment. The Energy Charter Treaty (1994) also recognizes the principle of permanent sovereignty over natural resources. Article 18(1) stated, "The Contracting Parties recognize state sovereignty and sovereign rights over energy resources. They reaffirm that these must be exercised following and subject to the rules of international law"(Farhad & Khalatbari, 2018).

The sovereignty of nations over their natural resources is not absolute and unconditional, and states must meet environmental reclamation standards in exploiting and using their natural resources. As mentioned earlier, the Stockholm Declaration on the Human Environment of the United Nations Conference on the Human Environment1972 and other documents point to the need to protect the environment. For example, General Assembly Resolution 37/7, World Charter for Nature, addresses sovereignty over natural resources and compliance with environmental requirements. Therefore, countries with oil resources should mention in oil contracts the need to meet environmental standards in the use of natural resources such as water, and they should not relinquish ownership of other natural resources under the pretext of gaining economic benefits because the right to natural resources belongs to all human generations(Hay, 2010).

Liability, Compensation, and Insurance

Oil operations play a negative role in the environment. Protecting the environment increases the costs for oil companies, which is not cost-effective; they have no incentive to fulfill environmental commitments. Also, there is no response to environmental degradation by oil companies in many developing countries with oil resources due to the lack of adequate laws to protect the environment and deal with polluters. Another problem for oil-rich countries is that if they adopt a strict approach to protecting the environment, international oil companies will not enter those countries to invest in and implement oil projects(Hatami & Karimiyan, 2014; Kuik, 2003).

Gradually, however, a strict doctrine of liability was adopted in the domestic laws of host countries and international treaties on environmental protection. The intervention of governments with oil resources to protect the environment, which is an integral part of human rights, is done through Environmental Regulation. One of the goals of government cooperation in combating environmental degradation is accepting legal principles on environmental protection. The "polluter pays" principle states that whoever is responsible for damage to the environment should bear the costs associated with it(Shiravi, 2016). This principle is expressed in the 16 Principles of the Rio Declaration on Environment and Development, which demonstrates, "The polluter would be a person, company, or other organization whose activities are generating that by-product. Payment should equal to the damage and be made to the person or persons that suffered harm". Today, many oil contracts adopt a strict liability doctrine. For example, according to Article 199 of the Civil Code of Qatar "Any person who commits an act that causes damage to another party shall be liable to indemnify such damage". This legal approach is reflected in Qatar's oil contracts. In a similar case in Abu Dhabi National Oil Company's Contract, it has been stipulated that "Except as specifically provided in this Contract, in no event, including the negligent act or omission on its part, shall either Party be liable to the other, whether under this Contract or otherwise in connection with it, in contract, tort, breach of statutory duty or otherwise, in respect of any indirect or consequential losses or expenses including if and to the extent that they might otherwise not constitute indirect or consequential losses or expenses, loss of anticipated profits, plant shut-down or reduced production, loss of power generation, blackouts or electrical shut-down or reduction, goodwill, use, market reputation, business receipts or contracts or commercial opportunities, whether or not foreseeable"(ADNOC, 2020; Mohamed & Al-Thukair, 2009).

One of the problems with liability for environmental damage is the ability of the polluter to pay for the damage. Unlike other contracts, oil contracts are complex, and if any of the obligations are violated, there are large costs to the offender. In some cases, these costs are so high that international oil companies cannot afford them. One solution to this problem is insurance. In oil contracts, oil companies are committed to ensuring environmental damage. Paragraph 3 of Article 30 of the 1997 Turkmenistan Production Sharing Contract states, "Contractor will obtain and maintain insurance covering clean-up costs for damage to the natural environment, including pollution of the air and water and surface and subsurface soils and waters contained within, under or over the Contract Area and other areas used in connection with activities conducted over or according to this agreement(Hatami & Karimiyan, 2014)." Paragraph 2 of Article 18 of the Contract or exploration and production of hydrocarbons at Morskoye field, Atyrau region, states: "Within 180 (one hundred eighty) days of the date of the contract's effectiveness, the contractor shall work out and submit for approval to the Competent authority a program of insurance against business risk, property and liability insurance concerned with exploration and production of hydrocarbons. Insurance shall be provided for property and liability risks related to environmental pollution including land pollution and costs of elimination of damage caused to the environment including land melioration and recovery"(Hey, 2001).

Environmental Impact Assessment

"Environmental Impact Assessments (EIAs) is an environmental decision support tool, which provides information on the likely impacts of development projects to those who decide as to whether the project should be authorized. The purpose of an EIA is to determine the potential environmental, social, and health effects of a proposed development so that those who make the decisions in developing the project and in authorizing the project are informed about the likely consequences of their decisions before they take those decisions and are thereby more accountable. It is intended to facilitate informed and transparent decision-making while seeking to avoid, reduce or mitigate potential adverse impacts by considering alternative options, sites, or processes. EIA has been regarded as both a science and an art, reflecting the technical aspects, such as impact identification and prediction, as well as the evaluation, management, and presentation of information". "Like any industry, oil and gas industry operation and activities have the potential to impact the environment, if its impacts are not adequately assessed and managed. The magnitude of the impact increases if the facility is located near sensitive receptors, such as drinking water sources, residential areas, or protected environments"(Hinkle & Rosencranz, 2008).

ENVIRONMENTAL CONDITIONS IN IRANIAN OIL CONTRACTS

To examine the environmental terms in Iran's oil contracts, the laws and legal documents must be reviewed, and then the oil contracts must be reviewed and analyzed.

Environmental Laws and Documents

As one of the countries with abundant oil and gas resources, the negative effects of oil activities in Iran are indisputable. Although there are many laws and legal documents on environmental protection in Iran, they have not been useful and have not taken a positive step towards environmental protection. Another problem is the multiplicity of laws and legal documents. However, there is still a lack of a comprehensive law covering all the principles of international environmental law, such as precaution. Article 45 of Iran's Constitution, which deals with the environment, states: " Public wealth and property, such as uncultivated or abandoned land, mineral deposits, seas, lakes, rivers, and other public waterways, mountains, valleys, forests, marshland, natural forests, unenclosed pastureland, legacies without heirs, property of undetermined ownership, and public property recovered from usurpers, shall be at the disposal of the Islamic government for it to utilize following the public interest. Law will specify detailed procedures for the utilization of each of the preceding items". It follows from this article that ownership of mines, including oil resources, is in the hands of the government. Government sovereignty over oil and gas resources has been used as a principle in the Iranian legal system in other laws and legal documents. Paragraph 1 of Article 14 of the Law on the Fourth Economic, Socio-Cultural Development Plan concludes exploration and development contracts conditional on maintaining the sovereignty and exercising state ownership of the country's oil and gas resources. This point is also emphasized in Article 2 of the 1987 Oil Law, and Oil Law Amendment Law approved in 2011. Article 48 of the Constitution states, "There must be no discrimination among the various provinces concerning the exploitation of natural resources, utilization of public revenues, and distribution of economic activities among the various provinces and regions of the country, thereby ensuring that every region has access to the necessary capital and facilities following its needs and capacity for growth." This article refers to the principle of environmental justice. Environmental justice (EJ) is the fair treatment and meaningful involvement of all people regardless of race, color, national origin, or income concerning the development, implementation, and enforcement of environmental laws, regulations, and policies"(Schlosberg, 2009; Nijar, 2013).

Thus, the right to the environment is an undeniable necessity. The third paragraph of Article 26 of the Sixth Five-Year Development Plan states: "The government is committed to allocating three percent (3%) of the revenues from crude oil and net gas condensate exports to one-third to oil-rich and gas-rich provinces and two-thirds to less developed regions and cities to implement development programs in the form of annual budgets approved by the council. Allocate planning and development to these provinces". Despite what has been said, protecting the environment of the areas where the oil operations are carried out is still not considered. Article 50 of the Constitution, which directly refers to the environment, states, "The preservation of the environment, in which the present and the future generations have a right to flourishing social existence, is regarded as a public duty in the Islamic Republic. Economic and other activities that inevitably involve pollution of the environment or cause irreparable damage to it are therefore forbidden". This principle points out that environmental protection is an obligation that the people and the government are committed to fulfilling. On the other hand, it can be inferred that all oil activities that cause irreparable pollution or irreparable damage to the environment are prohibited.

Moreover, Iranian legal instruments refer to some specific environmental standards. Environmental Impact Assessments (EIAs) are enshrined in other documents, such as the law of the Third Economic, Socio-Cultural Development Plan, and Sixth Five-Year Development Plan, and its implementation is mandatory. The Third Economic, Socio-Cultural Development Plan law also refers to the Polluter Pays Principle as one of the international environmental law principles. But other laws do not mention this important principle. Although environmental laws and documents have been enacted in Iran, comprehensive legislation still needs. This is because the law still does not mention principles such as the Polluter Pays Principle and the principle of precaution(Skjærseth, 2013; Smits et al., 2014).

Environmental Conditions in Iranian oil Contracts

Iran's legal system has to some extent addressed the issue of environmental protection in oil activities. The question is, do environmental conditions apply to Iran's oil contracts? To answer this question, oil contracts must be examined in detail.

Article 33 of The Iranian Buy Back refers to the contractor's obligations to comply with the requirements of Health, Safety, and Environment. "33.2. The contractor is under obligation to conduct the Development Operations and operating activities, if any (a) in an environmentally sensitive manner so that people today, and future generations, may benefit from the wiser stewardship of earth's resources; and (b) in a manner which ensures safety and health in the workplace for the benefit of the personnel and employees of each party, and its subcontractor(c) as well as the public at large" (Hatami & Karimiyan, 2014). In the latest model of Iran's oil contracts, called IPC, environmental commitments have been addressed. This type of contract fits the ESHIA Standards. Environmental, Social, and Health Impact Assessment (ESHIA) Plan requires multi-disciplinary teams to evaluate environmental, social, and health impacts and risks(McHugh et al., 2006).

Among the harmful environmental effects of oil activities in Iran, the pollution of the Persian Gulf, Respiratory diseases, damage to the marine ecosystem, Extinction of animal and plant species in the mangrove forest, and the accumulation of waste can be mentioned. Unfortunately, the focus on economic interests has reduced the government's rigidity in overseeing international oil companies' compliance with environmental commitments. Economic growth is short-term and dangerous, regardless of environmental considerations.

Years of experience in industrial activities have shown that environmental degradation can have far greater economic benefits. These harmful effects on the environment threaten the lives of today and the lives of future generations. Therefore, in addition to the adoption of domestic laws and international treaties on the environment, it is necessary to consider the environmental obligations of the parties to the oil contracts. Although oil companies incur high costs to meet their environmental commitments, compensating for the damage to the environment is higher. The environment cannot be endangered on the pretext that the contractual obligations are not economical. Norway is a good example for developing countries, which usually do not take strict environmental measures. Therefore, the time has come to make an incisive decision. Developing countries such as Iran must take a strict approach toward industrial operation, namely oil activities. This decision can prevent further environmental pollution.

CONCLUSION

This article examines contractual terms related to environmental issues in contracts and some oil laws and regulations. The main purpose of writing this study is to review the solutions offered in international agreements, analyze the current state of laws and agreements of the Islamic Republic of Iran, and present new solutions. Accordingly, the present study was conducted to analyze the content of international oil and gas laws and agreements.

It was observed that contrary to the laws and regulations of the oil and gas fields, in some operational areas, including South Pars, due to the acceleration and expansion of the scope of executive activities and the lack of proper health management. The environmental environment of the Rannap is as follows. Pollution and drought pollution in the sea in recent months in the city of Assaluyeh is a good example of these cases that have been mentioned. The discussion of the removal of fillers also highlights the need to accelerate the management of environmental pollutants in these areas. It is essential to center for monitoring of the environment in this area be established and take immediate action to implement the strategies presented in this article take and a report on the implementation of the requirements of laws and regulations and international standards contained within the given time (e.g., three months) to the environment Provide life.

Given the characteristics mentioned in this article, it is still necessary that oil countries, including the Islamic Republic of Iran, in addition to developing the necessary laws and regulations to monitor and control oil operations, implement a comprehensive monitoring system for the implementation of international oil contracts and apply conditions related to the environment.

REFERENCES

Abu Dhabi National Oil Company (ADNOC). (2020). *General Terms and Conditions, for the Sale of Crude Oil / Condensate and Liquefied Petroleum Gas,* P.31. ADNOC. https://www.adnoc.ae/-/media/adnoc-v2/files/adnoc_crude-and-lpg_gtcs_january-2020-edition-final_v1.ashx?la=en&hash=C955167 8CC5CBBBAB30DFE83A495800E8AD540A1

Amani, M. (2010). *International oil contract law*. Imam Sadiq University Press.

Berkhout, F., & Smith, A. (2003). Carbon flows between the EU and Eastern Europe: Baselines, scenarios and policy options. *International Environmental Agreement: Politics, Law and Economics*, *3*(3), 199–219. doi:10.1023/B:INEA.0000005624.46391.96

Dabiri, F., & Poorhashemi, S. A. (2009). A Study of the Principles and Concepts of International Environmental Law with a Look at Sustainable Development. *Journal of Environmental Science and Technology*, *11*(3), 220.

DiMento, J. F. (2021). *Book Review: Philosophies of Polar Law. International Environmental Agreements: Politics*. Law and Economics.

Farhad, D., & Khalatbari, Y. (2018). Achieving sustainable development from the perspective of international environmental law. *Journal of Human and Environment*, *16*(44).

Faruque, A. (2006). Validity and Efficacy of Stabilisation Clauses: Legal Protection vs. Functional Value. *J. Int'l Arb.*, *23*, 317.

Fuentes, X. (2002). International law-making in the field of sustainable development: The unequal competition between development and the environment. *International Environmental Agreement: Politics, Law and Economics*, *2*(2), 109–133. doi:10.1023/A:1020990026398

Gao, Z., & Kao, C. K. (1998). *Environmental regulation of oil and gas* (Vol. 11). Kluwer Law International BV.

Gray, J. S., Bakke, T., Beck, H. J., & Nilssen, I. (1999). Managing the environmental effects of the Norwegian oil and gas industry: From conflict to consensus. *Marine Pollution Bulletin*, *38*(7), 525–530. doi:10.1016/S0025-326X(99)00004-1

Habibi, M. H. (2005). *Environmental Law*. Tehran University Press.

Hatami, A., & Karimiyan, E. (2014). *Foreign Investment Law in Light of Investment Act and Contracts*. Teesa Publication.

Hay, J. (2010). How efficient can international compensation regimes be in pollution prevention? A discussion of the case of marine oil spills. *International Environmental Agreement: Politics, Law and Economics*, *10*(1), 29–44. doi:10.100710784-009-9096-8

Hey, E. (2001). The Climate Change Regime: An Enviro-Economic Problem and International Administrative Law in the Making. *International Environmental Agreement: Politics, Law and Economics*, *1*(1), 75–100. doi:10.1023/A:1010117910664

Hinkle, J., & Rosencranz, A. (2008). *Jon Birger Skjærseth and Tora Skodvin, Climate Change and the Oil Industry: Common Problem*. Varying Strategies.

Ismail, O. S., & Umukoro, G. E. (2012). Global impact of gas flaring. *Energy and Power Engineering*, *4*(4), 290–302. doi:10.4236/epe.2012.44039

Ismail, O. S., & Umukoro, G. E. (2016). Modelling combustion reactions for gas flaring and its resulting emissions. *Journal of King Saud University-Engineering Sciences*, 28(2), 130–140. doi:10.1016/j.jksues.2014.02.003

Kuik, O. (2003). Climate change policies, energy security and carbon dependency trade-offs for the European Union in the longer term. *International Environmental Agreement: Politics, Law and Economics*, 3(3), 221–242. doi:10.1023/B:INEA.0000005625.44125.54

Lindhout, P. E., & Van den Broek, B. (2014). The polluter pays principle: Guidelines for cost recovery and burden sharing in the case law of the European court of justice. *Utrecht Law Review*, 10(2), 46. doi:10.18352/ulr.268

McHugh, S., Maruca, S. D., Lilien, J., & Manning, A. (2006). Environmental, social, and health impact assessment (ESHIA) process. In SPE International Health, Safety & Environment Conference. OnePetro.

Mohamed, L., & Al-Thukair, A. A. (2009). Environmental Assessments in the Oil and Gas Industry. *Water Air and Soil Pollution Focus*, 9(1-2), 99–105. doi:10.100711267-008-9190-x

Nijar, G. S. (2013). The Nagoya–Kuala Lumpur Supplementary Protocol on Liability and Redress to the Cartagena Protocol on Biosafety: An analysis and implementation challenges. *International Environmental Agreement: Politics, Law and Economics*, 13(3), 271–290. doi:10.100710784-012-9187-9

Poorhashemi, S. A. (2013). *International Environmental Law*. Dadgostar Press.

Pūraitė, A. (2012). *Origin of Environmental Regulation*. Mykolas Romeries University Press, Faculty of Public Security. https://intranet.mruni.eu/upload/iblock/b7c/014_puraite.pdf

Ramazani, G. M. H. (2013). A Comparative Study of "Precautionary Principle" in Opinions and Decisions of Internationals Tribunals. *Public Law Journal*, 15(40), 143.

Sabour, M. R. (2015). *Alternative Energy*. Khaje Nasir Toosi University Press.

Schlosberg, D. (2009). *Defining environmental justice: Theories, movements, and nature*. Oxford University Press.

Shiravi, A.-H. (2016). *Oil and Gas Law*. Mizan Press.

Skjærseth, J. B. (2013). Governance by EU emissions trading: Resistance or innovation in the oil industry? *International Environmental Agreement: Politics, Law and Economics*, 13(1), 31–48. doi:10.100710784-012-9201-2

Smits, C. C., van Tatenhove, J. P., & van Leeuwen, J. (2014). Authority in Arctic governance: Changing spheres of authority in Greenlandic offshore oil and gas developments. *International Environmental Agreement: Politics, Law and Economics*, 14(4), 329–348. doi:10.100710784-014-9247-4

Sornarajah, M. (2006). A law for need or a law for greed?: Restoring the lost law in the international law of foreign investment. *International Environmental Agreement: Politics, Law and Economics*, 6(4), 329–357. doi:10.100710784-006-9016-0

Tienhaara, K. (2011). Foreign investment contracts in the oil & gas sector: A survey of environmentally relevant clauses. *Sustainable Development Law & Policy*, 11(3), 6.

Wawryk, A. S. (2002). Adoption of international environmental standards by transnational oil companies: Reducing the impact of oil operations in emerging economies. *Journal of Energy & Natural Resources Law*, *20*(4), 402–434. doi:10.1080/02646811.2002.11433308

Young, M. A. (2017). Energy transitions and trade law: Lessons from the reform of fisheries subsidies. *International Environmental Agreement: Politics, Law and Economics*, *17*(3), 371–390. doi:10.100710784-017-9360-2

ADDITIONAL READING

Ramazani, G. M. H. (2013). A Comparative Study of "Precautionary Principle" in Opinions and Decisions of Internationals Tribunals. *Public Law Journal*, *15*(40), 143.

Sabour, M. R. (2015). *Alternative Energy*. Khaje Nasir Toosi University Press.

Schlosberg, D. (2009). *Defining environmental justice: Theories, movements, and nature*. Oxford University Press.

Shiravi, A.-H. (2016). *Oil and Gas Law*. Mizan Press.

Skjærseth, J. B. (2013). Governance by EU emissions trading: Resistance or innovation in the oil industry? *International Environmental Agreement: Politics, Law and Economics*, *13*(1), 31–48. doi:10.100710784-012-9201-2

Smits, C. C., van Tatenhove, J. P., & van Leeuwen, J. (2014). Authority in Arctic governance: Changing spheres of authority in Greenlandic offshore oil and gas developments. *International Environmental Agreement: Politics, Law and Economics*, *14*(4), 329–348. doi:10.100710784-014-9247-4

Sornarajah, M. (2006). A law for need or a law for greed?: Restoring the lost law in the international law of foreign investment. *International Environmental Agreement: Politics, Law and Economics*, *6*(4), 329–357. doi:10.100710784-006-9016-0

Tienhaara, K. (2011). Foreign investment contracts in the oil & gas sector: A survey of environmentally relevant clauses. *Sustainable Development Law & Policy*, *11*(3), 6.

Wawryk, A. S. (2002). Adoption of international environmental standards by transnational oil companies: Reducing the impact of oil operations in emerging economies. *Journal of Energy & Natural Resources Law*, *20*(4), 402–434. doi:10.1080/02646811.2002.11433308

Young, M. A. (2017). Energy transitions and trade law: Lessons from the reform of fisheries subsidies. *International Environmental Agreement: Politics, Law and Economics*, *17*(3), 371–390. doi:10.100710784-017-9360-2

KEY TERMS AND DEFINITIONS

Equity: Defined by UNEP to include intergenerational equity - "the right of future generations to enjoy a fair level of the common patrimony" - and intragenerational equity - "the right of all people within the current generation to fair access to the current generation's entitlement to the Earth's natural resources" - environmental equity considers the present generation under an obligation to account for long-term impacts of activities and to act to sustain the global environment and resource base for future generations. Pollution control and resource management laws may be assessed against this principle.

Polluter pays principle: The polluter pays principle stands for the idea that "the environmental costs of economic activities, including the cost of preventing potential harm, should be internalized rather than imposed upon society at large." All issues related to responsibility for environmental remediation costs and compliance with pollution control regulations involve this principle.

Precautionary principle: One of the most commonly encountered and controversial principles of environmental law, the Rio Declaration formulated the precautionary principle: To protect the environment, the precautionary approach shall be widely applied by States according to their capabilities. Where there are threats of serious or irreversible damage, lack of complete scientific certainty shall not be used as a reason for postponing cost-effective measures to prevent environmental degradation. The principle may play a role in any debate over the need for environmental regulation.

Prevention: The concept of prevention can perhaps better be considered an overarching aim that gives rise to a multitude of legal mechanisms, including prior assessment of environmental harm, licensing or authorization that set out the conditions for operation and the consequences for violation of the conditions, as well as the adoption of strategies and policies. Emission limits and other product or process standards, the use of best available techniques, and similar techniques can all be seen as applications of the concept of prevention.

Public participation and transparency: identified as necessary conditions for "accountable governments,... industrial concerns," and organizations generally, public participation and transparency are presented by UNEP as requiring "effective protection of the human right to hold and express opinions and to seek, receive and impart ideas,... a right of access to appropriate, comprehensible and timely information held by governments and industrial concerns on economic and social policies regarding the sustainable use of natural resources and the protection of the environment, without imposing undue financial burdens upon the applicants and with adequate protection of privacy and business confidentiality," and "effective judicial and administrative proceedings." These principles are present in environmental impact assessment, laws requiring publication and access to relevant environmental data, and administrative procedures.

Transboundary responsibility: Defined in the international law context as an obligation to protect one's environment and prevent damage to neighboring environments, UNEP considers transboundary responsibility at the international level as a potential limitation on the sovereign state's rights. Laws that limit externalities imposed upon human health and the environment may be assessed against this principle.

Chapter 11
Environmental and Criminal Law

ABSTRACT

An overview of the evolution of environmental protection at a global level is the purpose of this chapter. At the beginning, the authors discuss the emergence of environmental concerns, followed by the second point, which addresses environmental protection at the international level. Third, the phenomenon of criminalizing harmful conduct against the environment is explored, with a special focus on the role of the European Union in this regard. The authors will review the most important social movements, international events with the most impact and the legislative evolution of community law in this area.

INTRODUCTION

The conception of the legal asset of the environment as an object of protection is relatively recent, since it did not appear until the post-industrial era, when technological progress and its impact led us to rethink the role of criminal laws, which until now only protected so-called property in "traditional" legal systems.

The so-called "risk society" leads us to a new understanding of society that will condition the actions of the criminal legislator. With each technological advance, new risks appear, so that "an increasingly industrialized society is an increasingly dangerous society"(Brickey, 1996). Thus, technological evolution implies the appearance of new forms of risk that will require criminal treatment.

As a result of the unstoppable economic development, the environment begins to deteriorate notably and resources begin to run out, directly affecting the lives and health of citizens. As a result of this, terms such as quality of life or respect for the environment appear. Public opinion is beginning to take an interest in the environment, producing the so-called "greening" of public opinion. "There is awareness that the imbalances in the biosphere caused by human activity endanger the development and survival of Humanity on the Planet". In other words, a clear awareness on the part of society of environmental problems originates, which is reflected in the emergence of the first environmental movements. Thus, in the 1970s, as a result of society's concern about the destruction of the environment that surrounds it, the first environmental groups began to emerge, advocating radical changes in environmental policies.

DOI: 10.4018/978-1-6684-4158-9.ch011

Social discontent over the governments' management of the environment/development binomial grew and soon manifested itself in citizen protests. An example is the Warren conflict, which occurred in 1982 in the United States. This movement began when a polluting waste dump was established near the town of Warren, North Carolina, a low-income area populated primarily by African-Americans. Faced with this action, which was described as an act of environmental racism, the population reacts and begins its protests, producing hundreds of arrests, which soon spread to the national territory, generating local conflicts in Los Angeles, New York, Houston or Chicago(Lazarus, 1994).

These phenomena became institutionalized and transformed in such a way that in the 1990s they became "state networks of the environmental justice movement". From October 24-27, 1991, 650 activists from around the globe gather for the First National People of Color Environmental Leadership Summit in Washington, DC. It broadens perspectives beyond racial protests against polluting activities and extends the phenomenon of environmental justice, which goes from being a movement for racial justice to a movement for justice for all. The Office of Environmental Justice and the National Environmental Justice Advisory Council were soon created, a phenomenon of entity creation that spreads internationally. For example, the national groups of Friends of the Earth International of England and Scotland are created, which promote the introduction of legislative elements and changes in public administration. In this way, the environmental justice movement is organized, resulting in institutions that will advocate for the right to a suitable environment for all citizens.

As we said, we are in a moment of great technological development, which is going to cause great environmental disasters. Thus, of those produced in the seventies, the industrial disaster of Séveso, in the region of Lombardy, Italy (1976), or the disaster at the Three Miles Island nuclear power plant, in the State of Pennsylvania, United States (1979). These environmental accidents caused by human action will occur continuously, highlighting in the following decades the Bhopal disaster, India (1984), the Chernobyl nuclear accident, Ukraine (1986), or the Fukushima I nuclear accident, Japan (2011).

Until then, social movements aimed at guaranteeing all citizens their right to a suitable environment. However, the events of this time reveal the imminent need to directly protect the environment. Thus, the social will to intervene in environmental problems is forged and what we can call "Political Ecology" is born, understood as the correct policy for the conservation of the environment, as environmental public management(Faure, 2004).

This change in perspective will be reflected in the international agenda, thus giving rise to a stage called the ecological era. The international community begins to consider the need to put limits on man's actions in his search for development" and, from then on, a series of international conventions will be held whose objective is the protection of the environment. environment.

PROTECTION OF THE ENVIRONMENT BY INTERNATIONAL INSTITUTIONS

The study of international environmental law "is essential to be able to undertake any other legal research on a specific aspect framed around the legal protection of the environment, because if we do not know what the principles and mandatory guidelines established at the international level are, we will not be able to understand the legal dimension of any other study carried out, both at the international level and at the national level, since the States must adapt their legal order to the international commitments assumed". Therefore, we will proceed to analyze the most important international events in the environmental field within the United Nations, with the aim of giving a vision of the evolution of international action on this issue.

As a precedent of international environmental law, we can highlight some regulations on certain matters related to the environment. Thus, there are some international treaties from the end of the 19th and 20th centuries on sectoral issues such as excessive fishing of certain species or the protection of birds useful for agriculture, among others. However, it was not a question of protecting the environment in itself, but rather certain related aspects insofar as they affected the development of human activity. That is, "the environment is protected not for its own benefit but for the benefit of man", adopting an anthropocentric position.

The birth of international environmental law is usually marked in the early 1970s, when Sweden proposes holding a meeting on environmental problems within the United Nations. Thus, following the Swedish initiative, the United Nations Conference on the Human Environment was held between June 5 and 16, 1972, which gave rise to the Stockholm Declaration, which "represents the first effort of a to protect the environment as a legal asset"(Cho, 2000).

This conference is an example of the awareness on the part of governments of the need to plan the fight against pollution with legal measures and adequate technical instruments. Proof of this is the great participation that there was in it, with representatives from 112 States attending.

In it, the need to preserve the environment for future generations is raised for the first time, principles are established that allow the synthesis between the environment and development and, for this, it is established that the States must be responsible for ensuring that activities carried out within its borders, jurisdiction and control do not cause harm to people, the natural environment or the environment of other States.

As a result of this first Conference, the Action Plan on the human environment makes a series of recommendations to the States to develop their policies, laying the foundations for all the actions carried out subsequently(Du Rées, 2001).

Since then, the environment has become an issue of international importance and will be discussed again on several occasions within the framework of the United Nations, as we will see below. Outside this area, international events related to environmental protection are also multiplying rapidly.

The Stockholm Conference marks a turning point in the approach to international environmental protection. From this moment on, the problem of environmental degradation is addressed from an integral point of view, because until then, as we have seen, the international conventions on the environment dealt with sectoral problems, a technique that proved ineffective in providing adequate protection. to the environment. Analyzing the Conventions of the time, we note that "indeed, all of them propose the protection and preservation of certain natural resources that are shared by the entire international community, regulating certain problems that affect everyone and not a certain area of the globe". In this way, the environmental issue begins to be treated with a global approach(Faure, 2016).

After a decade of the Stockholm Conference, the second United Nations Conference on the environmental problem is held. This is the United Nations Conference on Environment and Development or Earth Summit, held in Rio de Janeiro between June 1 and 15, 1992, attended by representatives of 176 States. It re-emphasizes the need to preserve the environment for future generations, and to treat environmental protection and economic development as a single issue.

As a result of the Conference we have, first of all, the Rio Declaration on Environment and Development, a document without legally binding force in which a series of principles are established by virtue of which future decisions will be made in relation to the compatibility of development and protection of the environment. In principle, the intended title was the Earth Charter, but it was finally changed at the insistence of developing countries, who considered that it did not refer to the objective of integrating environment and development(Faure, 2017).

Secondly, Agenda 21 results, a document that contains an action plan for sustainable development to be developed during the 21st century. Thus, it explains the main problems in environmental matters, establishes priorities, details the bases for action in each of the objectives and indicates the means of execution.

At the institutional level, the Commission on Sustainable Development is created, which is part of the UN Department of Economic and Social Affairs, to ensure the complementary action of the activities derived from the Conference.

The Rio Conference takes the global nature of the environment for granted, in the sense of dealing with environmental problems in a general way, not by compartments. However, this Conference is going to give a new meaning to the global nature of environmental problems, as it emphasizes the need for direct participation, not only of governments, but also of citizens. In this way, greater importance will be given to non-governmental organizations that, until now, attended international forums as mere observers.

It is important to highlight here the change of conception that occurs. While in 1972 the environmental problem was treated from an anthropocentric point of view, an ecocentric position is now adopted. That is to say, the Stockholm Conference, as its name indicates - United Nations Conference on the Human Environment - raises the protection of the environment insofar as its function is to serve man; however, the Rio Declaration treats man as "an integral part of Nature, together with other forms of life, with which it must be respected regardless of its use for man" (Faure, 2016).

Another ten years later, the UN organizes the World Summit on Sustainable Development, which was held from August 26 to September 4, 2002 in Johannesburg and in which 190 Heads of State or government participated. The fundamental objective of this event was to put a stop to environmental degradation, to which is added the fight against poverty.

The resulting Johannesburg Declaration on Sustainable Development is a renewal of the ecological commitment of the participating States, which extends to issues such as the eradication of poverty, the modification of unsustainable patterns of production and consumption, or health and sustainable development. Thus, most of the actions in environmental matters had already been indicated previously, this document being a reaffirmation of the commitments made in Agenda 21. New commitments are made in relation to reducing the number of people who lack access to safe water and to proper sanitation, as well as access to energy services.

At the same time, the Plan for the Implementation of the Decisions of the World Summit on Sustainable Development establishes the specific actions to be carried out to develop the commitments established in the Declaration(Delgado, 1985).

The major objectives achieved at this Summit are related to the eradication of poverty and health. However, in terms of environmental protection, it does not represent great progress, except for the agreements reached for the entry into force of the Kyoto Protocol. This establishes objectives for reducing the emission of gases into the atmosphere responsible for the greenhouse effect and climate change and for its entry into force, the ratification of at least 55 countries that add up to at least 55 percent of the emissions made was necessary. in the industrialized world. Thus, the commitment to ratify by China, Russia and Canada was achieved, with which the minimum required was already reached. In contrast, the United States, the world's largest polluter with 25 percent of emissions, refused to accept the Protocol.

In 2005, the XI Conference of the Parties to the United Nations Framework Convention on Climate Change -approved in 2002- and the I Conference of the Members of the Kyoto Protocol are held in Montreal. These conclude with the approval of the Montreal Action Plan, which represents an important advance in terms of environmental protection at the international level. In it, the signatory States undertake to negotiate a new instrument against climate change that allows actions beyond 2012, which is the deadline established by Kyoto. In addition, unlike what happened with this one, an agreement is reached without opposition from the United States.

One year later, from November 6 to 17, 2006, continuing with the thematic line of the previous ones, the XII Conference on Climate Change and the II Conference of the Members of the Kyoto Protocol are held in Nairobi.

In general, the results of both were modest, since "the summit culminated in the adoption of a limited number of agreements that, in general, do not refer to the main obstacles in the negotiations nor do they represent a qualitative leap in the reformulation of the regime on climate change for the post-Kyoto scenario"(Uhlmann, 2011).

More recently, the United Nations Conference on Sustainable Development (Rio+20) was held in Rio de Janeiro from June 20 to 22, 2012.

The resulting document, entitled "The future we want", is of an uncertain legal nature and does not devote all the necessary attention to the environmental issue. It concludes that it is essential to give a boost to the green economy, but does not establish specific guidelines to achieve these objectives. It also highlights the need for a change in the institutional framework for sustainable development, but, given the lack of consensus to create a specific Agency for it, only the United Nations Environment Program is reinforced in this area.

As we can see, environmental protection has suffered a slowdown in recent years. Perhaps the cause is the context of crisis, in which the economic resources of the States are less and, therefore, they are not willing to make firm commitments that involve additional spending. Given this, it is necessary to revive international interest in environmental protection, because only in this way will the conservation of the environment be guaranteed, which, in turn, means ensuring long-term human survival(Smith, 2012).

CRIMINAL LAW AS AN INSTRUMENT OF ENVIRONMENTAL PROTECTION

The measures and commitments related to the protection of the environment were shown to be ineffective, since the great ecological disasters continued to happen. Thus, given the ineffectiveness of the measures adopted to stop environmental deterioration, recourse to criminal law is proposed as the last resort of the legal system to protect the environment against the most serious attacks against it.

Within the international process of environmental protection, there is an initiative aimed at resorting to criminal law as a means to guarantee the conservation of the environment. In the Stockholm Conference of 1972, reference is made to criminal law as a way of preventing situations of extreme pollution. Between the sessions of the different working groups[36], one was held on the preconfiguration of a new crime of international scope, ecocide, understood as the disturbance or destruction, in whole or in part, of an ecosystem(Smith, 2012). Unfortunately, this proposal was not included in the final declarations. However, this initiative must be positively valued, since it represents the first international attempt to criminalize harmful conduct against the environment.

Apart from isolated projects such as the previous one, the main promoter of environmental criminal law in our environment has been and is the European Union (hereinafter, EU). The special situation of the EU as a supranational body with decision-making power and direct influence on the legislation of its Member States makes it a perfect setting to carry out a common environmental policy. The role that it plays is fundamental, since "the protection of the world's ecological heritage will largely depend on its ability to respond to supranational phenomena of environmental degradation".

As we will see, in recent decades, community bodies have carried out a gradual transformation of environmental protection legislation, now reaching the point of forcing their member states to follow a process of criminalizing conduct that threatens the environment. Thus, in the following pages we will review chronologically the community legislation on environmental protection, highlighting its evolution towards a common system of environmental criminal protection.

In the first moment of the communitarian legislative history we do not find direct reference to the environment in the founding Treaties of the European Communities(Smith, 2012). The protection of the environment was not, at that time, a priority or an objective at European level. However, this lack of express provision did not prevent the tendency to establish a common policy and provide the necessary legal bases to be able to legislate on environmental matters(Faure & Zhang, 2011).

Parallel to the international interest in environmental issues, in the 1970s the concern of the European institutions arose to create a common policy to protect the environment. Thus, after the celebration of the Stockholm Conference by the United Nations in June 1972, the Summit of Heads of State and Government of the European Community is held in Paris on October 19 and 20. In it, the protection of the environment is assumed as a mission of the EU, the basic principles of the environmental policy of the Community are established and the community institutions are urged to present an Environmental Action Program.

From then on, a line of action began to be devised, various community initiatives emerged and programs began to be developed in relation to the protection of the environment. However, these first initiatives referred only to the administrative regulation of certain activities and made reference to different sectorial elements of the environment, such as the atmosphere or waters, but not to the protection of the environment from a overall point of view(Fortney, 2002).

The first moment in which the use of criminal law as an instrument for protecting the environment is considered is with Resolution 28/1977/EU of the Council of Europe, on the Contribution of Criminal Law to the protection of the Environment. The same, in its third recital, refers to the need for the intervention of criminal law as a last resort "in the event that the remaining measures have not been respected, have not had an effect or have been inadequate(Bande, 2017)."

In it, for the first time, the Member States are recommended to criminalize polluting activities, both those carried out with intent and recklessness; the imposition of criminal sanctions, not only imprisonment and fines, but also other accessories such as the closure of polluting establishments or the disqualification of those responsible. Likewise, States are advised to review their criminal proceedings to adapt them to the singularities of environmental law, proposing the creation of specialized sections in the Prosecutor's Offices, as well as a special judicial registry of those convicted of pollution and the exclusion of the benefit of amnesty for serious infringements relating to the protection of the environment. As we can see, ambitious measures are contemplated that would involve the introduction of modifications, not only in criminal texts, but in the procedural system itself(Setness, 1996).

With this, we see how the European position regarding environmental crime changes. We have gone from initial disregard and disinterest to developing proposals aimed at establishing a common system for protecting the environment through the use of criminal law.

The recommendations made will be transferred to the internal legal systems of the States. In Spain, five years after the promulgation of the Constitution[43], LO 8/1983, of June 25, on the Urgent and Partial Reform of the Penal Code, is issued, which typifies the ecological crime for the first time in legislative history. Spanish in article 347 bis CP, within Title V, Chapter II (Crimes against public health and the Environment). However, we will have to wait a few years to see included in our legal system the procedural modifications referred to in Resolution 28/1977/EU of the Council of Europe, on the Contribution of Criminal Law to the Protection of the Environment.

Later, in the 1990s, the so-called "Group of specialists on the protection of the environment through criminal law" was constituted within the Council of Europe, which pursued a more complete protection of the environment using criminal law. as ultima ratio and establishes common guidelines to fight against environmental attacks. Within this body, the aim is to harmonize national legislation regarding offenses against the environment and, for this, it is recommended that the Member States modify their criminal legislation(Schroeder, 1993).

Seeking this objective of harmonization, Resolution 1/1990/EU of the Council of Europe is dictated, relative to the protection of the Environment through Criminal Law. This affects the recommendations already made, updating them and further specifying the measures to be adopted. Specifically, it recommends establishing a broader list of criminal types, configuring them as crimes of danger and their delimitation with administrative sanctions. In this way, express recommendations are made regarding the legislative technique to be used by national legislators.

In relation to the prosecution of environmental crime, the possibility of carrying it out is recommended, not only in the countries where the act was committed, but also in those that have suffered the consequences. In this way, not only are the Member States urged to include specific crimes against the environment in their internal legal systems, but the first lines of a cooperation system to combat this type of crime are outlined.

Formalizing the community commitments, the Spanish legislator includes the aspects indicated in LO 10/1995, of November 23, of the Penal Code, which represents an important step in the matter. Unlike the previous Code, which only had a criminal type dedicated to the ecological crime itself, the current one contains 13 articles and several common provisions, aimed at protecting natural resources and the Environment (articles 326 to 331) and flora and fauna. fauna (articles 332 to 337). This reform was well received by the doctrine, since it complies satisfactorily with the broad recommendations made by the Group of specialists(Faure, 2010).

On the other hand, as regards the EU's original law, in 1992, the issue of environmental protection was incorporated into the Maastricht Treaty or the EU Treaty (hereinafter, TEU). This one dedicates its Title XVI to the environment, consecrating it as community policy. From then on, the political-criminal references in the founding Treaties will increase.

Following this line, the Treaty of Amsterdam, in 1997, perfects the cooperation system. With this, the establishment of minimum standards by the Union on the constitutive elements of offenses and sanctions in the field of organized crime, terrorism and drug trafficking is allowed. In this way, the door of Community criminal law is opened, although at the moment the environment is not among the criminal matters on which the EU can legislate(Mistura, 2018).

Despite this, this Treaty will represent an advance in the process of European influence in environmental criminal matters, since it introduces the principle of sustainable development as one of the main objectives of the Community. Thus, the foundations of community policy on environmental matters are established, which will be: the principles of precaution and preventive action, the principle of correcting attacks and aggressions against the environment and the principle that the polluter pays.

The Treaty of Amsterdam, by giving the possibility to the European bodies to legislate in criminal matters, makes possible the approval of the Tampere Program in 1999, which establishes political-criminal guidelines of the EU with normative bases. It does include damage to the environment as one of the forms of crime that need specific treatment by the Union and is included in the priorities of European criminal policy(Lazarus, 1995).

On this basis, the community institutions can already legislate in environmental criminal matters, seeking the unification of the internal legislation of the Member States. Thus, after a conflict between the European Parliament and the Commission that leads us to the Court of Justice of the European Communities, Directive 2008/99/EU of the European Parliament and of the Council, of 11.19.2008, is finally issued, on the protection of the environment through criminal law. This represents a great strengthening of the European environmental criminal system and will decisively influence the legislative evolution of crimes against the environment in the internal legal system of all the EU Member States.

The third recital of Directive 2008/99/EU of the European Parliament and of the Council, of 11.19.2008, on the protection of the environment through criminal law, once again refers to the need to resort to criminal law to protect environment. In this sense, he tells us that "experience has shown that the existing sanction systems are not enough to achieve full compliance with the legislation for the protection of the environment. This compliance can and should be reinforced through the application of penal sanctions that reveal social disapproval of a qualitatively different nature than administrative sanctions or a compensation mechanism under civil law.

As a result, the text requires the criminalization of conduct that threatens various elements of the environment: the air, the soil or water, animals or plants, and even habitats and the ozone layer. Regarding activities, legislation is passed on the discharge, emissions, treatment or transfer of waste, the operation of dangerous installations, activities related to nuclear materials, activities harmful to flora or fauna and the commercialization of products that destroy the environment. ozone. As we can see, it is a complex norm that covers multiple aspects, reflecting an all-encompassing vision of environmental problems, which affects all elements of the environment.

The transposition of Directive 2008/99/EU of the European Parliament and of the Council, of 11.19.2008, regarding the protection of the environment through criminal law, brought about major changes in Spanish legislation. LO 5/2010, of June 22, which modifies LO 10/1995, of November 23, of the Penal Code, introduces important novelties in the Code, once again increasing the catalog of typical behaviors, as well as your penalty. In this way, we can again observe the correlation between the promulgation of community regulations and the legal reforms that have occurred in our internal legal system.

Following the lines of action of the EU, Spanish criminal legislation on environmental matters has evolved from a limited protection of the environment with the classification in a single precept - former article 347 bis CP - to dedicate a Chapter to "crimes against natural resources". natural resources and the environment", made up of seven precepts –arts. 325 to 331 CP- which contain a wide range of behaviors that can be considered criminal.

With all of the above, the EU has established itself as a promoter of ambitious environmental policies that encompass multiple aspects, achieving the criminalization and prosecution of many of the harmful activities against the environment. After the evolution that community law has experienced in environmental criminal matters, it is currently configured in a homogeneous way. The Member States already include in their internal legal systems criminal conduct, if not identical, with a great degree of similarity. In this way, sufficient and equal protection of the environment is guaranteed throughout the European territory(Mitsilegas et al., 2016).

CONCLUSION

The protection of the environment as a legal right does not begin until the post-industrial era, when, as a result of great development, the deterioration of the environment makes it clear that there is a need to intervene to guarantee the maintenance of optimal conditions for future generations. To this end, international institutions, especially the United Nations, are concerned with promoting cooperation in this matter and the signing of Treaties with which States commit to environmental problems.

The evolution of the decisions adopted at the international level show a clear ecocentric tendency, in the sense of protecting the environment more and more for the value it has, without there being a link between the protection granted to it and a direct benefit for the human being

As for the use of criminal law to protect the environment, in the United Nations it has timidly appeared. However, the role of the EU stands out here, which has promoted the creation of a Community environmental criminal law.

Thus, the situation of community legislation on environmental matters has changed notably, since it has evolved from an initial point of disregard for environmental issues to establish in its founding Treaties a firm legal basis on which it legislates in criminal matters. Thanks to this, a common penal system has been developed that provides a high level of protection to the environment and pursues a wide range of harmful activities against it.

As a consequence of the special political configuration of the EU, community initiatives have a direct impact on the internal legal system of the Member States, which have been forced to criminalize conduct that is harmful to the environment, as has been reflected in the successive reforms of the Spanish penal code. In this way, we observe that the appearance of international concern for the environment was much earlier than the national one in countries like ours, where environmental awareness arose after the criminalization of behaviors by community imperative.

REFERENCES

Bande, L. C. (2017). *Criminal law in Malawi*. Juta.

Brickey, K. F. (1996). Environmental crime at the crossroads: The intersection of environmental and criminal law theory. *Tul. L. Rev.*, *71*, 487.

Cho, B. S. (2000). Emergence of an international environmental criminal law. *UCLA J. Envtl. L. & Pol'y*, *19*(1), 11. doi:10.5070/L5191019216

Delgado, R. (1985). Rotten social background: Should the criminal law recognize a defense of severe environmental deprivation. *Law & Inequality*, *3*, 9.

Du Rées, H. (2001). Can criminal law protect the environment? *Journal of Scandinavian Studies in Criminology and Crime Prevention*, *2*(2), 109–126. doi:10.1080/140438501753737606

Faure, M. (2004). European environmental criminal law: do we really need it?. European Energy and Environmental Law Review, 13(1).

Faure, M. (2017). The development of environmental criminal law in the EU and its member states. *Review of European, Comparative & International Environmental Law*, *26*(2), 139–146. doi:10.1111/reel.12204

Faure, M. G. (2010). *Vague notions in environmental criminal law*.

Faure, M. G. (2016). The revolution in environmental criminal law in Europe. *Va. Envtl. LJ*, *35*, 321.

Faure, M. G. (2016). A paradigm shift in environmental criminal law. In *Fighting Environmental Crime in Europe and Beyond* (pp. 17–43). Palgrave Macmillan. doi:10.1057/978-1-349-95085-0_2

Faure, M. G., & Zhang, H. (2011). Environmental criminal law in China: A critical analysis. *Envtl. L. Rep. News & Analysis*, *41*, 10024.

Fortney, D. C. (2002). Thinking Outside the Black Box: Tailored Enforcement in Environmental Criminal Law. *Texas Law Review*, *81*, 1609.

Lazarus, R. J. (1994). Meeting the demands of integration in the evolution of environmental law: Reforming environmental criminal law. *Geological Journal*, *83*, 2407.

Lazarus, R. J. (1995). Mens rea in environmental criminal law: Reading supreme court tea leaves. *Fordham Envtl. LJ*, *7*, 861.

Mistura, A. (2018). Is There Space for Environmental Crimes under International Criminal Law: The Impact of the Office of the Prosecutor Policy Paper on Case Selection and Prioritization on the Current Legal Framework. *Colum. J. Envtl. L.*, *43*, 181.

Mitsilegas, V., Fitzmaurice, M., & Fasoli, E. (2016). The relationship between EU criminal law and environmental law. In *Research handbook on EU criminal law*. Edward Elgar Publishing. doi:10.4337/9781783473311.00024

Schroeder, C. H. (1993). Cool Analysis Versus Moral Outrage in the Development of Federal Environmental Criminal Law. *Wm. & Mary L. Rev.*, *35*, 251.

Setness, K. H. (1996). Statutory Interpretation of Clean Water Act Section 1319 (C)(2)(A)'s Knowledge Requirement: Reconciling the Needs of Environmental and Criminal Law. *Ecology Law Quarterly*, *23*(2), 447–494.

Smith, T. (2012). Creating a framework for the prosecution of environmental crimes in international criminal law. Ashgate Publishers.

Uhlmann, D. M. (2011). After the spill is gone: The Gulf of Mexico, environmental crime, and the criminal law. *Michigan Law Review*, *109*(8), 1413–1461.

ADDITIONAL READING

Faure, M. G. (2010). *Vague notions in environmental criminal law*.

Faure, M. G. (2016). A paradigm shift in environmental criminal law. In *Fighting Environmental Crime in Europe and Beyond* (pp. 17–43). Palgrave Macmillan. doi:10.1057/978-1-349-95085-0_2

Faure, M. G., & Zhang, H. (2011). Environmental criminal law in China: A critical analysis. *Envtl. L. Rep. News & Analysis*, *41*, 10024.

Fortney, D. C. (2002). Thinking Outside the Black Box: Tailored Enforcement in Environmental Criminal Law. *Texas Law Review*, *81*, 1609.

Lazarus, R. J. (1994). Meeting the demands of integration in the evolution of environmental law: Reforming environmental criminal law. *Geological Journal*, *83*, 2407.

Lazarus, R. J. (1995). Mens rea in environmental criminal law: Reading supreme court tea leaves. *Fordham Envtl. LJ*, *7*, 861.

Mistura, A. (2018). Is There Space for Environmental Crimes under International Criminal Law: The Impact of the Office of the Prosecutor Policy Paper on Case Selection and Prioritization on the Current Legal Framework. *Colum. J. Envtl. L.*, *43*, 181.

Mitsilegas, V., Fitzmaurice, M., & Fasoli, E. (2016). The relationship between EU criminal law and environmental law. In *Research handbook on EU criminal law*. Edward Elgar Publishing. doi:10.4337/9781783473311.00024

Schroeder, C. H. (1993). Cool Analysis Versus Moral Outrage in the Development of Federal Environmental Criminal Law. *Wm. & Mary L. Rev.*, *35*, 251.

Setness, K. H. (1996). Statutory Interpretation of Clean Water Act Section 1319 (C)(2)(A)'s Knowledge Requirement: Reconciling the Needs of Environmental and Criminal Law. *Ecology Law Quarterly*, *23*(2), 447–494.

Smith, T. (2012). Creating a framework for the prosecution of environmental crimes in international criminal law. Ashgate Publishers.

Uhlmann, D. M. (2011). After the spill is gone: The Gulf of Mexico, environmental crime, and the criminal law. *Michigan Law Review*, *109*(8), 1413–1461.

KEY TERMS AND DEFINITIONS

Equity: Defined by UNEP to include intergenerational equity - "the right of future generations to enjoy a fair level of the common patrimony" - and intragenerational equity - "the right of all people within the current generation to fair access to the current generation's entitlement to the Earth's natural resources" - environmental equity considers the present generation under an obligation to account for long-term impacts of activities and to act to sustain the global environment and resource base for future generations. Pollution control and resource management laws may be assessed against this principle.

Polluter pays principle: The polluter pays principle stands for the idea that "the environmental costs of economic activities, including the cost of preventing potential harm, should be internalized rather than imposed upon society at large." All issues related to responsibility for environmental remediation costs and compliance with pollution control regulations involve this principle.

Precautionary principle: One of the most commonly encountered and controversial principles of environmental law, the Rio Declaration formulated the precautionary principle: To protect the environment, the precautionary approach shall be widely applied by States according to their capabilities. Where there are threats of serious or irreversible damage, lack of complete scientific certainty shall not be used as a reason for postponing cost-effective measures to prevent environmental degradation. The principle may play a role in any debate over the need for environmental regulation.

Prevention: The concept of prevention can perhaps better be considered an overarching aim that gives rise to a multitude of legal mechanisms, including prior assessment of environmental harm, licensing or authorization that set out the conditions for operation and the consequences for violation of the conditions, as well as the adoption of strategies and policies. Emission limits and other product or process standards, the use of best available techniques, and similar techniques can all be seen as applications of the concept of prevention.

Public participation and transparency: identified as necessary conditions for "accountable governments,... industrial concerns," and organizations generally, public participation and transparency are presented by UNEP as requiring "effective protection of the human right to hold and express opinions and to seek, receive and impart ideas,... a right of access to appropriate, comprehensible and timely information held by governments and industrial concerns on economic and social policies regarding the sustainable use of natural resources and the protection of the environment, without imposing undue financial burdens upon the applicants and with adequate protection of privacy and business confidentiality," and "effective judicial and administrative proceedings." These principles are present in environmental impact assessment, laws requiring publication and access to relevant environmental data, and administrative procedures.

Transboundary responsibility: Defined in the international law context as an obligation to protect one's environment and prevent damage to neighboring environments, UNEP considers transboundary responsibility at the international level as a potential limitation on the sovereign state's rights. Laws that limit externalities imposed upon human health and the environment may be assessed against this principle.

Chapter 12
Environmental and Economic Law

ABSTRACT

In this chapter, the concept, the legal nature of the green economy, and its role in achieving sustainable development have been studied. While describing the actions of the international community to protect the environment, the analysis of the green economy and its vital role in achieving sustainable development has been discussed. Today, sustainable development can no longer be seen as a choice but as a commitment that all governmental and non-governmental actors must make every effort to achieve a sustainable economy to transition to a green economy. Paying attention to natural resources is valuable for any society, and short-term, medium-term, and long-term plans in the light of a green economy will bring growth and social welfare to the society.

INTRODUCTION

Environmental protection is a major concern of the international community. To protect non-renewable natural resources and respect the rights of current and future generations, the issue of sustainable development has been the focus of attention of states and international organizations. Sustainable development seeks to create a balance and compromise between the economic, social, and environmental dimensions. The traditional approach to brown economics and the maximum use of natural resources, regardless of the rights of future generations, leads to the destruction and pollution of the environment. In contrast, the green economy responds to the challenges facing the international community and facilitates the realization of sustainable development.

One of the most important consequences of population growth globally is energy consumption and increasing demand for energy resources. More than half of the world's population lives in cities. As a result, the demand for energy and its consumption in cities is much higher than in rural areas. So cities are responsible for emitting the most carbon dioxide and other pollutants harmful to the environment. Since the world's non-renewable resources are limited, clean and renewable resources must be sought. To address these issues internationally, sustainable development has been raised for more than a few

DOI: 10.4018/978-1-6684-4158-9.ch012

decades. Sustainable development is one of the most important issues discussed in international environmental law, and so far, both nationally and internationally, many steps have been taken to achieve sustainable development (Nanda, 1995).

The breadth and diversity of economic, social, and environmental dimensions require the concept of sustainable development so that different systems pay special attention to macro-level regulation. To this end, various programs and strategies have been developed by governments to achieve sustainable development. Specifically, the term "sustainable development" entered the international environmental law literature at the 1992 Rio de Janeiro Conference in Brazil. In other words, the main focus of the Second International Conference on the environment in 1992 was "Environment and Development" (Nanda & Pring, 2012). It speaks clearly and unequivocally on the issue of sustainable development.

The concept of "sustainable development" is re-emphasized in international conferences held after Rio 1992 and in the documents adopted by these conferences. Recognizing sustainable development in the upstream documents of national systems as a general policy indicates its importance. To implement the general policies of the system, programs and strategies should be developed in line with the main policy to achieve the lofty goals of sustainable development. The three dimensions of sustainable development and the need to balance between them have been repeatedly considered. The topic of "green economy" and its entry into the economic and legal literature can be studied and analyzed in this context.

The close connection between economic and environmental issues has led to the development of new approaches in international environmental law, one of the most valuable of which is the "green economy."

The transition from the traditional economic approach to the green economy is possible by observing the principle of fairness and environmental integration. In other words, traditional economics is based more on the unlimited use of natural resources and disregard for the rights of present and future generations. The effects of such an economy can be irreversible in practice. A change of approach in this area is inevitable and definite, and therefore the international community will have no choice but to use all international actors to deal with the effects. The transition to a green economy varies between governments, as the level of development of each of these countries differs from one another, and the natural and human capital of each will be involved. Considering the interdependence and inextricable dependence of the environment on major human issues such as ethics, politics, economics, development, etc., culture building and creating a systemic approach to the environment is important.

Challenges facing the international community, such as economic and financial crises, population growth, food insecurity, climate change, etc., prompted the United Nations to continue its work on sustainable development and monitor activity in this regard at the "Rio+20" conference held in 2012.

The accelerating pace of the country's development requires attention to all aspects of sustainable development. The green economy is one of the most important tools for achieving sustainable development. This perspective not only eliminates poverty but also increases public welfare, health and achieves social justice. With the familiarity of this concept, we can look at the performance of governments in this area. The main question of the present study revolves around what and why the green economy and its role in achieving sustainable development. This study aims to analyze the concept of the green economy and its relationship with sustainable development and measures taken at the international level. Therefore, while examining the green economy in the light of sustainable development, an attempt is made to analyze the recent actions of the international community regarding the green economy.

TRANSITION OF NEOCLASSICAL ECONOMICS TO ECO-FRIENDLY ECONOMICS

The Industrial Revolution has had many effects and consequences on human life. Economic growth resulting from the Industrial Revolution in several countries raised concerns about the capital distribution between economic agents and salaried workers. The resulting economic and social dimensions led to economic theories about two important factors of production, namely capital, and labor.

The rise of environmental issues in the 1970s raised concerns in the economic sector; Because, according to economic agents, strict observance of environmental standards in the production sector can lead to a reduction in economic growth. The gradual process of integrating economic, social, and environmental issues began with the Rio 1992 conference and is still in the hands of the Cardoles and some international organizations. Green economy is one of the most important manifestations of the integration of economic, social, and environmental dimensions and without it, sustainable development will not be achieved.

Regardless of which economic sector they belong to, most economists agree that immediate solutions must solve environmental problems.

Economists such as Pigou, Samuelson, and Quas have tried to fill in the gaps by proposing concepts such as side effects or external (external) effects, public property, and the right to pollute. Gradually, these concepts became one of the most important topics in environmental economics through the presentation of later economic theories.

The economic development of societies is inevitable, and in this way, the lack of attention to the environment causes instability. Comprehensiveness in consumer spending policies plays an important role in improving the efficiency of economic growth, creating employment and balanced development, and protecting the environment. The transition to a green economy is a solution that can address the international community's challenges in the environment. To move towards a green economy, one of the effective tools of green technology is applying the polluter-payer principle. Today, the polluter-payer is one of the most important principles of international environmental law. The origin of the polluter-payer goes back to the economic theory of the English economist Pigou. Pigou believes that the damage to the environment caused by the production of the product and the damage caused by its consumption is attributed to the person who caused it (Turner, 1992).

In 1974, the polluter-payer principle was developed to prevent pollution and to encourage countries in the European region to make appropriate and reasonable use of environmental resources under the auspices of the Economic Development Cooperation Organization. Article 16 of the Rio Declaration provides:

"National authorities should strive to internationalize the costs of environmental protection and the use of economic instruments, given that polluters should in principle pay for the elimination of pollution, and in the public interest without interfering with international trade and investment."

This principle has two important a priori functions to prevent the spread of pollution and a posteriori from compensating for the damage to the environment. Precautionary measures are more preventive and are imposed by the government and public authorities to monitor and control the quality of the environment.

One of the obvious examples is setting environmental standards. The use of market economic tools can be a great help in preserving the environment. With the help of these tools, economic agents are motivated to reduce pollution by complying with standards. Taxation is an important factor in influencing the internalization of externalities. This tax is imposed on producers of polluting products. On the other hand, raising the price of poor quality and polluting products by the economic broker to compensate for the loss caused by paying other taxes can not prevent the consumer from buying that product.

In this case, the producers of polluting goods are encouraged to maintain the quality standards of the goods to stay in the economic market.

Internalization of externalities means that the tax imposed is proportional to the amount of damage that comes from the production of that product to the environment. Economic logic dictates cost and benefit so that the economic agent seeks to reduce environmental costs to reduce the cost of commodity prices due to environmental taxes by pursuing environmental and technological standards. These taxes result from the simultaneous intervention of two key principles of the environment: the principle of prevention and the principle of polluter-payer. The main purpose of these taxes is to reduce pollution. Of course, this tax rate is very important because there will be no incentives for the polluter to reduce pollution if it is too low. Now, if the tax rate is too high, as a result, prices will rise, and costs will go very high. In other words, with no increase in prices, shareholder profits fall, and workers' wages fall. On the other hand, consumers reduce pollution by incurring taxes on goods and services and incur additional costs.

Whatever has been said about the importance of the a priori function will not be hindered until the posterior function is discussed. The posterior function of the pollutant-payer principle is achievable in the form of repair tools. The anterior function is based more on preventive measures to prevent environmental damage, while the posterior function of the original pollutant-payer is more compensatory. Following the general principle of law, no damage should be left without compensation. Major damage to the environment must be repaired and compensated.

Therefore, to compensate for the damages to the environment, the rules of civil liability law must be used. Most international rules and regulations in the field of civil liability are based on pure liability. The Lugano Convention on Civil Liability for Damages Caused by Dangerous Activities to the Environment emphasizes the appropriateness of establishing an objective liability regime in the light of the originator of the polluter-payer (Turner, 1992). The negligence of the cause of the damage is not as it is in the theory of liability based on fault and existence. Therefore, as mentioned before, the damages must be compensated, and the compensation cost will be borne by the person who caused the damages.

A priori functions and the application of economic hardships can play an important role in internalizing the costs of side effects and external damages to the environment through the polluter-payer axis. In addition to implementing this principle in the legal system of any country in the field of environment, it can be in the service of sustainable development and integration of its three dimensions.

Integration of Environmental Dimensions in the Framework of Green Economy

Sustainable development with economic, social, and ecological dimensions is achieved. Compromise between economic and environmental issues requires us to move away from the traditional development model and take a serious approach to introduce green environmental economics. The traditional development model is based on the maximum use of natural resources regardless of the rights of current and future generations. Undoubtedly, such an approach has adverse effects on the environment. Changing production methods and modifying consumption patterns is an effective step towards the optimal use of natural resources.

Meeting the challenges of the environment at the global level requires us to move towards a green economy. The transition to a green and ecological economy must reduce greenhouse gas emissions, protect natural resources, and achieve social and individual justice to combat inequality.

Reducing greenhouse gas emissions is one of the most important goals of the green economy. The real transition to a green and ecological economy requires a reduction in these gases. The Framework Convention on Climate Change is one of the most important international instruments set up to prevent greenhouse gas emissions in 1992. The most important goal of this convention is to stabilize the number of emissions. To that end, States Parties to Annex 1 to the Convention were required to return their greenhouse gas emissions to their current level by 1990. To take effective measures to reduce greenhouse gases, the protocol was developed. The need for these measures is considered important because the increase in greenhouse gases is one of the problems of the international community. Damages caused by climate change will have various consequences, such as rising sea levels, tsunamis, problems with environmental migrants, and so on. Under the Kyoto Protocol, industrialized nations pledged to reduce their greenhouse gas emissions by an average of 5.2 percent compared to the same states' emissions in 1990 between 2008 and 2012 (Von Stein, 2008). The Kyoto Protocol recognizes flexible mechanisms. Like most other similar conventions, the United Nations Framework Convention on Climate Change set out general commitments. As noted above, the convention's commitment to stabilizing greenhouse gas emissions was recognized, but the implementation of general commitments was left to the definition of specific mechanisms for achieving their goals. One of the special techniques of international environmental law is the use of framework conventions. The Framework Conventions set out the general principles and commitments relating to the cooperation of States within a given area but leave the manner of cooperation and its details to the Protocols of Supplementary Agreements. Of course, it should be noted that the relationship between the Framework Convention (the original document) and the protocol or the Additional Agreements (subsequent documents) is that each of these documents is independent of the other. In other words, in terms of treaty law, each of these instruments is in the form of an international treaty, but with the difference that only two parties to the Framework Convention (the original document) can be a party to the protocol or a party to the agreement. On the other hand, it should be noted that membership in the Framework Convention (the main document) does not necessarily mean membership in the Protocol or Additional Agreements (subsequent documents) (Palmer, 1992).

Therefore, the Kyoto Protocol seeks to define and define the limits of the obligations contained in the Framework Convention for the Reduction of Greenhouse Gas Emissions. To that end, the Kyoto Protocol paved the way for the three main tasks of the publishing, joint implementation, and development business.

A) the mechanism of the publishing trade or the trade of the publishing right; This mechanism allows member states to commit to reducing greenhouse gas emissions by exchanging their allocated emission rights to fulfill their obligations. Thus, a country that reduces its emissions more than necessary in fulfilling its international obligations will then have the power to transfer its excess emissions to another country (Von Stein, 2008).

B) joint implementation mechanism allows member states to obtain emission reduction units from other developed countries' implemented greenhouse gas emission reduction plans. Under the Kyoto Protocol, either side of the plan, namely the host government and the investing government, is required to reduce greenhouse gas emissions. In other words, a State Party may define and implement a plan to reduce emissions in the other State Party to fulfill part of its commitment to reduce emissions. Therefore, Government A can acquire emission reduction units by implementing the plan.

C) Clean development mechanism; The clean development mechanism is the same as the joint project or project-based implementation. In Sazekar Clean Development, plans are implemented to reduce greenhouse gas emissions. The fundamental difference between a clean development mechanism and a shared implementation mechanism is that the host government is developing. In other words, this structure operates between acceding states and developing governments.

To help industrialized countries fulfill their obligations at a lower cost, they can implement their emission reduction plans at a lower cost in a developing country and return a certificate of emission reduction from the developing country (The host state). On the other hand, implementing this mechanism contributes significantly to realizing sustainable development in developing countries.

Thus, the Sazikar of clean development plays an important role in fulfilling the commitment of governments to reduce the volume of greenhouse gas emissions and serves the sustainable development and economic development of developing countries.

One of the important and necessary criteria for accepting plans related to clean development is the criterion of redundancy and sustainable development. The redundancy criterion means that the project must lead to a reduction in emissions, in addition to what happens in the absence of project activity. In other words, greenhouse gas emissions should be less than the amount emitted without this design (Dessus et al., 1999).

But the second criterion, which is also mentioned at the beginning of Article 12 of the Protocol, is of great importance. The purpose of the Clean Development Mechanism is to assist the members not listed in Annex 1 in achieving sustainable development, assisting in the convention's ultimate goal, and assisting the members listed in annex I to meet their mitigation and mitigation obligations.

Therefore, it can be inferred that the clean development mechanism is efficient and is based on win-win logic; Because each party to the project (investor country and host country) is in their interest. With the investment of developed countries in PAC development plans, these countries can effectively fulfill their reduction commitments and receive credit for their reduction. Developing countries will also benefit from increased investment flows in clean development projects, and greenhouse gas emissions in those countries will be reduced, thus helping to achieve sustainable development. That is why the two countries have acknowledged in the documents of the Rio + 20 conference that the policies of the green economy in the field of sustainable development and eradication of poverty should strengthen international cooperation, including financing, capacity building, and transfer of funds to developing countries. (Barbier, 2012). To further clarify the issue of the green economy, we will continue to develop it at the "Rio + 20" conference.

Formation of a Green Economy in the Framework of the Rio 2012 Conference

The Rio + 20 Conference, like the 1992 Rio Conference, was very important because most countries and international organizations attended it. In addition to reviewing the steps taken in previous periods of sustainable development, this conference addresses the international community's challenges in terms of sustainable development. Paying attention to the record of actions taken by the international community in this field makes it possible to draw the future perspective of sustainable development and establish a framework for progress in the coming years. As for the participants in this conference, it should be noted that in addition to the two members of the United Nations and international institutions, non-governmental organizations also had the right to participate in this conference. The final document of the conference was entitled "The Future We Want." In its four sections, the document addressed issues related to a) renewal of political commitments, b) a green economy in the context of sustainable development and poverty eradication, c) the institutional framework for sustainable development, d) the framework for action, and follow-up.

In the introduction of this document, the issues of eliminating all forms of poverty and stabilization and economic growth for all, achieving the Millennium Development Goals, strengthening cooperation and reviewing current and emerging challenges and focusing on human development, as well as renewing commitment to sustainable development and filling the gap Existing points. Governments are determined to make every effort to achieve the internationally agreed development goals and the Millennium Development Goals by 2015.

The final document of the Rio+20 conference relies on three main points in the renewal of political commitments.

A) Emphasis on Rio principles and previous action plans related to the environment and sustainable development. The document confirms and reaffirms the complete list of international instruments drafted before the Rio 2012 Conference. Most of the documents cited were declarations, action plans, or strategies, which have no legal obligation in the strict sense of the word. At the same time, it refers to treaties such as the United Nations Framework Convention on Climate Change, the Convention on Biological Diversity, and the United Nations Convention on Desertification and emphasizes the importance of these three development documents. Encourages all States Parties, in addition to fully complying with their obligations under the principles and regulations outlined in those instruments, to take effective and specific measures at all levels and to strengthen international cooperation.

B) Strengthen integration, implementation, and coherence: Assess the progress made and the gaps in the implementation of the documents of the big meetings on sustainable development and the face of existing and new problems. Progress in sustainable development and poverty alleviation has not been equal, and governments have insisted on pursuing previously accepted commitments. It was emphasized to fill the gap between the level of development between developing and developed countries and create appropriate conditions and strengthen international cooperation, especially in finance and lending, trade, and technology transfer at the national and international levels.

In addition to integrating the three dimensions of sustainable development equally, it should point to the multiple financial, economic, food, and energy crises that have affected the capacity of different countries, especially developing countries. For example, more than one billion people in the world are still in extreme poverty, and about 14% of the world's population is suffering from malnutrition, which should add to public health problems and infectious diseases, which are still a concern (Barbier, 2012).

There are other problems in achieving sustainable development, including the special situation of vulnerable countries, especially in Africa, less developed countries without land and enclosure, middle-income countries, or at war. Of course, it should be noted that for each of the above challenges before the Rio + 20 conference, a special action plan or strategy has been allocated, which the Rio Conference final document reaffirms and reaffirms.

C) Major groups and other influential parties. The last point that has been considered in the renewal of political commitments is the role of major and influential groups. For this reason, the beginning of this section deals with the issue of public participation. Participation means that everyone's voice is heard in decision-making centers. Actual participation, the ability and opportunity to influence the decision-making process, and the fact that copper is an active and ongoing tool of engagement (King et al., 1998). Considering the needs and requirements of stakeholders, make informed choices, and in this regard, these groups can protect their rights. The participatory approach causes the involvement of groups that are directly and indirectly affected by the decisions made; in other words, they feel that the success and failure of the approved projects are due to their actions. The more citizens and stakeholders feel belonging, the more actively involved in the decision-making process.

The public's role and participation in sustainable development are considered important because their right to access information and administrative and judicial authorities is necessary to promote sustainable development. The provisions of Article 10 of the 1992 Rio Declaration on the principle of partnership are once again mentioned, and its observance in the field of sustainable development is taken into account. From this point of view, access to information and participation in decision-making and decision-making regarding policies and programs for sustainable development and their design and implementation at all levels is the international community's concern. Thus, major and influential groups such as women, youth, non-governmental organizations, natives, workers, trade unions, farmers, fishers, etc., are counted in sustainable development, and the active role of civil society in sustainable development is emphasized.

As noted above, the green economy is one of the issues that are already on the agenda of some governments and international institutions today because this economy can serve the realization of sustainable development. This economy plays an important role in eliminating poverty and eradicating it while paying attention to the values of nature and the environment. If we look at the implementation plan of the Johannesburg World Summit on Sustainable Development, the same three basic elements of sustainable development, namely economic growth, social development, and environmental protection, have been considered, and the fight against poverty, change and reform Naturally, they have been considered necessary for economic and social development.

The main difference between the final document of the Rio + 20 conference and its previous documents on sustainable development is that the recent ratification document has paid special attention to the green economy. But not all countries agree on this issue, and even some countries did not have a proper approach to the economic concept. Developing countries' pessimism and concern about the green economy stemmed from seeing it as a slowdown in their economies.

The speech of the President of Bolivia at the Rio + 20 conference is proof of such a claim. Bolivian President Morales called the green economy a "new colonialism" imposed by rich countries on developing countries. In their view, northern countries want to intervene in developing countries' national policies under the pretext of protecting the environment.

This issue is also addressed in the UN Summit, 2012 trade relations between developed and developing countries. The position of developing countries seems a bit ambiguous, and they do not have a very favorable view on the liberalization of international trade because this liberalization puts them in direct competition with large industrial countries. At the same time, these countries view the barriers and restrictions imposed on international trade for ecological reasons as a reduction in national competitiveness and a barrier to access to the markets of developed countries. These countries also see these restrictions as a new form of protection at the service of developed countries to implement their trade policy strategies.

In contrast, industrialized nations see restrictions on international trade as necessary to reduce environmental problems. However, these countries do not agree on environmental policies and strategies. Some countries, such as the United States, oppose some environmental measures and instruments, such as the failure to ratify multilateral environmental agreements, such as the Convention on Biological Diversity and the Kyoto Protocol, for economic and commercial interests.

"60 percent of greenhouse gases are produced by 20 percent of the richest countries, while 20 percent of the poorest countries produce less than one percent of these gases," Ecuadorian President Rafael Cora said at the Rio 2012 conference. At the urging of world leaders to protect the land, the Indian government declared that "the green economy is considered a trade-in nature and a crime against the earth and humanity."

The green economy may create problems for developing countries in the short run, but in the long run, it can go a long way in creating jobs, capacity, and empowerment to improve social integration and eradicate poverty. On the other hand, the greenness of this economy and the reasonable and fair use of resources will help its sustainability and renewability, in which case the third element of sustainable development, which is environmental protection, will be strengthened.

Therefore, the green economy is seen in the context of sustainable development and the elimination of poverty. Each country has various approaches, perspectives, models, and tools to achieve sustainable development in its three dimensions concerning the situation and national priorities. A green economy should be seen as an opportunity and not a threat; Realizing this economy is one of the valuable tools for achieving sustainable development and can be a solution for policymaking.

The final document of the Rio + 20 conference sets out the conditions for green economy promotion policies in the context of sustainable development and poverty eradication. It first refers to the compliance of these policies with the Rio Principles, Agenda 12, the Johannesburg Plan, and then lists a long list of conditions for the implementation of these policies as follows:

A) Compliance with international law; B) Respect for the sovereignty of each country over its natural resources, taking into account the circumstances, goals, responsibilities, and national priorities, as well as the margin of maneuver of each country's decision-making concerning the three dimensions of sustainable development; C) by giving the first role to governments and the participation of all relevant parties, including civil society and emphasis on the environment and institutions that are properly employed at all levels; D) Promoting sustainable and non-exclusive economic growth, facilitating innovation, providing facilities, privileges, and practical tools for all, and ensuring respect for human rights for all; E) Paying attention to the needs of developing countries, especially countries with a special situation; F) Strengthening international cooperation, including financial resources, capacity building, and technology transfer for the benefit of developing countries; G) refraining from imposing unjustified conditions on public development and financial assistance; H) non-arbitrary or unjustified discrimination, covert restrictions on international exchanges, refusal to take unilateral measures to resolve major ecological problems beyond the jurisdiction of the exporting state, and consideration of the basis for measures to combat transboundary or global environmental problems. International; I) Participate in filling technological gaps between developed countries and while developing and reducing the technological dependence of developing countries through all possible means; J) Improving the well-being of indigenous peoples and communities, other local and traditional populations, and ethnic minorities, recognizing and relying on their identity, culture, and interests, and refraining from endangering their cultural heritage, practices, and traditional knowledge by preserving and respecting Non-commercial approaches that address poverty alleviation; K) Improving the well-being of women, children, youth, the disabled, miners and small farmers, fishers and small and medium-sized enterprises, and improving the livelihoods and independence of indigenous and vulnerable groups, especially in developing countries; L) Advocating for all capacities offered to men and women and ensuring participation in them; M) Promoting productive activities to eliminate poverty in developing countries; N) Responding to concerns related to inequality and promoting social inclusion, especially minimal social support; O) Improving sustainable production and consumption methods; and P) Pursue the efforts made to implement equitable and non-monopolistic development approaches to combat poverty and inequality.

Most conditions for green development promotion policies are somehow identified in the first and second periods of sustainable development documents. Some of these conditions have a general aspect, such as the first condition that the compliance of policies to promote a green economy in sustainable development and the elimination of poverty following international law.

To further clarify the dimensions of the issue in the next section, the concept and legal nature of economics will be analyzed.

CONCEPT AND LEGAL NATURE OF THE ECO-FRIENDLY ECONOMY

The formation of international environmental law has led the international community to turn its attention to various environmental issues. One of the topics that entered the economic literature in the 1970s was green products, green economics, or green growth (Steward, 2011).

Green economics is commonly used instead of brown economics; When fossil and non-renewable resources are used indefinitely, the economy is brown. Observance of environmental regulations in all stages of production causes the product to be considered green. To further explain the issue, we will analyze the green economy.

So far, various definitions and interpretations of the green economy have been made. The green economy can be studied from the economic point of view in micro and macroeconomics. At the micro-level, "green economy" refers to an economy in which factories and small businesses produce green products. From a macroeconomic point of view, it is a bit more complicated; This is because the production of green products alone is not enough, but the production of these products should be compared with the use of non-renewable resources. In other words, the share of green products in a country's GDP should be separated, for example, from the share of products from that country's oil resources. Therefore, the evaluation of the share of each of these products in GDP will determine the position of the green economy in the final analysis in terms of macroeconomics.

According to the European Union, a green economy is considered to "lead to the growth and creation of jobs and the elimination of poverty through investment in nature for the long-term survival of the earth." The European Union has paid more attention to economic growth factors. It emphasizes less carbon production, the efficient and sensible use of resources, and the promotion of sustainable consumption and production methods. The European Commission pays special attention to market instruments such as tariffs, customs duties, and environmental subsidies for implementing these measures because these mechanisms are considered a means to achieve its economic and environmental objectives.

The United Nations Environment Program (UNEP) considers the green economy to be an economy that "promotes human well-being and social justice, as well as a significant reduction in environmental risks and resource scarcity." It is carbon. According to UNEP, the green economy seeks to strengthen convergence as much as possible through the three dimensions of sustainable development. The development of the economic system is aimed at strengthening and improving the natural capital of the earth, increasing economic productivity, and reducing social inequalities (PNUE Report, 2011).

In economic analysis, cost and benefit are effective factors in the continuity of activity and competition in the economic field. The high economic costs of green products create numerous barriers for producers. In other words, in the short term, the production of these products leads to the reluctance of firms and factories in developing countries, and for this reason, they do not approve of the application of environmental standards in the production process. For example, these countries believe that carbon dioxide emissions by industrialized countries cause the most pollution. They are at the beginning of the path of development and industrialization and are hardly willing to accept the economic costs associated with meeting environmental standards.

The green economy is recognized as the key to sustainable development. In other words, the realization of sustainable development depends on the transition to a green economy. Debating the legal nature of the green economy makes it somewhat difficult to disagree on the nature of sustainable development, as there are many differences between the legal teachings of international law on "sustainable development." In his independent opinion, Judge Trinidad argued that "sustainable development" was the key to "sustainable development." The concept of "is mere, and this concept is, in fact, a principle, with normative value" (Weeramantry, 1997).

Achieving sustainable development is impossible without a green economy. The transition to a green economy is one of the requirements for sustainable development. Now, if we approve of sustainable development as a rule or a customary principle, this could mean that the green economy, as one of its components, will have a customary aspect.

It seems that when legal scholars so much debate the "principle" of sustainable development, one cannot hope for the secularity of a green economy.

Even though since the 1990s, numerous environmental documents have repeatedly stated their subject and goal as sustainable development, and also the constitutions of many countries have directly or indirectly addressed sustainable development; The International Court of Justice only speaks of the "concept of sustainable development" (Nagymaros, 1997) and is somewhat cautious, still refusing to recognize sustainable development as a "principle." However, it is reasonable to believe that sustainable development can be considered a general policy that should be explained and defined strategies to achieve that policy.

After a brief understanding of the green economy and its legal nature, it is necessary to examine the goals and players of the green economy.

GOALS AND ACTORS OF THE ECO-FRIENDLY ECONOMY

As mentioned earlier, protecting the environment faces many challenges. To meet the new challenges in the field of environment, the international community seeks to solve some environmental problems with a green economy plan. The goals of the green economy can be summarized as follows: a) Paying attention to greenhouse gas emissions; B) Preservation of natural resources; c) Attention to the goals of social justice and fairness. Undoubtedly, achieving these goals requires the cooperation of all actors in the international arena.

International law is witnessing the entry of many actors into the international arena. Other states and international organizations are not the only actors in this field. In various ways, new players intervene in forming, implementing, and monitoring international rules and norms. International environmental law is no exception.

An important role should be assigned to the primary subjects of international law. Governments' cooperation and adherence to international commitments in all areas, including the green economy, will be crucial. Although the green economy is not considered a legally binding obligation, we must go through a transition to a green economy to achieve sustainable development. For example, one of the most important goals of the green economy is to reduce carbon emissions. If the countries with the highest carbon emissions do not cooperate, how can we hope for the desired goals in the green economy?

Implementing green economy promotion policies will ensure the transition to sustainable development. Of course, the implementation of such a policy requires the cooperation of all. The production of green products can be considered with the credibility of producers and consumers because each of them, in turn, plays an important role in achieving the desired goals in the green economy. Producers will contribute to the production of green products by observing environmental rules and standards, and consumers will complete another link in this chain by purchasing such products. The production of green products is due to the observance of standards and the use of clean energies. The use of renewable resources leads to the preservation of natural resources, and from this perspective, sustainable development is guaranteed while producing green products. Due to the high cost of green products, the consumer should choose green products in choosing between these products and products that do not meet environmental standards. Of course, it is not an easy task because the consumer must have achieved that degree of social awareness to pay the extra price of green products without any hesitation.

Therefore, the role of the government in this regard is vital because, on the one hand, it must monitor the production of green products and compliance with environmental standards, and on the other hand, it must promote and promote the consumption of green products so that consumers tend to buy such products. It seems that governments should focus most of their activities on the second part because by changing the attitude and taste of consumers and their desire to buy green products, producers will automatically tend to produce green products. The success of green products in society and the high demand for such products cause the producer to tend to produce these products because otherwise, they will have to leave the economic circle. Educating and educating consumers about green marketing will lead to the production of a green product. Otherwise, these concepts will lose their meaning and concept.

By purchasing green products, part of the production costs of such products should be paid by the consumer, and governments should support the other part. Discounts and tax exemptions for producers of such products can be a good and effective incentive for the continued production of green products and support for producers. Therefore, the role of the government in this regard will be significant and can be effective in achieving sustainable development. The final document of the 2012 conference considers the implementation of green economy policies by countries that seek to use them as a common commitment to sustainable development and emphasizes that each country can choose (Barbier, 2012).

The second player in this field is international organizations. The integration of economic, trade, and environmental issues and regulations in this field can greatly help improve the state of the environment. The formation of the green economy in the form of the United Nations can be examined in the framework of the World Bank and the United Nations Environment Program (UNEP). The turning point in the World Bank's focus on environmental issues dates back to the 1980s. Under pressure from non-governmental organizations in the bank's major shareholder countries, environmental issues, especially environmental impact, were addressed. Between 1984 and 1994, the World Bank drafted about 400 documents under the headings "Supportive Policies," "Operational Guidelines," "Strategic Lines," "Pollution Prevention and Reduction Book," and "Good Practice Guide." These documents cover various topics such as environmental assessment, ecological habitats, indigenous populations, forests, pesticides, dams, forced relocation, water management, and energy (Cioffo et al., 2016).

The World Bank also invests in environmental projects, But this investment never exceeds 10% of the total investment. In general, the observance of environmental policies by the bank has a limited aspect. The World Bank classifies projects in terms of environmental impact into three groups. Only high-risk designs are considered in detail in local consultation. In the 1990s, 186 projects underwent in-depth environmental assessments or only about 12 percent of high-risk Group A projects. More than half of the bank's operations were found to have no negative impact on the environment.

In some cases, the World Bank has come under fire. Many problems have arisen in projects such as large dams and industrial forest exploitation in Brazil or Indonesia because these measures have led to deforestation (Cioffo et al., 2016). In 2012, the World Bank released a report entitled "Green and Sustainable Growth Necessary." In this report, the World Bank encourages governments to move towards growth policies. A recent report by the World Bank (Rapport de la Banque Mondiale, 2012) highlights the following:

A) "Green growth of efficient growth" is essential and is necessary to achieve sustainable development.
B) Political obstacles, highly static behaviors and norms, and the lack of financial instruments are the main obstacles to green growth. Focus on policies and investments for the next 5 to 10 years. The goal is to exit unsustainable growth and avoid the harmful effects of public health.
C) Progress occurs through interdisciplinary solutions that consider economic, political, sociological, and psychological aspects.
D) Green growth is not a one-dimensional and rigid concept and is based on various strategies based on the countries according to the situation, priorities, and local resources. All rich or developing countries can grow without slowing down.

The United Nations Environment Program report entitled "Towards a Green Economy: For Sustainable Development" (Erickson et al., 2020) is a pioneering explanation of the green economy. Therefore, it is appropriate to briefly mention the role of UNEP in the formation of the green economy and the provisions contained in this report.

The recent UNEP report is important because: First, it specifically examines the issue of green economics. Second, it presents the executive strategies considered by governments.

Thirdly, this document guides the discussions in Rio + 20 about the green economy. The results of the UNEP report can be summarized as follows:

1. Investing only 2% of the world's gross domestic product in 10 key sectors will make it possible to move to an economy with low carbon emissions and reasonable use of resources.
2. A very green economy leads to growth in natural capital and gross domestic product growth.
3. Green economy leads to investment in natural capital and increases its value.
4. Green economy plays a role in the fight against poverty.
5. The transition to a green economy over time creates new jobs.

6. Prioritizing public spending and investment in areas that facilitate ecological change in economic sectors is essential.
7. Green economy can lead to growth and job creation like a brown economy and bring the best results in the medium and long term by granting and offering the most environmental and social benefits.

Economic actors, local councils, trade unions, farmers, citizens, and civil society as a whole can play a variety of roles. Economic actors play a key role in inefficient production and use of resources, waste reduction, etc. It will not be possible to move towards a green, ecological and fair economy without the active cooperation of non-governmental actors.

After analyzing the issue, we will deal with the appropriate conditions for the transition to the green economy and the measures taken in this field in the international arena.

CONDITIONS FOR THE TRANSITION TO A ECO-FRIENDLY ECONOMY

Undoubtedly, regulation in removing obstacles to green investment is one of the cases that can provide a good platform for the green economy. By setting minimum standards, it can counteract the instability of some economic activity. The development of product technical standards and production process standards at the national and international levels are all within this framework. At the international level, the improvement of energy status and its efficiency through various measures to reduce carbon dioxide emissions within the framework of the Kyoto Protocol is one of the clear examples of effective tools to achieve environmental goals. Encouraging green investment in various sectors of the economy is another important factor that can facilitate the transition from a green economy to a green economy.

Subsidies are economic tools that have different effects on the economies of countries and create an artificial market. One of the most important negative effects of subsidies is the distortion of prices. As a result, distorted subsidies artificially reduce the cost of doing business in an unstable environment. One of the major environmental problems globally is the subsidies given to the production and consumption of fossil fuels. Subsidies in 2008 were estimated at $ 557 billion for fossil fuels and $ 100 billion for the production of these fuels. Subsidies of this nature will cause great damage to the environment and be a major obstacle to the development of renewable energy technologies. Therefore, the gradual elimination of subsidies related to the production and consumption of fossil fuels can reduce greenhouse gas emissions by 6.9% (Erickson et al., 2020).

Of course, it should be noted that subsidies are not limited to the energy sector but can also be achieved in sectors such as agriculture, transportation, and fishing, etc. Subsidies to the new, clean and renewable energy sector can not be considered subsidies for fossil fuels, as these subsidies support green investment and prevent greenhouse gas emissions. Taxes are another economic tool that provides the achievement of maximum social welfare. Revenues from taxation play an important role in achieving sustainable development and ensuring social justice (Krass et al., 2013). Taxes on production and consumption are considered as an appropriate response to protect the environment. Of course, such an idea was first proposed by a British economist named Pigou. The internalization of the damage to the environment resulting from the production and consumption of goods and services forms the basis of the Pigou theory(Aall, 2014).

Damages related to the production of goods and damages resulting from their consumption are attributed to the person who caused it. Therefore, these damages must be integrated into the price of the desired goods and services. The polluter must pay a compensatory tax according to the amount of pollution entering the environment. The origin of the "polluter-payer," which is considered one of the progressive principles of international environmental law, should be sought in Pigou's theory and the subsequent actions taken in the 1970s within the framework of the Organization for Economic Co-operation (Turner, 1992).

Negative (external) side effects of producing a product or service can cause a lot of damage to the environment without affecting the price of the product or service. For example, pesticides and chemical fertilizers will have many negative effects on soil and groundwater, while these damages are not reflected in the price of agricultural products. In this case, the consumer of the relevant products benefits from the market price, i.e., in fact, the real price of the product is not reflected. In other words, calculating the cost of environmental damage reflects the real price (Luppi et al., 2012).

CONCLUSION

The international community has taken appropriate measures since the 1990s to achieve sustainable development. The 1992 Rio Conference marked a turning point in the development of sustainable development. Achieving the three economic, social, and environmentally sustainable development dimensions requires a brown economy to a green economy. The green economy seeks to counter the unlimited use of fossil and non-renewable resources. The Framework Convention on Climate Change and the Kyoto Protocol are tangible effects of reducing greenhouse gas emissions. By compiling these documents, the international community is working to achieve the desired goals with the participation of developed and developing governments. Unfortunately, despite the flexible forecasts under the Kyoto Protocol, it does not offer a bright and promising outlook for controlling greenhouse gas emissions. This frustration stems more from the reliance of developing countries, especially China and India, on preferential rights and the non-accession of the United States to the Kyoto Protocol, as each group (developing and developing) emits more than a quarter of greenhouse gases. As can be seen, half of the main causes of greenhouse gas emissions, or in other words, the causes of climate change, are outside the control of the Kyoto Protocol.

In 2012, the Rio + 20 Conference, with its final document entitled "The Future We Want," tried to re-emphasize the issue of sustainable development, citing documents from previous conferences and meetings and paying special attention to the green economy. Despite the impressive expectations of most observers of the tangible changes to the existing framework of the international environmental protection system, the establishment of a "High Commissioner for Future Generations" and the "upgrading of the United Nations Environment Program" to a specialized international organization. To explain the specific goals and create a kind of "roadmap for the green economy" that they had in mind, but the results of the 2012 conference were not very promising. Of course, it should be noted that no definition was given due to some ambiguities in the concept of the green economy in the document. At the same time, the definition of the green economy is important for many reasons. A clear definition of the green economy will be effective for implementing effective accounting and the formulation and interpretation of future economic and trade agreements at the WTO level. Finally, it must be acknowledged that sustainable development is a commitment and a necessity and cannot be seen as a choice. In this case, the transition to a green economy will be inevitable.

REFERENCES

Aall, C. (2014). Sustainable tourism in practice: Promoting or perverting the quest for a sustainable development? *Sustainability*, *6*(5), 2562–2583. doi:10.3390u6052562

Barbier, E. B. (2012). The green economy post Rio+ 20. *Science*, *338*(6109), 887–888. doi:10.1126cience.1227360 PMID:23161980

Cioffo, G. D., Ansoms, A., & Murison, J. (2016). Modernising agriculture through a 'new' Green Revolution: The limits of the Crop Intensification Programme in Rwanda. *Review of African Political Economy*, *43*(148), 277–293. doi:10.1080/03056244.2016.1181053

Dessus, B., Thomas, J. P., & Tillerson, K. (1999). *MDP and priorities in African development; MDP et priorites du developpement en Afrique*. Liaison Energie-Francophonie.

Erickson, P., van Asselt, H., Koplow, D., Lazarus, M., Newell, P., Oreskes, N., & Supran, G. (2020). Why fossil fuel producer subsidies matter. *Nature*, *578*(7793), E1–E4. doi:10.103841586-019-1920-x PMID:32025022

King, C. S., Feltey, K. M., & Susel, B. O. N. (1998). The question of participation: Toward authentic public participation in public administration. *Public Administration Review*, *58*(4), 317–326. doi:10.2307/977561

Krass, D., Nedorezov, T., & Ovchinnikov, A. (2013). Environmental taxes and the choice of green technology. *Production and Operations Management*, *22*(5), 1035–1055. doi:10.1111/poms.12023

Luppi, B., Parisi, F., & Rajagopalan, S. (2012). The rise and fall of the polluter-pays principle in developing countries. *International Review of Law and Economics*, *32*(1), 135–144. doi:10.1016/j.irle.2011.10.002

Nanda, V. (1995). *International environmental law & policy*. Brill Nijhoff.

Nanda, V., & Pring, G. R. (2012). *International environmental law and policy for the 21st century*. Martinus Nijhoff Publishers.

Palmer, G. (1992). New ways to make international environmental law. *The American Journal of International Law*, *86*(2), 259–283. doi:10.2307/2203234

Steward, F. (2012). Transformative innovation policy to meet the challenge of climate change: Socio-technical networks aligned with consumption and end-use as new transition arenas for a low-carbon society or green economy. *Technology Analysis and Strategic Management*, *24*(4), 331–343. doi:10.10 80/09537325.2012.663959

Turner, R. K. (1992). *Environmental policy: An economic approach to the polluter pays principle*. CSERGE.

Von Stein, J. (2008). The international law and politics of climate change: Ratification of the United Nations Framework Convention and the Kyoto Protocol. *The Journal of Conflict Resolution*, *52*(2), 243–268. doi:10.1177/0022002707313692

ADDITIONAL READING

Dessus, B., Thomas, J. P., & Tillerson, K. (1999). *MDP and priorities in African development; MDP et priorites du developpement en Afrique*. Liaison Energie-Francophonie.

Erickson, P., van Asselt, H., Koplow, D., Lazarus, M., Newell, P., Oreskes, N., & Supran, G. (2020). Why fossil fuel producer subsidies matter. *Nature*, *578*(7793), E1–E4. doi:10.103841586-019-1920-x PMID:32025022

King, C. S., Feltey, K. M., & Susel, B. O. N. (1998). The question of participation: Toward authentic public participation in public administration. *Public Administration Review*, *58*(4), 317–326. doi:10.2307/977561

Krass, D., Nedorezov, T., & Ovchinnikov, A. (2013). Environmental taxes and the choice of green technology. *Production and Operations Management*, *22*(5), 1035–1055. doi:10.1111/poms.12023

Luppi, B., Parisi, F., & Rajagopalan, S. (2012). The rise and fall of the polluter-pays principle in developing countries. *International Review of Law and Economics*, *32*(1), 135–144. doi:10.1016/j.irle.2011.10.002

Nanda, V. (1995). *International environmental law & policy*. Brill Nijhoff.

Nanda, V., & Pring, G. R. (2012). *International environmental law and policy for the 21st century*. Martinus Nijhoff Publishers.

Palmer, G. (1992). New ways to make international environmental law. *The American Journal of International Law*, *86*(2), 259–283. doi:10.2307/2203234

KEY TERMS AND DEFINITIONS

Equity: Defined by UNEP to include intergenerational equity - "the right of future generations to enjoy a fair level of the common patrimony" - and intragenerational equity - "the right of all people within the current generation to fair access to the current generation's entitlement to the Earth's natural resources" - environmental equity considers the present generation under an obligation to account for long-term impacts of activities and to act to sustain the global environment and resource base for future generations. Pollution control and resource management laws may be assessed against this principle.

Polluter pays principle: The polluter pays principle stands for the idea that "the environmental costs of economic activities, including the cost of preventing potential harm, should be internalized rather than imposed upon society at large." All issues related to responsibility for environmental remediation costs and compliance with pollution control regulations involve this principle.

Precautionary principle: One of the most commonly encountered and controversial principles of environmental law, the Rio Declaration formulated the precautionary principle: To protect the environment, the precautionary approach shall be widely applied by States according to their capabilities. Where there are threats of serious or irreversible damage, lack of complete scientific certainty shall not be used as a reason for postponing cost-effective measures to prevent environmental degradation. The principle may play a role in any debate over the need for environmental regulation.

Prevention: The concept of prevention can perhaps better be considered an overarching aim that gives rise to a multitude of legal mechanisms, including prior assessment of environmental harm, licensing or authorization that set out the conditions for operation and the consequences for violation of the conditions, as well as the adoption of strategies and policies. Emission limits and other product or process standards, the use of best available techniques, and similar techniques can all be seen as applications of the concept of prevention.

Public participation and transparency: identified as necessary conditions for "accountable governments,... industrial concerns," and organizations generally, public participation and transparency are presented by UNEP as requiring "effective protection of the human right to hold and express opinions and to seek, receive and impart ideas,... a right of access to appropriate, comprehensible and timely information held by governments and industrial concerns on economic and social policies regarding the sustainable use of natural resources and the protection of the environment, without imposing undue financial burdens upon the applicants and with adequate protection of privacy and business confidentiality," and "effective judicial and administrative proceedings." These principles are present in environmental impact assessment, laws requiring publication and access to relevant environmental data, and administrative procedures.

Transboundary responsibility: Defined in the international law context as an obligation to protect one's environment and prevent damage to neighboring environments, UNEP considers transboundary responsibility at the international level as a potential limitation on the sovereign state's rights. Laws that limit externalities imposed upon human health and the environment may be assessed against this principle.

Chapter 13
Environmental and Theological Law

ABSTRACT

In this chapter, some aspects of environmental law and theology will be addressed. Accordingly, first the authors will conceptually analyze the doctrinal foundations of the right to the environment, and then discuss the general doctrinal principles of this right from the perspective of Theological law. The basic presumption of this article is based on the grounds that the right to the environment can not achieve their goals regardless of religious and metaphysical status to ensure a healthy environment.

INTRODUCTION

The destructive activities of human beings cause the crisis of climate change, deforestation, air pollution, water pollution, extinction of biological species, etc., and endanger human life if they are not reduced or not controlled. Therefore, moral, religious, and philosophical issues related to the environment have been formed for many years, and scientists have proposed various opinions and theories in this field and have proposed solutions to control human behavior towards the environment.

In a paper entitled "The Historical Roots of Our Ecological Crisis" (White, 1967), one historian writes that the roots of this crisis lie in the Judeo-Christian way of thinking that man has the right to rule over nature. In this view, nature is alien and merely a source of exploitation, and the history of this humane and authoritarian attitude goes back to the "journey of Genesis", especially to verse 28 of its first chapter: "And God made them (Adam and Eve) He blessed them and told them: "Be fruitful and multiply, fill the earth and dominate it, rule over the fish of the sea, the birds of the sky and all the animals that live on the earth" (Penn, 2003).

DOI: 10.4018/978-1-6684-4158-9.ch013

White writes that the pattern of ownership separates human beings from nature and places them in the form of a kind of duality as soul-matter, or soul-body, according to which the soul must rule over matter and man must rule over nature. While the pagans were in perfect harmony with the earth before the Judeo-Christian tradition, they thought that spirits nested in animals, trees and streams. According to White, in the current Mazan, under the influence of the Christian duality, wild animals and the whole organic and plant world are considered dead.

White sees panpsychicism as opposed to the anthropocentrism of Christianity. In the worldview of non-Abrahamic religions, it is disgusting to harm trees, plants and animals unless there is a reason and they have already performed certain rituals because they carry the sacred. But Christianity, by eliminating the animism of these religions, caused the human spirit to ignore the feelings of natural objects and exploit them.

On the other hand, Thomas Berry and some other Christian scholars consider the environmental crisis to be influenced by crises in the field of spirituality and morality. He believes that it is this boundless materialism that has led to the aggression and plunder of nature, and in fact the separation of religion from secular life may be one of the reasons for this crisis. Therefore, according to this group of researchers, a philosophical and religious understanding of ourselves as natural creatures is essential for us humans (Bourdeau, 2004).

Some have considered science and the spread of technology as the most important causes of environmental crisis. Simultaneously with the evolution of civilization and its manifestations from the thirteenth and fourteenth centuries, attention to nature has been abandoned. Descartes believed that nature is composed of tangible qualities such as size and weight and lacks intrinsic and immeasurable values such as beauty. People like Galileo, Newton, and Kepler have founded a science that emphasizes experience, observation, and the partisan view or theism, in which nature is no longer respected and sanctified.

Francis Bacon also considers the acquisition of knowledge to gain power so that he can occupy the world and nature (Pedersen, 2018). Like Kant, he speaks of the domination of nature and the ability to change and adapt nature to the needs and desires determined by man. Because of this thinking, the value of nature is raised and that nature has no value and is made only from raw materials for consumption in connection with human ideas. Therefore, some believe that the general and important difference between technical and ethical results today can be eliminated only when the study of science is supervised and directed to the findings and activities that are ultimately beneficial to humans and the environment(Dallal, 2010).

The environmental crisis is one of the problems that has arisen for human society. Religions have proposed solutions to preserve the environment and nature. New economic development was due to human activities, especially after the Industrial Revolution, which was carried out at the cost of destroying the environment. As a result of improper use of the environment, crises such as climate change, deforestation, soil erosion, habitat degradation, extinction of various wildlife species, biodiversity loss, radioactivity pollution, ozone depletion, acid rain, Rising global temperatures, chemical pollution, water shortages, floods, vegetation loss and desertification have occurred. Therefore, people concerned about the environment went to find solutions to solve environmental crises. New concepts such as environmental theology were introduced in this regard. This theology seeks to take advantage of the religious tradition and find solutions to the environmental crisis (Gottlieb, 2006). In other words, environmental theology expresses man's relationship to the environment by defining the relationship between God and the world. Christian theologians sought to extract moral advice for the protection of the environment by re-examining religious principles and church tradition. Therefore, it can be argued that Christian environmental theology is related to a concept such as environmental ethics. This is why issues such as environmental ethics, which provide comprehensive and systematic reasons for the need for moral relations between humans and the environment (Saniotis, 2012) were considered. With this approach in mind, this chapter seeks to examine Christian theology and its solutions to the challenges posed by the environmental crisis. Questions are also raised about this. Can Christianity offer solutions to this crisis or should other religions also be used to introduce solutions? Can the environment be considered important in texts related to environmental theology?

In Islam, the protection of the environment and nature is one of the teachings intended for Muslims. "The environment, while at the disposal of man as the divine caliph, has been described as a sign of divine power and greatness, which this 'caliph' undertakes to preserve and protect as he deserves" (Gada, 2014). God placed nature and the environment in the conquest of man. One of the verses of the Holy Qur'an mentions in this regard: If man recognizes that property correctly and adjusts it properly, he has in fact conquered that being. God commanded heaven and earth to follow man.

Some of the chapters of the Qur'an, including Spider, Thunder, Light, Toor, Najm, Naml, Qamar Hadid, Shams, Earthquake, Elephant and Night are called elements of nature. In the Holy Quran, it is not enough to name only a part of the suras as elements of nature; Rather, there are many parts of the verses of the Qur'an in this regard. For example, the word "water" has been used more than six hundred times, the word "earth" more than five hundred times and the word "sky" about thirty hundred times in the Holy Qur'an. In addition, we can refer to the narrations of the Holy Prophet of Islam (PBUH) who say, "Whenever you see spring, remember the Day of Judgment a lot."

Despite the above, it should be noted that nature is at the disposal of man on loan. Man is committed to protecting nature, developing it and preventing its destruction. Thus, just as the use of the environment is a right and authority given to man by the particular kind of creation he has been given, there are also obligations and responsibilities in return for it. In the Islamic legal system, there are rules and some harm to the environment that not only prevent humans from destroying the environment; Rather, they oblige the loser to compensate for all the damages caused.

EASTERN RELIGIONS

Buddhism

Most Buddhists believe people need to live simply and respect the cycle and balance in nature so everything can continue for future generations. To some Buddhists, living 'skilfully ' means to live without producing waste. The whole world benefits from avoiding needless exploitation. More generally, Buddhism can go hand in hand with environmental education in terms of promoting harmonious living between all living creatures and the environment and fostering environmentally friendly attributes such as selflessness, thriftiness, loving-kindness, social responsibility, and compassion.

Dr. Suzuki, a renowned scholar of Zen Buddhism, attributes the causes of Western man's spiritual turmoil to subjectivism, scientism, and alienation from nature. "The Buddha Says," "It preserves that the infinite seal must also flourish in all living beings." The Buddha taught that karma was cetana – action was intention – and that the intentional quality of actions determines their results: whether they lead to well-being or to suffering. ... Buddhist tradition indeed sees the 'law of karmavipaka' (as it is commonly called as a law of nature). The Dalai Lama said he regularly emphasises on the importance of maintaining a sense of the oneness of humanity, the idea that every human being is a part of us. "The threat of global warming and climate change is not limited by national boundaries; it affects us all(Cooper & James, 2017).

Hinduism

Human beings have no dominion over other creatures. They are forbidden to exploit nature; instead they are advised to seek peace and live in harmony with nature. The Hindu religion demands veneration, respect and obedience to maintain and protect the harmonious unity of God and nature. Hinduism teaches that the 5 significant elements (space, air, fire, water, and Earth) that constitute the environment are all derived from prakriti, the primal energy. Each of these elements has its own life and form; together, the elements are interconnected and interdependent. Most Hindus understand 'environment' to mean the natural world – everything around us that is part of the Earth and nature. ... Many Hindus believe that nature cannot be destroyed without humans also being destroyed, because we need the natural world in order to survive, and also because every atman is a part of Brahman.

In Hinduism, although human beings are at the top of the evolutionary pyramid, they are not a separate factor from the planet and other creatures. "If there is a flower or fruit tree in a village, that place becomes valuable for worship and respect," says the Mahabharata.

Seven recommendations for the realization of environmental ethics: rejection of human domination over nature, coexistence with the environment, prevention of domination of nature and waste of resources, responsibility for dealing with nature, protection of the environment, sanctification of the environment due to its oneness with God and Prohibition of self-forgetfulness is expressed due to environmental crises. In different religions, environmental laws are mentioned and considered in different ways. The Gita says: Let the human community protect the environment for their own survival and the biodiversity around. The tenth chapter of the Gita-Bibhuti Yoga (the yoga of divine manifestations) is a thought-provoking one (Box 1)(Framarin, 2014).

Zoroastrianism

The obsession of the Zoroastrian Iranians in keeping water, soil, air and fire clean has been famous among the Greeks. Herodotus and Xenophon wrote about it and wrote: The Iranians do not throw anything dirty in the water and take care that the soil is not clean.

In the religion of Zoroaster, everything given by God is pure and lovable. There is a big difference between the religions that consider matter and body as evil and the Zoroastrians who consider them as divine forgiveness and worthy of protection. It is man who should not use matter in the wrong way.

God has created this joyful world, with all its beauties, for the benefit and enjoyment of human beings. All people, as trustees, must fight to preserve it. Respecting the natural elements is acknowledging the greatness of God and religiosity. Darius the Great writes in the inscription of Biston: Ahuramazda created the world, created our wages, created happiness and created happiness for the people.

Zoroastrian Iranian culture is a culture of happiness. The culture of mourning is a non-Iranian culture. In one of the prayers, Zoroastrians pray in all four directions of the earth. This means that God is everywhere. It does not have a special house and it is the house of God everywhere, and the fact that all the blessings of God are everywhere should be respected and supported.

The reason why Zoroastrians look at light (light, sun, moon or fire) in prayer is that light is a symbol of God, and wherever there is light, it is wisdom, it is truth, it is love, it is reach, and God is there. Before Ashura, the Iranians worshiped the lord of all kinds (the gods of the heavens) and paid ransom to them. Zarathustra considered the worship of various gods to be incorrect and called on people to worship the one God - Ahur al-Mazda. After Zarathustra, the clerics who were not present to give up imaginary gods at all, introduced some of them to the religion in the name of god or angel. At this stage, the gods or angels are the creatures of Ahuramazda, and in order to be the source of the work, they assigned a task to each of them, and generally became supporters of natural elements and phenomena, and became Ahuramazda's collaborators. If we understand the history of the origin of the gods and their duties in religious literature, we will see the basis of thought as beautiful. Understanding the duties of the gods emphasizes the importance of the environment and the prevention of pollution in the Zoroastrian tradition. Among Anahita and Aban are the guardian angels of water, Azar and Niriusang, the guardian angels of fire, Zamyad, the guardian angel of earth and soil, and Vivo, the guardian angel of wind and air. Angels also protect the sun, moon, stars, and rain, as well as guardianship of purity, righteousness, and covenant. Some post-Islamic Greek and Iranian writers, out of ignorance or intent, have misinterpreted the Zoroastrians' extraordinary respect for natural elements, one or two of which are not without merit. Zoroastrians have a great respect for fire, because light is a symbol of Ahura. Zoroastrians refrain from polluting the fire - pouring unclean things and wicked animals - on the fire, to prevent air pollution. The issue of environmental pollution today has complexly attracted the minds of scientists, citizens and statesmen. In my youth, I heard some biased people say: fire burns and is not polluted! This is true, but as a result of burning dirty and toxic substances, the air is polluted and public health is endangered. This is not fire worship! Nowhere in the Avesta does God appear as fire. While "Yahweh", the Jewish god, speaks to the Jewish prophet from inside Mount Sinai. In Islam, too, God is known as the light of the heavens and the earth. In all religions, light is a sign of divinity, being given the relation of fire.

Another issue is the use of crypts for the dead. The crypt, or as the Persians call it, the Tower of Silence, has become a fantasy subject for foreign writers. The "crypt" is one of the ways to get rid of the dead body. Burial was also common among Zoroastrians.

The cave was built on top of a mountain far from the city, where the carnivores simply ate the dead meat in a short time, and then the guards of the cave poured the remaining bones into a well with sharp water to disinfect and destroy it.

It is natural to use this method. That is, animal meat is eaten by birds. It also does not pollute the environment because where there is a lot of carnivore, dead meat is eaten in a few minutes. In contrast, burial is fraught with difficulties. One that pollutes the soil and the earth. Another is the loss of land that can be used for agriculture and development. Of course, with the development of cities, the use of crypts becomes impractical. The results of putting the dead in the crypt are similar to the results of "burning the dead body", that is, the dead body disappears with the same speed and cleanliness. As we have said, the importance of Zoroastrianism for the preservation and protection of the environment has led some scholars to call Zoroastrianism the first religion of environmental protection.

The religion of Zoroaster is the first religion to pay attention to the environment and ecology, in which everything that God has given is pure and lovable. In the book of Avesta it is stated: We glorify the waters of the spring / And the waters of the passages / The interconnections of the road / And the intersections of the road / We glorify the flowing mountains / The aquatic lakes / And the profitable fields of wheat / And we praise the Creator / the earth and the sky / and the agile wind created by Mazda / and the Alborz mountain / the earth and all good things (Avesta-Yashtha).

In this religion, all the blessings of God, wherever they are, should be honored and supported. At this stage, the gods or angels are the creatures of Ahura Mazda, and in order to be the source of the work, they assigned certain tasks to each of them, they generally supported the elements and natural phenomena. "Includes in the Zoroastrian tradition. Including Anahita and Aban, the guardian angels of water, Azar and Niriosang, the guardian angels of fire, Zamia, the guardian angel of the earth and Vivo, the guardian angel of wind and air, and also take care of purity and correctness, etc(Kistaubayev, 2020)."

ABRAHAMIC REGILION

Judaism

In the Jewish religious texts, doctrines related to the environment are implicitly mentioned, because these doctrines are part of the category of the relationship between God, man and the world, and each of these cases will be mentioned here.

God in Judaism

In Judaism, God is the only fact that how he is known is not so important; Rather, more emphasis is placed on the divine attributes. Judaism is based on two pillars: belief in one God and the selection of the children of Israel as evangelists of this belief. The one God of Israel is a living God whose creative power is always at work in the world of creation, and this power was first revealed with the creation of the world (Epstein, 1945).

Moreover, it is said that God is in everything; However, he should not be considered one with the whole world of creation and with any part of it. Of course, this understanding of God is almost recent in evolution and has been common among mystics since the Middle Ages, meaning that God is separate from the world of creation but is deeply connected to it (Epstein, 1945).

In the shadow of Shekhina (the divine presence), the Jews in exile felt God everywhere, and the rabbis said that Shekhina would go from one synagogue to another in Israel, or that she would stand at the synagogue. And blesses the footsteps of the Jews who go to the synagogue (Epstein, 1945).

Zabhali tribes have an all-divine tone and teach that the world has been blessed by God. Whereas according to the Talmud, God is "our Father in heaven", a father who is always close to his children on earth, and this contradicts the doctrine of all divinity, which equates God with nature (Epstein, 1945).

The realization of the divine purpose in Judaism is related to the idea of the kingdom of God, but this kingdom is not in heaven and any other life; Rather, it is built here on earth and under divine guidance and with human hands. This sovereignty of God is realized with the advent of the Savior Christ (Epstein, 1945).

Man in Judaism

In Genesis (27: 1) it is stated that God created man in his own image. In some interpretations, humans are considered superior to animals, and animals superior to plants, and Bratz plants superior to inanimate objects. This hierarchy has two consequences: the responsibility of higher ranks over lower ones and the right of precedence of higher ranks over lower ones. (Epstein, 1945) Therefore, man is of central importance among all creatures, and his creation as God, the command to man to fill the earth, dominate it and rule over all creatures, evokes a human-centered structure (Epstein, 1945). Of course, the idea that man has a higher status as the axis of the universe depends on his responsibilities and duties to God and the world (Anterman, 2007). The Talmud expands the biblical view of man and introduces him as a friend and co-worker of God. In order to achieve the goal of creation, human cooperation is necessary and God has created him as his partner. Man is a special agent to achieve goals that go beyond the boundaries of the natural world through the natural world (Epstein, 1945). The process of creation is not over with the creation of man and the world; Rather, this task was entrusted to man to lay the groundwork for the evolution of what has been created. God is the Creator. Therefore, man must be a creator and the basis of his cooperation with God is in obedience to him (Epstein, 1945).

But in spite of this superior dignity for man, the land is not in the possession of man to exploit it; Rather, he is only the guardian to preserve the earth for the benefit of his contemporaries and for future generations. The Talmud is also a special approach to one's duty to (Yoreh, 2014) according to the Torah, the human soul and body are inseparable, therefore, all human attention should be focused on life in this world and not life after death. The command to worship and serve God is only for this world and the life of this world (Swartz, 1996). So the axis of Jewish life is based on the Torah, the law and the duties of this world.

Nature in Judaism

According to the Bible, the creation of the world was done by God, and the first verses of the Bible tell the story of creation. The Bible is full of awareness and appreciation of nature, from the "Genesis Journey" to the "Marmara". Nature bears witness to God in creation (Isaiah 26:40; Amos 8: 5; Job 41:38) Contrary to the corrupt worldviews of the ancient religions, which regarded things in nature as divinity, the Bible clearly He considers nature to be the work of God and does not consider it a part of God in any way. (Ivry, 1997)

The material world is based on the "journey of Genesis". Evil enters the world through the free action and choice of people, not during the process of creation. It is in later periods that the duality of body and soul is introduced and this world is recognized as an evil substance. It's God's nature, Psalm 1:24 states, "The earth and all that is in it and the world and all that lives in it belong to God." God does not leave nature to man; It simply gives him permission to use it. Man does not own the earth, the earth is for God and all of His creatures, many rulings in Judaism emphasize that man needs God's permission to use creation (Watling, 2009) "Remember that the earth "It belongs to God and you can not sell it forever. You are God's guest, you can only use the product of the earth." (Leviticus, 23:25) Thus, in Judaism, the world is centered on God, and it is the place of manifestation of the meaning and value of God's creation (Watling, 2009).

Christianity

The serious criticism of some thinkers that the theism of the Abrahamic religions is the root of the world's environmental problems has aroused the concern and attention of theologians and led to the establishment of new studies and findings under the title of environmental theology. In this theology, God and His action are examined from an environmental point of view, and God's relationship with the world can be examined under different models. Christian theology has not accepted the model of God's enmity with nature and has taken various approaches to reject it. In the defense approach, some theologians have tried to change the face of the issue and introduce causes other than religion as the cause of the environmental catastrophe. However, this approach has failed to provide a specific model and answer the question of what is the relationship between God the Father and the planet and the environment. Also, in the extra-Christian approach, theologians believe that the Christian view of God's teaching should be reconsidered. And he did not consider it as God Almighty and separate from nature. Proponents of this approach, who have often made the Eastern version their top priority, even some believe that God is nothing but nature, and that nature is God. This approach is essentially unrelated to Christian teaching and has, in practice, led to a change in the religion's beliefs about creation. In addition, these models are incapable of explaining the goodness of nature. Thus, many Christian theologians in the Third Approach believe that religious texts and Christian theology are sufficiently grounded in environmental accusations, and to clarify this fact without the need to change the position of Christian theism, one only needs to re-read and focus on the sacred texts. Is. If we look at the holy books, including the Bible, with the lens of green, it is inferred that matter has no inherent evil and that man is no more than a trustee. In the Christian approach, on the one hand, the good nature of God and His good order are emphasized, and on the other hand, by predicting the realization of the divine kingdom, a healthy environment is one of the signs of the realization of the divine kingdom and promise.

In the meantime, by examining the proposed models, it can be concluded that the doctrine of creation from non-existence is the most important and safe achievement of environmental theology in the view of religion, according to which, on the one hand, matter and the natural realm of the universe are evil. Mobrasat, on the other hand, emphasizes and obliges the responsibility of human beings in its protection. In other words, in the model of good ending, the evil of matter (albeit transversely) is accepted; That is, God acts as a savior who saves nature and the environment, which has been afflicted by original sin, and empties it of depravity, while, according to the creation doctrine of non-existence, there is essentially no theological root. It is inconceivable to know matter as evil and its inferiority, and the whole universe is considered good even after its descent. Based on the model of good creation, in addition to theological products and results, a series of moral principles are inferred, the most important of which is the principle of benevolence or service, which emphasizes the responsibility of man to protect other creatures. According to this moral principle, since man does not own the earth and his dominion over the earth has never been absolute, man is only a servant who has been asked to treat creatures in a holy and wise manner and when he is unable to do his duty earth is damaged(Watling, 2009).

Islam

Attention to the environment is evident in the Holy Quran, the narrations and biography of the Prophet (PBUH) and the Imams.

Environment in the Holy Quran

In the Holy Quran, nature and its resources have been conquered 33 times, and nature has been revived eighteen times. Also words like (sea) forty-one times, (mountains) thirty-nine times, (tree) twenty-seven times, (existing) twelve times, (earth) more than five hundred times, (sky) about thirty hundred times And (sun) have been used 33 times. Also, chapters of the Holy Quran with names such as Tin, Fajr, Najm, Naml, Ankabut, Dohi, Adiyat, Dhariyat, Shams, Lil, Qamar, Hadid, Raad, Noor, Toor, Shams, Zalzal,

Lil, Fil, Baqara, Nahl, Dokhan, Borouj and have been named and God has sworn to some of them such as Teen, Fajr, Shams and Lil.

What is seen in the verses of the Qur'an is that the real and main owner of everything is God. (Verse 126 of Surah An-Nisa ') Nature is conquered by man by the command of God. Verses 65 of Surah Al-Hajj, 36 of Surah Al-Baqarah, 16 of Surah Al-Mu'min, 2 of Luqman, 13 of Surah Al-Jathiya, 32 and 33 of Surah Ibrahim and 14 of Surah An-Nahl are among the verses that refer to this subject. But one thing to note is that this privilege that God has given to man to conquer nature is only a test for him. (Verse 165 of Surah An'am and 130 of Surah A'raf) Therefore, man should always refrain from abusing this right and privilege. Verses 2 of Surah Al-Furqan, 3 of Surah Al-Aali, 8 of Surah Al-Raad, 49 of Surah Al-Qamar, 21 of Surah Al-Hijr and 70 of Surah Al-Hajj advise human beings to use the environment correctly and to avoid destroying the environment. Misuse of nature is an example of ingratitude for the privilege bestowed on man. In this regard, the eighty-sixth verse of the remaining surah states, "Do not waste the pure resources that God has given you and do not exceed your limits, because the aggressors will not be subject to the love of God." Therefore, it can be said that man has a commitment to use this conquered nature correctly, and to destroy it means to stay away from God's mercy and grace.

Man should always thank his God for these blessings, because these blessings are signs of the power of God (verse 101 of Surah Yunus, verse 53 of Surah Fussilat, verse 9 of Surah Saba, verse 44 of Surah Ankabut and verse 5 of Surah Jathiya) . Man should also think about the system of creation according to nature (verse 99 of Surah An'am, verse 33 of Surah Yasin, verse 101 of Surah Yunus, verses 3, 4 and 13 of Surah Jathiya).

Another issue that the verses of the Holy Quran address is the right of all human beings to a healthy environment. Verse 25 of Surah Al-Baqarah states in this regard, "And He is the God who created for you (mankind) all that is in the earth." The tenth verse of Surah Ar-Rahman also states, "And God made the earth for all(Khalid, 2002)."

Environment in Narrations

In the narrations issued by the Holy Prophet of Islam (PBUH) and the Imams, the importance of the environment in Islam can also be deduced. Trees are very important in the environment. The Prophet of Islam (PBUH) forbade Muslims from cutting down trees and burning agricultural products. (Sheikh Klini, 1407). In another narration attributed to Imam Ali, he considers not harming the trees and leaving them as signs of piety (Nahj al-Balaghah, p. 16). According to one of the narrations, planting a tree is considered as a current charity and strengthens the family (Khalid, 2002).

Protecting water resources and preventing their pollution are also among the commandments that are reflected in the narrations. For example, the Prophet Muhammad (pbuh) forbade Muslims from bathing in stagnant water. (Khalid, 2002) Muslims should also dig drinking water wells at a distance from the sewage well. (Khalid, 2002) Not only the protection of water resources, narrations also emphasize the need not to pollute the air. Imam Sadegh (AS) states in a narration that "the enjoyment of life depends on the three factors of clean air, abundant water and pleasant and fertile land" (Khalid, 2002).

In addition to the above teachings, which called on man to protect the environment and to avoid polluting water resources, air, and cutting down trees, we can mention one of the commands to cultivate landless lands that are ownerless. This will lead to the development of human society. The construction of land without owners is one of the recommendations that can be taken from some narrations. For example, the Prophet (PBUH) ordered the revitalization of the dead lands that have no owner. (Khalid, 2002)

A healthy environment affects the quality of human life. There are narrations from Imams in this regard that point to the positive impact of the environment on human life. For example, Imam Kadhim says that "looking at lush green plants brightens the eyes and increases their brightness" (Khalid, 2002). In another narration, Imam Sadegh says, "Looking at colorful flowers, beautiful blossoms, and lush trees gives man so much pleasure that no pleasure can be equated with it."

RELIGIOUS SOLUTION TO THE ENVIRONMENTAL CRISIS

Judaism Theological Approaches

Nature and the environment In Judaism, it is a sign of renewed allegiance or the completion of God's authority and a manifestation of love and compassion. If the commands of God are obeyed, clean air, heavy rain, fleshy and leathery cattle, fertile soil, etc. will be for man; And if God's commands are not obeyed, there will be earthquakes, infertility, plant pests, roof collapse, floods, storms, famines, etc. (Motloch, 2000). Judaism, then, both in the Bible and beyond, emphasizes the relationship between people and nature, the prosperity of the people depends on their obedience to God, and the prosperity of the earth depends on the social justice and moral integrity of the people of that land (Penn, 2003).

As man tried to dominate the earth, he became aware of his limitations and that God is the king of the earth. Therefore, every year Sabbath and other celebrations thanked him and took refuge in him. They realized that they had to treat the earth well, not just rest it; Rather, respect it, keep water sources clean, create green areas near urban areas, regulate sewage and waste disposal, and prevent animal harm (Swartz, 1996).

The Law of Sabbath is one of the most important laws emphasized in the Bible. What is understood from these laws is that the land that is being exploited should be treated with love and care. The treatment of the land must be based on justice and the avoidance of exploitation. You should also not use it for one day after six days. "And God blessed the seventh day, and sanctified it, because on that day God gave up all that he had made to build" (Genesis 2: 2).

The divine doctrine of Bal Tashhit emphasizes the protection of all that is created. According to this law, the destruction of what can be useful to human beings is forbidden. This prohibition also covers animal and plant life and emphasizes the prohibition of extravagance (Watling, 2009).

Zaar baaliei hayyim law prohibits harm to animals. This law is a sign of Hallachis' interest in animal welfare, and the Babylonian rabbinic law states that a Jew must feed his animal before doing his work (409, Ber, TB).

The year of Sabbath or the year of fall clearly shows the idea of nature protection and Ibn Maimun considers the reason for the year of fall to increase the yield of the land and its revitalization. Gather, but dedicate the land to God during the seventh year, and do not sow anything in it. For the earth, it is a year of rest "(Dallal, 2010).

The prohibition against cutting down trees is stated in the Bible as "When you besiege a city for a long time, do not cut down fruit trees, eat their fruit, but do not cut down trees. Trees are not your enemies" (Deuteronomy 19: 1). And in Holakha's interpretation, this doctrine has been extended to the subject of tree irrigation (Gottlieb, 2006). There is an Eid for tree planting called Ilanot, which is held between January and February in Iran to celebrate and sanctify the trees and to awaken nature again. In the past, it was customary to plant a tree as soon as each baby was born, and the child himself was later forced to irrigate the tree so that he could use it at his wedding years later (Ivry, 1997).

In Shulhan Arukh, it is commanded that "a person who damages a tree puts his life in danger" (Bourdeau, 2004).

One of the laws in Judaism is the Kashurt or Kosher law (halal food) which restricts production and consumption, and the laws of purity and cleanliness of food (for example, eating meat only from non-predatory animals, Slaughtered in a painless way and not mixed with milk is permissible (insist on the balance of the sacred relationship between man and nature (Watling, 2009). There are also special environmental laws. For example, there are several divine laws prohibiting the contamination of running water, and even the Talmudic law forbids drinking water that has not been covered for a long time (Bourdeau, 2004). Because they may have been infected by insects or other items. The Talmud also places great emphasis on cleanliness and even considers it necessary to wash one's hands before eating or doing certain things. In traditional Judaism, a person who did not observe cleanliness, hygiene, and related laws was rejected. The Talmud states that "the cleanliness of clothing and body parts is sacred, the impurity of clothing brings madness" (Epstein, 1945). Judaism emphasizes maintaining personal health. There are also rules for annoying odors. For example, it is emphasized that the place where animal carcasses are dumped or the cemetery and tannery should be away from cities so that people are safe from its bad smell. Mishna forbids placing the threshing machine up to 50 cubits away from cities so as not to cause damage or eye irritation (Epstein, 1945).

Christian Theological Approaches

Christian Theological Approaches and Models in Resolving the Environmental Crisis

As mentioned, there were criticisms of the Abrahamic religions' approach to the environment. Therefore, Christian theologians read the texts and expressed their opinions. A number of commentators defended Christian theology, responded to the criticism, and dismissed the allegations. On the contrary, some theorists accepted the criticism and tried to find a solution from the Eastern and Native American religions. These views, in a way, showed the ineffectiveness of conventional Christian theology in the face of environmental crises. In addition to the above views, some views were raised which, by accepting Christianity's disregard for the environment, sought solutions from within Christian texts. This view, unlike the second view, sought to provide an intra-religious solution that could meet environmental challenges.

Defensive Approach

This approach states that religious tradition is not the only component that forms the culture of societies, so it can not be said that religious tradition is the historical root of environmental crises and no historical or scientific evidence can be provided for it. In this regard, we can refer to the opinion of one of the researchers that he states that factors such as technology, urbanization, democracy, increasing personal wealth and aggressive attitude towards the environment are related to environmental crises and can not be the only tradition. He considered Judeo-Christian to be the only cause of environmental crises. In this regard, this researcher raises questions. Are people who do not follow the Christian tradition less prone to excessive consumption, extravagance and waste than Christians? If the environmental crisis is a religious problem, why are other parts of the world facing it as well? (Saleh & Hasani, 2017)

"Double" also believes that influential factors such as economic, social and political factors in the production of new science and economics in the emergence of environmental crises should not be ignored. According to him, most of the main roots of the current environmental crisis should be sought in the world of secular nationalism and non-Christianity, scientism and liberalism of the sixteenth to nineteenth centuries. (Saleh & Hasani, 2017) This approach does not provide an answer to solve environmental crises; It only provides an answer for critics.

Extro-Christian Approach

While accepting criticisms of environmental Christian theology, a group of Christian theologians offered religious models. It should be noted that these models are mostly derived from Eastern traditions and religions.

Intro-Christianity Approach

This approach, while acknowledging the church system's neglect of the environment, sees the environmental crisis only as the result of the actions of religious people, not the inefficiency of the Christian religion and teachings. That is why it seeks to re-read divine sources to give religion a special look at the environment.

Islamic Theological Approaches

The religion of Islam, in addition to the issues that are interpreted as beliefs, includes rules and regulations that are referred to as Sharia. The Islamic legal system contains rules and regulations that Shiite and Sunni jurists have deduced from verses of the Holy Qur'an, hadiths and other sources such as reason and consensus, and have described them as instructions for Muslims.

The destruction of the environment is one of the harms identified in the Islamic legal system, and rules have been laid down for criminalization and compensation. Among these rules, we can mention the rule of no harm, loss and causation.

No Harm Principle

The no-harm rule is one of the most important rules of Islamic law, which is interpreted as the most important legal rule in the Islamic legal system regarding the civil liability of individuals. This jurisprudential rule is based on a narration of the Prophet (PBUH) who says "no harm and no harm in Islam"(Khalid, 2002). There are many opinions about this jurisprudential rule. According to one of the views that the author also believes in, this jurisprudential rule has not only ruled on negation; It also proves the verdict. This view means that according to the no-harm rule, harm to others is prohibited, and if harm occurs, the harmer is obliged to compensate. According to this rule, all losses must be compensated. One of the disadvantages is the damage caused to the environment by human activities.

Loss Principle

According to this rule, any person who causes damage to the environment is obliged to compensate for the damage caused by his action. It should be noted that the loser, as soon as the damage is caused, is obliged to compensate the damages and it does not matter if he did it intentionally or unintentionally. The theory of waste can play a significant and appropriate role in the system of liability for environmental damage. According to this rule, any person who destroys the environment is a guarantor and must compensate the damages. Harm to nature and the environment leads to the realization of civil liability for the victim, whether proven guilty or not.

Causation Principle

According to this theory, anyone who causes damage or loss of property to others must compensate it. Sometimes people do not directly pollute the environment. It may cause direct damage to nature. According to this rule, if there is damage to the environment, other persons become aware of the occurrence of this damage, are obliged to inform the competent authorities such as ministries and organizations related to the environment and the judicial system. If people do not do this, according to the rule of causation, it can be considered responsible for compensating the damage to the environment.

Transgression Principle

Among the theories related to civil liability in Islamic law is the theory of abuse. This rule is more applicable to "trustee" civil liability and to trust contracts such as deposits, power of attorney, etc., which hold persons such as the depositor and the lawyer liable if the loss or defect of the property is documented in their misconduct. This rule can also be invoked in the area of liability for environmental damage. For example, if the owner of agricultural land lights a fire on his land and this fire spreads to other agricultural lands and leads to the destruction of agricultural products and livestock, according to this theory, civil liability will be realized, because It is permissible to light a fire only if one can control it, otherwise if it spreads to the property of others, it is an example of encroachment.

Respect for Property Principle

This theory is based on the famous narration "The sanctuary of the Muslim is the sanctuary of the blood" (Khalid, 2002). The Messenger of God (pbuh) also states in a narration that insulting a believer is a sin, killing him is disbelief and his disobedience to sins and respect for his property is like respect for his blood (Khalid, 2002). The environment in Islamic law is one of the property that needs to be protected and any damage to it has led to the obligation of the person who caused the damage to compensate.

ENVIRONMENTAL IMPACTS OF MODERNIZATION

In the modern era, the view of nature is a purely instrumental and calculating view of how to make the most of it. By raising the level of man's scientific consciousness, he considers himself not a member of nature; Rather, he knows the only effective and decisive being in it. The environment is defined and organized according to human will. Therefore, the result, unlike in the past, is no longer a composite order; It is independent and its own. More than ever, nature becomes passive and intervenes without feeling the need to understand language and the laws that govern it. "What is important is the effectiveness of the method. The world and nature become objects that can be calculated, measurable and predictable; that is, ideas are reduced to numbers" (Mol, 2006).

The unity and integrity of nature is set aside. It is assumed that the universe and each unit as a whole is composed of its own elements and components. Cognition can be obtained by studying these components and the properties of the behavior of phenomena, and there is no other way but to analyze to simplify complex elements. Since the component is the representative of the whole and knowing each component is simple and easy, so through this cognition can be achieved the whole complex. Therefore, the earth, like any other phenomenon, is reduced to its constituent parts as much as possible. The various branches of art and science follow a specific path, each without a horizontal connection of meaning and one piece. In this way, the single soul, including the subtle sciences, is destroyed. Of course, along with empiricism, its mechanistic role should not be overlooked. Life, like a mechanical device, is reduced to mere physical, chemical, and mechanical laws. By knowing the form and function of each, one can achieve knowledge of the whole form and function. What helps man to know these components is to know the causal relationship between them. In other words, every disability is one hundred percent due to its cause, and the causal relationship is an algebraic and inevitable relationship.

It was because of this insight that the interconnected and organic structure that governs nature falls apart. Over time, many habitats and living species are severely damaged, and natural resources such as water, soil and air are polluted, and forests and pastures are destroyed, causing the earth's temperature to change, the ozone layer to be perforated, and too much to enter the dangerous and carcinogenic UV rays and dozens of other complications and diseases.

Instead of focusing on environmental quality management and the richness of human experience, many policymakers are greedy for new styles. These people think that their job is to design the project, not to design a phenomenological whole. They often look at the project as a manifestation of form or design, not a creative response to its context. Instead of designing a suitable environment and nature, they design the building and the site. It is often mistaken to think of design as the creation of form, not the creation or discovery of a new experience of nature (Motloch, 2000) or the instrumental use of nature to create institutions and ways of thinking and acting that, willingly or unwillingly, individual relationships and Socially affects human beings. The human ideal is defined in uniting the rest of human society in order to force nature to obey the human will under the guidance of science.

DISCUSSION

In the last decade, due to the persistence of environmental crises (despite the advancement of science and technology), the world view of religion has changed and has been considered as an effective factor in resolving environmental crises. Many environmentalists now believe that policies Not only should the environment be comprehensive, but such policies, in implementation, require religious cultural support that underpins environmental protection, because proponents of this theory believe that the historical role of religion in protecting the environment confirms the loss of Spiritual connection with nature has made it difficult to protect it. As a result, by emphasizing the need for the presence of a religious base to resolve environmental crises, the presence of the religious factor in environmental discussions is considered necessary.

In 1967, Lynn White, a professor of history at the University of California, Berkeley, wrote an article entitled The Historical Roots of Our Ecological Crisis, claiming that the roots of the crisis lay in the Jewish-Christian way of thinking that man has the right to rule over nature. In this view, nature is alien and merely a source of exploitation, and the background of this humane and authoritarian attitude goes back to (Genesis), especially to verse 28 of its first chapter: "And their God (Adam). And he blessed Eve and said to them: Be fruitful and multiply, fill the earth and subdue it, and rule over the fish of the sea and the birds of the air and all the animals that live on the earth. Islam has a special view on the environment. One of the religious advices that can prevent environmental crises is environmental ethics based on theology and monotheism and self-knowledge and self-preservation, which is also called ecology of environmental ethics. Environmental efforts should be ecological And be based on the religion, beliefs, values and beliefs of the region. Different religions, depending on their laws and beliefs, can be a means to a better understanding of nature.

CONCLUSION

In the modern age, which is the result of the extensive social and cultural changes of the European Renaissance, the ideas of scientists such as Descartes and Bacon played an important role. If in the past man was mainly in search of the general truths of the universe, he came to earth in the Renaissance and returned to nature and society and sought reality with the power and ability of science. After knowing the world and society and in accordance with his method and technique, he dominated nature and society, and this domination became more and more day by day. In this period, what was considered more than anything else was reason and rationality, which itself became the source of the formation of the relationship between subject and object, and consequently the hegemonic relationship of man with man and man with nature. In this sense, the origin of any social decision arose from the necessity of exercising rational sovereignty and rational justification. In this approach, man came out of the passive state and found a unique position in the study and prediction of things and, above all, paid attention to physical and material needs. The undisputed dominance of modern man has led to environmental crises such as climate change, water and sea pollution, and animal and plant ecosystem change. Thus, it can be said that modern man did not use scientific achievements properly and instead of responding to his needs by properly exploiting nature, he destroyed the environment under the pretext of economic development, not only his life; It has also endangered the lives of future generations.

Due to the divinity of different religions and its conformity with human nature, the rules and regulations of the divine religions for the healthy life of human beings in a clean and unpolluted environment have always been proposed. The title should be stated as a fundamental factor in the way of living better, which in this way will prevent man from destruction and corruption in nature. Many thinkers have also come to the conclusion that in the present age it is only religious teachings that have the power to curb human haste in the age of scientism and materialism.

REFERENCES

Bourdeau, P. (2004). The man− nature relationship and environmental ethics. *Journal of Environmental Radioactivity, 72*(1-2), 9–15. doi:10.1016/S0265-931X(03)00180-2 PMID:15162850

Cooper, D. E., & James, S. P. (2017). *Buddhism, virtue and environment*. Routledge. doi:10.4324/9781315261195

Dallal, A. (2010). *Islam, science, and the challenge of history*. Yale University Press.

Epstein, I. (1945). *Judaism*. Epworth Press.

Framarin, C. (2014). *Hinduism and environmental ethics: law, literature, and philosophy*. Routledge. doi:10.4324/9781315852522

Gada, M. Y. (2014). Environmental ethics in Islam: Principles and perspectives. *World Journal of Islamic History and Civilization, 4*(4), 130–138.

Gottlieb, R. S. (Ed.). (2006). *The Oxford handbook of religion and ecology*. OUP USA. doi:10.1093/oxfordhb/9780195178722.001.0001

Ivry, A. (1997). Jewish philosophy. *Bulletin de Philosophie Medievale, 39*, 45–48. doi:10.1484/J.BPM.3.534

Kistaubayev, S. U. (2020). Zoroastrianism And The Expression Of The Relationship Between Nature And Man In Islam. *Theoretical & Applied Science, 84*(4), 624–626. doi:10.15863/TAS.2020.04.84.104

Mol, A. P. (2006). Environment and modernity in transitional China: Frontiers of ecological modernization. *Development and Change, 37*(1), 29–56. doi:10.1111/j.0012-155X.2006.00468.x

Motloch, J. L. (2000). *Introduction to landscape design*. John Wiley & Sons.

Pedersen, K. P. (2018). Environmental ethics in interreligious perspective. In *Explorations in global ethics* (pp. 253–290). Routledge. doi:10.4324/9780429500626-12

Penn, D. J. (2003). The evolutionary roots of our environmental problems: Toward a Darwinian ecology. *The Quarterly Review of Biology, 78*(3), 275–301. doi:10.1086/377051 PMID:14528621

Saleh, S. M. H., & Hasani, S. A. (2017). Critical study of the models of environmental Theology in the new Christianity. *Religions and Mysticism, 50*(1), 103–127.

Saniotis, A. (2012). Muslims and ecology: Fostering Islamic environmental ethics. *Contemporary Islam, 6*(2), 155–171. doi:10.100711562-011-0173-8

Swartz, D. (1996). Jews, Jewish texts, and nature: A brief history. *This sacred earth: Religion, nature, environment,* 87-103.

Watling, T. (2009). *Ecological imaginations in the world religions: An ethnographic analysis*. A&C Black.

White, L. Jr. (1967). The historical roots of our ecologic crisis. *Science, 155*(3767), 1203–1207. doi:10.1126cience.155.3767.1203 PMID:17847526

White, L. (2004). The historical roots of our ecological crisis. *This sacred earth: religion, nature, environment,* (179), 192.

Yoreh, T. S. (2014). *The Jewish Prohibition Against Wastefulness: The Evolution of an Environmental Ethic.*

ADDITIONAL READING

Pedersen, K. P. (2018). Environmental ethics in interreligious perspective. In *Explorations in global ethics* (pp. 253–290). Routledge. doi:10.4324/9780429500626-12

Penn, D. J. (2003). The evolutionary roots of our environmental problems: Toward a Darwinian ecology. *The Quarterly Review of Biology*, *78*(3), 275–301. doi:10.1086/377051 PMID:14528621

Saleh, S. M. H., & Hasani, S. A. (2017). Critical study of the models of environmental Theology in the new Christianity. *Religions and Mysticism*, *50*(1), 103–127.

Saniotis, A. (2012). Muslims and ecology: Fostering Islamic environmental ethics. *Contemporary Islam*, *6*(2), 155–171. doi:10.100711562-011-0173-8

Swartz, D. (1996). Jews, Jewish texts, and nature: A brief history. This sacred earth: Religion, nature, environment, 87-103.

Watling, T. (2009). *Ecological imaginations in the world religions: An ethnographic analysis.* A&C Black.

White, L. Jr. (1967). The historical roots of our ecologic crisis. *Science*, *155*(3767), 1203–1207. doi:10.1126cience.155.3767.1203 PMID:17847526

White, L. (2004). The historical roots of our ecological crisis. *This sacred earth: religion, nature, environment,* (179), 192.

Yoreh, T. S. (2014). *The Jewish Prohibition Against Wastefulness: The Evolution of an Environmental Ethic.*

KEY TERMS AND DEFINITIONS

Equity: Defined by UNEP to include intergenerational equity - "the right of future generations to enjoy a fair level of the common patrimony" - and intragenerational equity - "the right of all people within the current generation to fair access to the current generation's entitlement to the Earth's natural resources" - environmental equity considers the present generation under an obligation to account for long-term impacts of activities and to act to sustain the global environment and resource base for future generations. Pollution control and resource management laws may be assessed against this principle.

Polluter pays principle: The polluter pays principle stands for the idea that "the environmental costs of economic activities, including the cost of preventing potential harm, should be internalized rather than imposed upon society at large." All issues related to responsibility for environmental remediation costs and compliance with pollution control regulations involve this principle.

Precautionary principle: One of the most commonly encountered and controversial principles of environmental law, the Rio Declaration formulated the precautionary principle: To protect the environment, the precautionary approach shall be widely applied by States according to their capabilities. Where there are threats of serious or irreversible damage, lack of complete scientific certainty shall not be used as a reason for postponing cost-effective measures to prevent environmental degradation. The principle may play a role in any debate over the need for environmental regulation.

Prevention: The concept of prevention can perhaps better be considered an overarching aim that gives rise to a multitude of legal mechanisms, including prior assessment of environmental harm, licensing or authorization that set out the conditions for operation and the consequences for violation of the conditions, as well as the adoption of strategies and policies. Emission limits and other product or process standards, the use of best available techniques, and similar techniques can all be seen as applications of the concept of prevention.

Public participation and transparency: identified as necessary conditions for "accountable governments,... industrial concerns," and organizations generally, public participation and transparency are presented by UNEP as requiring "effective protection of the human right to hold and express opinions and to seek, receive and impart ideas,... a right of access to appropriate, comprehensible and timely information held by governments and industrial concerns on economic and social policies regarding the sustainable use of natural resources and the protection of the environment, without imposing undue financial burdens upon the applicants and with adequate protection of privacy and business confidentiality," and "effective judicial and administrative proceedings." These principles are present in environmental impact assessment, laws requiring publication and access to relevant environmental data, and administrative procedures.

Transboundary responsibility: Defined in the international law context as an obligation to protect one's environment and prevent damage to neighboring environments, UNEP considers transboundary responsibility at the international level as a potential limitation on the sovereign state's rights. Laws that limit externalities imposed upon human health and the environment may be assessed against this principle.

Chapter 14
Environmental and Energy Law

ABSTRACT

Energy law is basically aimed at making energy accessible at a reasonable cost, while environmental law in energy matters focuses on avoiding pollution when producing, making accessible and/or consuming energy. Perceived in this way, they constitute found paradigms, in permanent friction. Given that the human being cannot live or develop without using energy, it is that in this chapter, the authors seek to investigate aspects of encounter between both paradigms for a more adequate relationship between them, considering them with their sights set on the goal of sustainable development. In the first term, the authors will consider central aspects of the vision of energy law and then attend to the vision of environmental law in energy matters.

INTRODUCTION

Energy Law

From the point of view of Energy Law, energy is the engine of the production of goods and services in all economic sectors and the center of social development, so regulation focuses on these aspects, generally linked to costs, accessibility, technology . From the social and economic point of view, energy is considered a primary or derived natural resource.

Although we will consider different energy sources, any form of energy, once converted, is equivalent to any other, since it can be measured in identical units.

As the supply of energy requires multiple processes (production, conversion, transformation, distribution), public and private entities usually enter into competition in order to control energy sources, to influence the preferences of users, to access resources (material, financial, human and others).

The value of energy lies in its ability to heat, light and make things move. Energy has been considered to be the vector of human evolution and development. Each country uses various energy sources to different degrees and in combinations that often reflect its endowment of natural resources. The combination of extraction and conversion technologies and processes by which energy services (electricity, heat) are provided to end users is known as the energy system, and the individual formula as the energy mix. The

DOI: 10.4018/978-1-6684-4158-9.ch014

depletion of finite energy resources (coal, crude oil, natural gas, uranium) creates the need to invest in alternative sources or import, the latter is a real drain on developing countries, in addition to being a network of supply vulnerable to changes beyond the control of national planners. It is well known that an energy crisis necessarily entails a socio-economic crisis(Arney, 2010).

Developing nations are increasingly concerned with the issue of "security" of energy supply. Energy must be priced appropriately to cover the total cost of supply, although access to it should not be limited based solely on economic considerations. It is necessary to adjust prices based on "ad hoc" models so that all consumers can take advantage of the benefits of energy. This is not only necessary to determine the prices to be paid by the different social classes but also by the different commercial sectors, since energy prices directly influence the competitiveness of goods and services both locally and internationally.

Environmental Law

From the point of view of Environmental Law, many organizations and political positions call for a reduction in energy consumption and consumption of natural resources. Already in the 1960s, ecologists have pointed out the need to change behavior towards nature (change in the mode of civilization and means of development). Since the 1970s, talk has begun of "limits to growth" (Bradbrook, 2011), a report requested by the Club of Rome from specialists in system dynamics at MIT shortly before the oil crisis, not only due to the scarcity of resources such as coal, oil or gas, but due to the ecological footprint of the human being. Although it is often said that subsistence economies leave little mark, this inevitably depends on the number of individuals and the habitat. Some authors point out the hypocrisy of those who blame the deterioration of the planet on oil companies when they, although encouraged by commercial benefits, extract resources to satisfy consumers, who also seek to obtain them at the lowest possible cost. A change in lifestyle models is advocated, as Pope Francis has done in his encyclical Laudato Si.

Governments must act on horseback of the economic-financial interests of energy and those of the preservation of the environment, since they must protect public health and the natural environment; In addition, they must take care that the benefits of long-term public investments in infrastructure development are not reduced or distorted by businessmen seeking short-term profits.

Energy production, energy markets, and energy use are generating serious and visible environmental problems in our times, since energy-related activities represent 84.3% of the planet's anthropogenic GHG, more than produce millions of tons per year of fly ash, bottom ash, boiler slag, which contaminate the soil and water, if not, negatively modify the soil and subsoil with still unsuspected consequences. For example, the fracking15 system for obtaining shale gas and oil has been increasing in most of the countries of the world despite its multiple risks. Among the negative environmental effects of fracking, the following stand out: decrease in the availability of water for other uses; contamination of water sources; soil contamination with toxic waste; carcinogenic effects, mutative effects, affectation of the endocrine and nervous systems -ia- in areas close to wells, storage and transport areas due to polluting gas emissions (in particular, methane, sulfur dioxide, nitrogen oxides, volatile organics) caused by the high number of trucks and drilling equipment; emergence of allergies; acceleration of global climate change. In addition, the magnitude of the undertaking implies the use of large vehicles and platforms that damage roads, pollute per se, affect the quality of life of the residents and fauna and flora in the vicinity with their noise and land use.

In the context of the environment, consumption also presents many problems, particularly in areas such as the inefficient use of energy, the use of substandard fuels and technologies, and the weakness of many environmental standards. The provision of cleaner fuels and more efficient technologies to consumers must be accompanied by effective environmental policies and regulations and the corresponding mechanisms to enforce them.

One of the most effective ways to promote greater sustainability in production and consumption is to properly account for the costs of environmental impacts in energy costs, which in addition to encouraging consumers to use energy rationally and carefully, provides a direct incentive for developers and entrepreneurs to invest in technologies that mitigate impacts, manage emissions and waste more effectively, and use energy more efficiently(Bosselman, 2011).

Attentive to the environmental effect of some energy sources, we indicate below the CO_2 emission rates of the different forms of energy generation, indicating them in increasing order of emissions:

- Nuclear
- Wind
- Hydroelectric
- Photovoltaic
- Biomass
- Coal (carbon capture and storage)
- Gas (carbon capture and storage)
- Gas
- Petroleum
- Coal
- Lignite

In the preceding list, for example, while nuclear and wind power do not exceed 20 grams of carbon per kWh, oil reaches 340 and lignite reaches 475.

The production of biofuels, widespread in the last decade, like all sources, has positive and negative environmental aspects that each government must weigh based on the availability of its natural resources. The most widely used biofuel is biodiesel obtained from vegetable oil (eg soybean, sunflower, palm) or animal fats. Another widely used bioethanol is obtained from the fermentation of raw materials rich in sucrose, starch, or cellulose (eg sugar cane, corn grains, grass, straw, forest residues). As biofuel acts as a substitute for fossil fuel, it is considered environmentally friendly, however, the increase in the area of crop production for the generation of biodiesel and bioethanol seriously affects biodiversity(Dincer, 1999).

Sources other than those listed above are also used, but this has not been done on a large scale so far. These include: geothermal (interior heat of the earth), "blue energy" (difference in salt concentration between fresh water from a river and sea water), tidal or wave energy (movement of sea waves), kinetic (use of the moment of acceleration in the movement), biomass (plants transform the radiant energy of the sun into chemical energy through photosynthesis, and part of that chemical energy is stored in the form of organic matter; the chemical energy of biomass can be recovered by burning it directly or transforming it into fuel), ocean thermal gradient or tidal wave (takes advantage of the temperature differences between the warmer upper waters and the lower ones that are increasingly colder depending on the depth), thermoelectricity (when two different metals at different temperatures are brought into contact form a bimetallic union, between both sides of the union an electromotive force is generated), fission and nuclear fusion, etc.

The ITER (International Thermonuclear Experimental Reactor) is a large-scale scientific experiment of most of the important countries of the Northern Hemisphere under the auspices of the IAEA, which seeks to produce energy commercially through nuclear fusion. ITER is being built in Cadarache (France) at a cost of 14 billion euros, making it the fifth most expensive project in history, after the Apollo Program, the International Space Station, the Manhattan Project and the development of the GPS system. It should generate its first nuclear plasma in November 2020 and be fully operational in March 2027(Jura, 2012).

RELATIONS BETWEEN ENERGY LAW AND ENVIRONMENTAL LAW

As Aagaard points out, traditionally, energy and environmental regulations have managed their inter-relationships in a state of conflict, imposing negative constraints on each other. Thus, the federal energy regulatory commissions in different countries have had to comply with the requirements of national and international environmental standards as an obstacle to their economic achievements, while the environmental protection agencies have had to foresee specific requirements and exceptions when dealing with issues related to energy, due to its condition as a basic factor in the economy and in the quality of life of the peoples(Kotzé, 2014).

However, the current requirements point out the advantages of the convergent and synergistic relationships between both regulatory activities, which also extends to other linked fields in which the action overlaps (eg transport, storage). The perception of the need for convergence between both regulatory sectors becomes increasingly evident.

Currently, political alignments are sought that simultaneously support the objectives of the energy law (low costs) and those of environmental law (preservation of the environment), taking advantage of opportunities for positive synergy.

It must be recognized that for much of the 20th century, energy policy benefited from economies of scale in the energy sector, in which increased energy production led to lower energy prices. Low energy costs depended on increased energy use (especially coal and other fossil fuels), while higher energy use implied increased environmental impacts.

All energy production necessarily has direct and indirect effects on the environment. Minimizing these effects requires permanent institutional adjustments in the form of policies and regulations to guide the development of the energy system. Power supply planning helps determine the resources to be exploited; it also influences the direction of investments and the orientation of technological development. In other words, it is from the environment that policies and regulations related to energy must be formulated.

Environmental law has attempted to reduce environmental damage from energy-related activities, with a focus on preventing pollution and avoiding damage to natural resources. This is expensive. For example, the cost alone of installing a pollution control system at a single coal-fired power plant can be more than $200 million. Thus, paradoxically, environmental regulations often increase the costs of production and aggravate the negative economic effects of energy use(Khajehpour et al., 2017).

If the two fields are kept separate (Energy vs. Environment), with negative models, limiting each other, even when they are effective. By their very design, they involve obvious irrational wear.

The first link between both paradigms was given based on the need to guarantee the health of the population, in a similar way to what was given in environmental law in general and its evolution. At the same time, given the economic difficulties of the many developing countries, at the international level, environmentally friendly energy efficiency programs and projects have been assumed through, ia, the World Bank, the Inter-American Development Bank, the Global Environment Facility (GEF).

CASE OF MENA OIL AND GAS TRADE

The way of human life and its interaction with the environment before the Industrial Revolution and the economic growth of societies was such that there was not much pressure on ecological potential and natural resources. After the Industrial Revolution, more access to natural resources, especially fossil fuels, became more important. In the twentieth century, after World War II(Bradbrook, 2011), policymakers and planners turned their attention to the fact that any economic growth would inevitably be due to the availability of natural resources. It is relevant, and in all these years, no one thought that these resources are limited and finite and are one of the most important causes of environmental degradation. Thus the obvious link between energy use and the rights of nature was ignored for years. The focus of energy law was on providing energy resources at affordable prices, while environmental law focused on environmental protection. In general, the emergence of energy law and environmental protection law in the international community has not happened all at once but has gradually formed and evolved. For years, the obvious link between energy production and consumption rights and environmental protection has been largely ignored, and the main focus of energy law has been to ensure affordable energy supply and focus on environmental law, nature protection, and a clear demarcation between energy law and environmental law was also not enforced. The first link between energy law and environmental law was indirect to reduce energy consumption and increase energy efficiency, which led to the oil shock of the year 1971, followed by 1973 and 1977(Khajehpour et al., 2017), when oil prices rose sharply due to events in the Middle East. Since then, many oil and its derivatives consumers have changed their minds, examining the possibility of alternative sources of fossil fuels and ways to reduce energy consumption. Thus, energy law was implicitly and indirectly related to environmental law. Increasing energy demand, declining fossil fuel resources, and the negative effects of different energy levels on the environment are important issues in the modern world. Therefore, national and international laws and regulations should be geared towards integrating energy law, environmental law, and issues(Wildermuth, 2011). Economics tended to protect all components of the environment, including air, water, and soil, from the negative effects of different energy levels from exploration and extraction, transmission to consumption. In other words, energy efficiency and environmental policies separately Invincibility are related. Because energy production and consumption have environmental effects and this issue connects the challenges of energy production and consumption, sustainable energy sources, and the protection of the natural environment(Wildermuth, 2011).

Conflict Between Energy Law and Environmental Law

International energy law is based on the principles of governance, contracts based on mutual interests, reciprocity, and acceptance of states as key actors, which is a distance from international environmental law, which is based on the rules of international obligations of states, international rules and the role of non-state actors. It has a lot. In this regard, the following is a review and analysis of general legal principles and laws and regulations regarding energy and the environment to identify conflicts between energy law and environmental law.

One of these differences is the difference in legal sources. Since international energy law and international environmental law are both branches of public international law, both of these fields of law are based on the sources in Article 38 of the Statute of the International Court of Justice. However, due to the novelty of international environmental law and these sources, there are other sources such as international rules of procedure, international obligations of governments, and resolutions of international organizations and UN Security Council resolutions. And it can be argued that they do not necessarily have much application in energy law. Therefore, the requirements and requirements of these two disciplines are different(Bradbrook, 2011).

On the other hand, international energy law is binding resources and bilateral or multilateral agreements that guarantee its implementation. But in international environmental law, the role of non-binding sources is more important than binding sources and greatly impacts the development of this field of law. The legal effects of the non-binding Stockholm Declaration 1972, the 1982 Universal Declaration of Nature, the 1992 Rio Declaration, and the Johannesburg Universal Declaration 2002 were far more than many binding documents, such as international conventions(Wildermuth, 2011).

Another difference between international energy law and international environmental law is the issue of actors. In international energy law, governments are the main actors and their subjects. In contrast, in international environmental law, the role of non-governmental actors (international non-governmental organizations, world public opinion, non-governmental groups, etc.) is recognized, and They attach great importance to the development of legal rules, protection, management, and environmental monitoring. For example, the ratification of the 1998 Aarhus Convention on Access to Information, Public Participation in Decision-Making, and Access to Courts on Environmental Issues reflects the tendency of international environmental law to recognize the role of soft actors in the development of the legal field .

In terms of international cooperation, the difference between these two legal disciplines is serious. In fact, the principle of cooperation is a common international obligation and one of the basic principles of the Charter of the United Nations and is a feature of contemporary international law. The principle of cooperation under the UN Charter is a binding principle. In the field of international law, whenever we talk about energy law, we mainly consider the legal aspects of production, transfer, sale, rights, and obligations of investors and consumers in the field of energy, energy security. Therefore, the principle of cooperation in international law Energy is formed in the direction of mutual economic benefits for the parties or around it. In addition, the Energy Charter Treaty (ECT) is the first multilateral treaty in the world to address the issue of energy(Bosselman, 2011). This treaty has many innovations in various fields, from issues related to investment and energy trade to transit and increasing energy efficiency and environmental protection with respect to the principle of sovereignty of governments and the concept of sustainable development, intergenerational justice, the principle of prevention refers to the principle of cooperation and the principle of payment by the polluter. Also, the nature of environmental issues is cross-border, and its solution is beyond the power of one country, and since the environment does not know the territory, so protecting the environment and combating environmental hazards is beyond the capacity of one or more countries, so the international community needs cooperation. International measures to control, prevent and reduce negative environmental effects. Therefore, the principle of cooperation can play an important role in international environmental law. Also, the basis of the principle of cooperation in international environmental law is the rule of Argamens.

Therefore, the basis of international environmental law, unlike energy law, which is based on mutual interests, is the rule of Argamens. Participate in international conferences and, ultimately, cooperate in critical situations by concluding treaties, conventions, and protocols. The principle of joint but different liability is also one of the basic concepts and principles in international environmental law in the field of cooperation in combating environmental challenges(Jura, 2012).

The issue of international responsibility of states is also another point of contention in these two fields of law. The international responsibility of states in international energy law is based on subjective and objective responsibility because, in this type of law, mutual interests are considered by the parties. Whereas the new trend in international environmental law has shifted to pure liability, under which no damages are left without compensation; Therefore, a new perspective has opened up in the international responsibility law of states. The establishment of environmental insurance, formation of joint national and international funds, environmental assistance, etc., are examples of the new trend of international environmental law in the field of compensation for environmental damage. On the other hand, according to customary international law, governments must cooperate internationally in the field of environmental protection, and due to the problems facing international responsibility, governments must protect their environment with a precautionary approach. In this context, the need to apply the principles of cooperation, prevention, and precaution in international environmental law is doubled. Hence, along with the classical legal systems based on subjective and objective responsibility, the tendency towards pure responsibility is also progressing and developing. Increasing the conclusion of international treaties and conventions on environmental compensation in various issues such as marine pollution such as the Civil Liability Convention on Oil Pollution at Sea 1972, as well as nuclear issues, air and ... and the establishment of various mechanisms for environmental compensation through The establishment of funds and the promotion of "insurance" contracts indicate the development of the theory of pure liability in international environmental law(Bradbrook, 2011).

Can it be concluded that just as soft resources (statements, declarations, and agendas, etc.) and soft actors (international non-governmental organizations) are recognized in international environmental law, so is "soft responsibility" as the foundations of the regime? Recognize the international responsibility of governments for environmental damage?

Similarities and Interactions Between International Energy and Environmental Law

Despite the conflicts between energy law and environmental law mentioned earlier, there are similarities and interactions between energy law and environmental law. First, both are subject to international law. Accordingly, one of the main obstacles to integrating international energy and environmental law is the reluctance of governments to delegate or limit sovereignty to the benefit of environmental organizations at the international level. The political structure of government always tends to focus and has no interest in delegating it to other centers of power and decision-making. On the other hand, the conflict of interests between the main actors of international law (states) regarding environmental protection, which arises from the national sovereignty of states, challenges the development and expansion of international energy and environmental law. This conflict of interests includes political interests, Economic, commercial, etc. The conflict between the two is often seen(Lazarus, 2008). So that energy law focuses on two main goals; Maximizing economic benefits from resources and limiting the monopoly power to ensure competitive pricing, and in the field of international law, whenever energy law is discussed, it mainly examines the legal aspects of production, transfer, sale, rights, and obligations of investors and consumers. In the field of energy, energy security is considered. However, one of the most important principles since the Stockholm Declaration of 1972, so far in international environmental declarations, is the principle of sovereignty, which redefines the sovereignty of states and redefines the concept of sovereignty in law The International Environmentalists rely on the concept of "rational use" of land. Still, the shadow of absolute sovereignty continues to weigh on international environmental rights and energy rights, and governments continue to seek to increase their monopoly power over energy and the environment(Nouzha, 2000).

Second, energy law and environmental law were developed as two separate issues with separate goals, and there is a less systematic connection between the two disciplines. Environmental rules, regulations, and laws often provide an urgent response to environmental problems or take immediate action depending on what happened. Finally, energy law and environmental law focus on short-term planning. Hence, their response is short-lived. Therefore, energy and environmental laws, regulations, and policies must be on a long-term horizon and in a more comprehensive framework. Energy law and environmental law both have their complexity due to their different components. For example, energy law is different energy levels from exploration and extraction, transfer and finally consumption, each of which has its legal regime. The environment also includes all aspects of human life such as water, air and soil, and everything on earth and beyond the atmosphere. Because environmental issues are very broad and several conventions have been developed for each. However, less attention has been paid to developing a comprehensive and integrated convention on energy and the environment(Lazarus, 2008).

Also, the right to energy, the right to a healthy environment, environmental human rights, the common heritage of humanity, the rights of future generations, energy security, economic and trade issues, and ultimately the sustainable development of commonalities between energy law and environmental law in the world. And numerous UN statements have emphasized the recognition of sustainable development.

The final document of the Rio + 20 conference, "The Future We Want," encourages governments to use the provisions of this document to achieve the three dimensions of sustainable development (economic, social, and environmental).

Achieving these goals depends on the will of governments so that in their domestic law to provide a suitable basis for achieving sustainable development. It should be noted that with the end of the Millennium Development Goals in 2015, sustainable development goals will replace those goals, and in formulating New international instruments in this field, environmental protection will be introduced as part of the process of sustainable development. The relationship between the environment and other areas of human life and international relations such as trade and investment will also be considered(Tertrais, 2020).

Therefore, the principle of sustainable development is a new field that simultaneously considers both politics and culture and emphasizes the prosperity of the economy, trade, and industry, supports the environment and coexistence with nature and the common heritage of humanity, and Considers the rights of future generations. Sustainable development is also expected to have a major application in international environmental law, and ultimately, as the United Nations promises, the world will soon see the transformation of international environmental law into international sustainable development law(Tertrais, 2020).

In terms of regional agreements, the EU strategy to reconcile the conflict between energy rights and the right to the environment is significant. The union has proposed solutions that the member states of the union should use. One of these strategies is energy efficiency and reducing energy consumption. Energy efficiency includes the cycle of production, transmission, distribution, and consumption of energy, the use of which can greatly help reduce greenhouse gas emissions, air pollution, destructive effects on water and land, habitat destruction, and biodiversity. Measures to increase the share of sustainable renewable energy sources in the energy mix can also reduce environmental and climatic pressures compared to other forms of energy. These measures can also lead to the efficiency of other energy sources. Today, the importance and necessity of using renewable energy are clear to everyone. The European Union considers the use of renewable energy as important in the face of climate change and global warming and sees its use as crucial in securing energy resources and their diversity. Therefore, renewable energy has become one of the most important issues in political issues and legal development over the past few years. According to the EU, renewable energy sources are sources of energy that, unlike conventional sources such as oil, natural gas, coal, and uranium, are constantly being rebuilt due to their natural forces. Renewable energy sources are widely available in large areas of the EU, from the sunny south to water-rich north, from vast forests to windswept coasts to western Europe. This is an obvious difference compared to conventional energy sources, most of which must enter the EU from insecure areas. Today, renewable energy sources play an important role in the security of the EU's energy supply[8], and renewable energy can reduce the EU's dependence on energy imports. In addition, unlike fossil fuels that exploit climate damage - carbon monoxide - renewables can be converted to electricity, heating/cooling and fuel, without the effects of environmental damage(Outka, 2012).

Final Words

Because of the above, the gradual development of international environmental law owes more than anything to the quantitative and qualitative expansion of international treaties, international judgments and procedures, and the expansion of soft law. International energy law is also largely based on bilateral or multilateral agreements, contracts, and instruments such as the Energy Charter Treaty. However, the growing risks, threats, and widespread environmental degradation due to fossil fuels are a cause for concern. There have been calls for international energy and environmental law to be ineffective in addressing these challenges. In this context, a fundamental review of treaty law, the attitude of judges of the International Court of Justice, reform of the structure of the UN program for the environment, identification of facilitating mechanisms to achieve sustainable development, recognition of the role of non-governmental actors can be solutions Legal and international executive energy and environmental law(Outka, 2012).

To fill the legal and enforcement gaps in international energy and environmental law in the face of new and emerging environmental challenges, the "principle of effectiveness" should be pursued with comprehensive regulations that can define a link between Established sustainable development, environmental human rights, energy rights, and the economic system. Also, "comprehensiveness and integration" in the formulation, adoption, and implementation of international rules and regulations are important perspectives for the future of international energy and environmental law. In this approach, in addition to the three mentioned relationships, in all plans, programs, policies, and political, economic, cultural, and social decisions in the domestic and international realm of environmental protection as the axis of development. Therefore, the reconstruction and modernization of international energy and environmental law to provide a favorable legal framework for sustainable development concerning countries' geographical, economic, and social characteristics seem necessary(Lazarus, 2008).

In regional agreements, the legal model used in the EU regarding the interaction of energy law and environmental law can be a model to be studied in this field. This model is based on energy efficiency, including the cycle of production, transmission, distribution, and energy consumption, and reduction of energy consumption. On the other hand, it is based on increasing the share of sustainable renewable energy sources in the energy mix. It can reduce the destructive effects of the environment. Accordingly, the use of renewable energy is important in the face of climate change and global warming, but also its utilization is important in ensuring the security of energy resources and their vital diversity.

CONCLUSION

Various specialized papers have pointed out the inconvenience of receiving direct foreign investment in oil, gas and mining areas that include fossil oils, recalling that, when it comes to business, everything is usually included except energy and weapons, since there are no friendly disputes in these areas, manifesting themselves as the most contentious and forcibly renegotiated. Rightly, Susan Maples wonders if the lack of an adequate link between energy, environment and human rights does not herald even greater litigation and conflict in the future unless these three aspects are considered synergistically together. For our part, we understand that any energy proposal by public or private entrepreneurs must be accompanied by environmental and social considerations, based on a pre-established structure. Citizen participation and the obligation to provide a registered response to citizen questions should be foreseen, even when these objections do not per se have sufficient entity to stop a project.

We understand that, in the relationship between the energy and environmental paradigms, it is necessary to:

- break the archetype of energy policy that attends only to the benefits of the economy of scale (more production and consumption, lower cost) and replace it with that of the more rational use of the environment;
- include the environmental cost in the cost of energy, taking into account the valuation principle "from the cradle to the grave";
- promote the development of investments in technology aimed at more efficient production, management and use of energy;
- provide "Guides" to the development of the energy system from Environmental Law, including aspects related to supply planning to determine the resources to be exploited and their modalities -particularly when determining the energy system and the energy matrix-, the direction of investments, the orientation of technological development;
- prepare regulations from Environmental Law (together with specialized technicians in Energy) in emergency matters in the face of natural or human-caused catastrophes linked to the production, conversion, transformation and distribution of energy, both in terms of prevention regulations as ex post action protocols;
- establish joint, regular and continuous sessions between the state areas that deal with energy, environment and social impact for the purpose of permanent inter-feeding, with the presence of pre-designated people based on their scientific-academic training in the reference areas;
- establish training programs for citizens, using all available means, aimed at the rational and efficient use of energy at all levels;
- Implement virtual control models of energy expenditure in public spaces and institutions.

In this way we can say that the energy and environmental paradigms, by finding meeting points, will be in a position to build the bases for the achievement of the Millennium Sustainable Development Goals, in particular, its goal 7, by incorporating the principles of sustainable development in national policies and programs and reduce the loss of environmental resources.

REFERENCES

Arney, C. (2010). Hot, Flat, And Crowded: Why We Need A Green Revolution-And How It Can Renew America. *Mathematics and Computer Education*, *44*(2), 180.

Bosselman, F. (2011). A Brief history of energy law in United States law schools: An introduction to the symposium. Chi.-. *Kent L. Rev.*, *86*, 3.

Bradbrook, A. J. (2011). Creating law for next generation energy technologies. *Geo. Wash. J. Energy & Envtl. L.*, *2*, 17.

Dincer, I. (1999). Environmental impacts of energy. *Energy Policy*, *27*(14), 845–854. doi:10.1016/S0301-4215(99)00068-3

Jura, C. (2012). Considerations on a New Branch of Public International Law–Energy International Law. *Procedia: Social and Behavioral Sciences*, *62*, 801–805. doi:10.1016/j.sbspro.2012.09.135

Khajehpour, H., Ahmady, M. A., Hosseini, S. A., Mashayekhi, A. N., & Maleki, A. (2017). Environmental policy-making for Persian Gulf oil pollution: A future study based on system dynamics modeling. *Energy Sources. Part B, Economics, Planning, and Policy*, *12*(1), 17–23. doi:10.1080/15567249.2015.1004382

Kotzé, L. J. (2014). Rethinking global environmental law and governance in the Anthropocene. *Journal of Energy & Natural Resources Law*, *32*(2), 121–156. doi:10.1080/02646811.2014.11435355

Lazarus, R. J. (2008). *The making of environmental law*. University of Chicago Press.

Nouzha, C. (2000). Réflexions sur la contribution de la Cour Internationale de Justice à la protection des ressources naturelles. *Revue juridique de l'Environnement, 25*(3), 391-420.

Outka, U. (2012). Environmental law and fossil fuels: Barriers to renewable energy. *Vand. L. Rev.*, *65*, 1679.

Tertrais, B. (2020). French Nuclear Deterrence Policy, Forces, And Future: A Handbook. Fondation pour la recherche stratégique. *Recherches & Documents*, *4*, 12.

Wildermuth, A. J. (2011). The Next Step: The Integration of Energy Law and Environmental Law. *Utah Envtl. L. Rev.*, *31*, 369.

ADDITIONAL READING

Dincer, I. (1999). Environmental impacts of energy. *Energy Policy*, *27*(14), 845–854. doi:10.1016/S0301-4215(99)00068-3

Jura, C. (2012). Considerations on a New Branch of Public International Law–Energy International Law. *Procedia: Social and Behavioral Sciences*, *62*, 801–805. doi:10.1016/j.sbspro.2012.09.135

Khajehpour, H., Ahmady, M. A., Hosseini, S. A., Mashayekhi, A. N., & Maleki, A. (2017). Environmental policy-making for Persian Gulf oil pollution: A future study based on system dynamics modeling. *Energy Sources. Part B, Economics, Planning, and Policy*, *12*(1), 17–23. doi:10.1080/15567249.2015.1004382

Kotzé, L. J. (2014). Rethinking global environmental law and governance in the Anthropocene. *Journal of Energy & Natural Resources Law*, *32*(2), 121–156. doi:10.1080/02646811.2014.11435355

Lazarus, R. J. (2008). *The making of environmental law*. University of Chicago Press.

Nouzha, C. (2000). Réflexions sur la contribution de la Cour Internationale de Justice à la protection des ressources naturelles. *Revue juridique de l'Environnement, 25*(3), 391-420.

Outka, U. (2012). Environmental law and fossil fuels: Barriers to renewable energy. *Vand. L. Rev.*, *65*, 1679.

Tertrais, B. (2020). French Nuclear Deterrence Policy, Forces, And Future: A Handbook. Fondation pour la recherche stratégique. *Recherches & Documents*, *4*, 12.

Wildermuth, A. J. (2011). The Next Step: The Integration of Energy Law and Environmental Law. *Utah Envtl. L. Rev.*, *31*, 369.

KEY TERMS AND DEFINITIONS

Equity: Defined by UNEP to include intergenerational equity - "the right of future generations to enjoy a fair level of the common patrimony" - and intragenerational equity - "the right of all people within the current generation to fair access to the current generation's entitlement to the Earth's natural resources" - environmental equity considers the present generation under an obligation to account for long-term impacts of activities and to act to sustain the global environment and resource base for future generations. Pollution control and resource management laws may be assessed against this principle.

Polluter pays principle: The polluter pays principle stands for the idea that "the environmental costs of economic activities, including the cost of preventing potential harm, should be internalized rather than imposed upon society at large." All issues related to responsibility for environmental remediation costs and compliance with pollution control regulations involve this principle.

Precautionary principle: One of the most commonly encountered and controversial principles of environmental law, the Rio Declaration formulated the precautionary principle: To protect the environment, the precautionary approach shall be widely applied by States according to their capabilities. Where there are threats of serious or irreversible damage, lack of complete scientific certainty shall not be used as a reason for postponing cost-effective measures to prevent environmental degradation. The principle may play a role in any debate over the need for environmental regulation.

Prevention: The concept of prevention can perhaps better be considered an overarching aim that gives rise to a multitude of legal mechanisms, including prior assessment of environmental harm, licensing or authorization that set out the conditions for operation and the consequences for violation of the conditions, as well as the adoption of strategies and policies. Emission limits and other product or process standards, the use of best available techniques, and similar techniques can all be seen as applications of the concept of prevention.

Public participation and transparency: identified as necessary conditions for "accountable governments,... industrial concerns," and organizations generally, public participation and transparency are presented by UNEP as requiring "effective protection of the human right to hold and express opinions and to seek, receive and impart ideas,... a right of access to appropriate, comprehensible and timely information held by governments and industrial concerns on economic and social policies regarding the sustainable use of natural resources and the protection of the environment, without imposing undue financial burdens upon the applicants and with adequate protection of privacy and business confidentiality," and "effective judicial and administrative proceedings." These principles are present in environmental impact assessment, laws requiring publication and access to relevant environmental data, and administrative procedures.

Transboundary responsibility: Defined in the international law context as an obligation to protect one's environment and prevent damage to neighboring environments, UNEP considers transboundary responsibility at the international level as a potential limitation on the sovereign state's rights. Laws that limit externalities imposed upon human health and the environment may be assessed against this principle.

Chapter 15
Environmental and Islamic Law and Jurisprudence

ABSTRACT

Although there is no independent discipline of Islamic environmental law, Islamic law is proclaimed as a source of the legal system in the constitutions of Muslim countries. Thus, we can find in Islamic law a theoretical and practical foundation for environmental law. All sources of Islamic law can be used for this purpose as they all have a potential ecological application. This chapter explores the sources of Islamic law in order to find avenues to Islamic environmental law. In this regard, it analyzes its potential sources and paradigms. Then according to results of these analysis general Islamic principles introduced with a comparative view over the globally accepted environmental law. These general principles can be used in those countries with Islamic jurisprudence as a reference to improve their environment conservation lawmaking.

INTRODUCTION

In the 1960s, the world witnessed a huge public awareness regarding environmental issues(Tortell, 2020). This awakening is due to the factors such as the publication of Rachel Carson's book entitled "Silent Spring" which is related to the use of Dichlorodiphenyltrichloroethane (DDT) Pesticide, and the International Convention on Trade in Endangered Species was established in 1963, the United Nations Environmental Conventions, the Nuclear Non-Proliferation Law in 1968, etc. In the 1970s, earth-loving and green peace groups were formed(Meyer, 2021). After three decades of the activity of these groups and other non-governmental organizations in North America and Europe, especially in the development of legislation and standardization for corporate social and environmental issues, it found a global concept for the first time in the mid-1970s, and in fact the first official world meeting on environmental protection was held in Stockholm in 1972 (Trivedi, 2019). In 1992, the Earth Summit on sustainable development was held in Rio de Janeiro, Brazil (by the United Nations) and eventually led to the issuance of the Rio Declaration by the leaders of Rio. This declaration was a treaty framework on climate change, a treaty framework on biodiversity, the establishment of the World Bank's environmental aid

DOI: 10.4018/978-1-6684-4158-9.ch015

fund, the equipping of global environmental facilities, and 21 programs for implementation (Abdillah et al. 2020, Miller et al. 2020).

It has been more than thirty years since discussions related to religion and the environment have received attention in scientific and specialized circles. Many environmentalists now believe that environmental policies should not only be comprehensive, but such policies need the support of religious and moral culture in their implementation, which are the basis of environmental protection (Abdillah et al. 2020). In such a situation, examining religion's view of the environment becomes doubly important. In order to understand the place of the environment in the system of religious teachings, it is possible to consider a fourfold division of all human relationships, the relationship between man and God, the relationship with oneself, the relationship between man and other humans, and the relationship between man and nature (Smith & Veldman, 2020). How humans interact with nature and natural resources, water, soil, air, forest, animals, etc., is defined in the fourth field of human relations, and the rulings of these relations are explained in the jurisprudential view of this field (Soleymani, 2019, Hancock, 2019).

Today, many countries have taken steps toward sustainable development. The requirement of any sustainable development is to meet the needs of the current generation, without harming the ability of future generations to meet their own needs. Therefore, in order to continue inclusive and sustainable development, it is necessary to make adjustments in macro and microeconomic and cultural policies, and in the field of legislation, solid legal principles with an executive guarantee of all conditions and education related to environmental protection should become universal and customary, as it has become popular among the general public, and education, which is the introduction of culture, should provide the general public with the basis for continuous compliance with environmental laws. Of course, the basic and strong factor that is very influential in culture building is the dos and don'ts of religion and the religious and jurisprudential orders of Islam, which is the most important factor of the executive guarantee of Muslims (Hancock, 2019). Regarding the recognition of the capacities that can be relied upon in Islamic jurisprudence for the protection of the environment, it is necessary to address three different areas of public responsibility, sovereignty, general and special rules and principles, and then explain the approach and principles codified in Islamic jurisprudence and law (Badran, 2021). One of the important considerations in the Islamic jurisprudential view of the environment is that although man, as the noblest of creations, and from the jurisprudential point of view, according to the purity of the environment, has permission to use the environment, but due to the limitations, the benefit of this right is limited. The conflict with the rights of others has ended (Jamil, 2021).

Problems caused by humans in nature cannot be solved only by technology. It is necessary to change human behavior and this change should be based on environmental ethics. In fact, the relationship between man and nature should be re-examined. Environmental ethics is a branch of applied ethics that expresses the moral responsibility and ethical drive to protect the environment. Environmental ethics is based on the idea that ethics should be expanded to include the relationship between humans and nature. Another characteristic of environmental ethics is the dynamic nature of ethics[1], which tries to adapt itself to the conditions of time and place. The common boundaries of ethics and jurisprudence can be found in such cases where both of them firstly try to change and correct behaviors and secondly, they try to align themselves with the rapid changes of time and circumstances, but the executive guarantee of having jurisprudential orders and deriving the law from Jurisprudence always defines jurisprudence in different situations and draws different expectations from it (Badran 2021).

Now, the main question of this chapter regarding the interpretation of jurisprudence rules is that "can jurisprudence rules define the requirements for preserving the environment?" According to the approach of this research to base jurisprudence rules related to environmental protection. As the first step, key terms and the unique features of jurisprudence rules for recognizing the requirements of environmental protection are explained(Yusuf, 2020). In the second step, there is a brief overview of jurisprudential sources in the field of environment (according to holy Quran, tradition, reason and consensus) and in the third step, the relevant jurisprudential evidence in the field of environmental protection are analyzed. The importance of this study is that the environmental protection from the perspective of jurisprudence rules have high enforcement and binding guarantee in Islamic law-based countries, which can be effective in the reform of bioethics (Hancock, 2019).

RULES OF JURISPRUDENCE

Ahl al-Bayt (AS) paid attention and care in inculcating and teaching the principles to their special companions and left many jurisprudential rules for their followers, which are often expressed in general terms (Al-Kulayni 2008). A number of these rules have been discussed in the books of jurisprudential rules, but with this definition, there are still many other rules in the hadiths and possibly the words of the jurists that have not been extracted. Knowing the sources of jurisprudence rules makes us able to extract and discover new rules and familiarizes us with the method of obtaining common jurisprudence rules (Ameli 1989). In order to know the sources of jurisprudence rules, it is necessary to pay attention to its nature and its difference from the jurisprudential issue(Haq et al., 2020). Basically, the jurisprudence rule has two elements, one of which is common to the jurisprudential issue and the other is different from it: firstly, in the jurisprudential rules we are dealing with an issue that needs a proper ruling, and secondly, this issue is comprehensive(Hancock, 2019). But in jurisprudence, there is a minor issue that seeks its own ruling (Ameli 1991). In terms of the first element, the sources of jurisprudence are not different from jurisprudential issues, and in general, the sources are the book, Sunnah, consensus and reason in Shia jurisprudence, and in addition to them, Qiyas, Istihsan, etc. in Sunni jurisprudence. But from the second point of view, which is one of the coordinates and characteristics of jurisprudence, the sources of jurisprudence are somewhat different from the jurisprudence problem, in other words, in addition to referring to the mentioned sources, we need a special mechanism and method to understand and discover the rule(Marandi, 2020). In the term, jurisprudential rules are the general orders of jurisprudence that are in various jurisprudence disciplines. A jurisprudential rule is a rule that includes a general shari'a rule, and from its application, minor shari'a rules are obtained, which are examples of the general rule (Ghaemi Khargh, 2022).

JURISPRUDENTIAL SOURCES

Holy Quran

With regard to several environmental concepts in the Holy Qur'an, which has been raised in relation to the environment of human life and has analyzed issues in each of them, nature and the natural environment can be considered as a heritage and trust of God for man so that by following certain rules, take advantage of it and pay attention to its preservation and development (Haq et al., 2020). From the minimum directive and moral dos and don'ts or obligatory rules of liking and disliking, the guarantee of collective rights and the benefit of all is manifested. By reflecting on the verses, one can obtain the mandatory and legal rules and mandatory rules regarding the destruction of nature and its illegal use and the obligation to preserve the environment. About 198 verses include environmental themes and 444 verses refer to the earth from which 90 verse is related to the environment protection. Some of these verses have a positive and commanding aspect, some point out the quality of formation and the importance of the earth in the livelihood of the people, and the contents of some verses consider the harming the environment as an strictly forbidden or unlawful act (Yusuf, 2020).

Tradition, a Review of Narrations

Several narrations from Infallible Imam[2] advised Muslims to protect and conserve environment or banned them from harming it. This is mainly because of this fact that the process of life in this world has tied the fate of humans together, and one cannot live a complete and healthy life without being indifferent to the disruption of the life cycle and the responsibilities of life in the environment. The jurisprudential necessity of caring for the environment can be carefully obtained in the narrations that describe the quality of interaction with nature (water, soil, air), animals, plants, trees, etc. (Siyavooshi et al., 2019). What can be obtained from the sum of the Islamic narrations is that these teachings are not only used in the field of a person's action and determination of his duty regarding the personal possession. But also narration teaching can be extended to the field of statutory law(Begum et al., 2021).

Logic, Reason, and Ration

Logic in islam is divided into theoretical and practical ration in terms of what it understands. The meaning of the rule of reason in the environmental issue is a part of the evidence of practical reason that has been taken into consideration in the later books of principles, and it is based on the theological rule of rational good and badness, as well as the rule of connection between the rule of reason and Sharia. From an intellectual point of view, any human action that disrupts the balance and equilibrium of the environment and causes disruption in the ecosystem and the destruction of human beings is wrong. On the other hand, human effort to preserve the environment and restore it, is right(Soleymani, 2019). The main pillar of this reason is the issue of preserving the system and the survival of the human species, which is considered one of the well-known cases, and the opinions of all intellectuals have been settled on it, and the Shariah, as the creator of reason, agrees with it(Hancock, 2019).

ISLAM AND PUBLIC RESPONSIBILITY TOWARD ENVIRONMENTAL PROTECTION

The Islamic system mentions observing and caring for each other in society as enjoining the good and forbidding the evil, and without a doubt, protecting and caring for the environment is part of social life, and for this reason, it is subject to these arguments (Javadi Amoli, 2020).

About quranic verses of Ale-Imran/104 and Araf/199 it can be said(Siyavooshi et al., 2019):

The word (custom) means beautiful traditions and customs that are known by the intellectuals of the society. On the other hand, (customary) (known) are rare and unusual behaviors that are not understood by social reason, which is called (denial) in the term. In general, what is known as good and acceptable in the eyes of reason and Sharia, are called (customs) in the term of the Qur'an, which a person should continuously adopt in the society and invite others to it (Maghrebi, 2020).

The Qur'anic order (and other known commands) and other verses and narraions in this context emphasize that it is not enough to do good and righteous deeds, but one should invite others in this way so that righteous deeds spread in the society, become common and in spread everywhere(Soleymani, 2019).

If a person is not equipped with the power of faith and science, and on the other hand, the supervision of other responsible people over him becomes weak and lax, that person easily deviates towards corruption, which can lead to a crisis in the environment and human life. He faces many problems. Therefore, inviting people to environmental normative behaviors in the form of (enjoining good and forbidding evil) has an environmental nature, which in fact guarantees the implementation of individual and social duties towards the environment and is considered as their soul (Maghrebi, 2020).

According to divine teachings, standing against corruption is considered the secret of the survival of religion and humanity. The Holy Quran considers the destruction of the past nations to be the result of the wise not dealing with the corruption of the society: "Why were there not among the generations before you a remnant [of the wise] who might forbid corruption in the earth, except a few of those whom We delivered from among them? Those who were wrongdoers pursued that in which they had been granted affluence, and they were a guilty lot(Javadi Amoli, 2020).

It should be said that what is meant is evolutionary corruption and not the thing that gets corrupted in the course of time; Because humans do not have a hand in some of the corruptions, rather, it means legislative corruptions, that is, corruptions that occur by the hands of humans. One of the most tangible corruptions in the blessings of God occurs when humans seize nature to fulfill their desires; A destructive occupation without considering the rules of nature. Of course, the fight against corruption requires appropriate tools (Majlisi Esfahani, 2020).

Therefore, it is necessary that the issue of enjoining good and forbidding evil, which is a social responsibility and is emphasized by Islam, should be taken seriously in relation to environmental issues in order to prevent corruption and deviation and to promote noble values and social health. It should be considered one of the most important duties and examples of enjoining good and forbidding evil(Begum et al., 2021). Therefore, it is necessary to analyze the issue and the jurists should understand the scope of each person's privacy. The 50th article of the Iranian constitution is the most explicit ruling of the constitutional legislator in this field (Makarem Shirazi, 2020):

In the Iranian Legal system which is based on Islamic Fiqh, protecting the environment in which today's generation and future generations should have a growing social life is considered a public duty. Therefore, economic and other activities that are associated with environmental pollution or irreparable destruction are prohibited(Javadi Amoli, 2020).

ISLAM AND GOVERNMENTAL RESPONSIBILITY BEFORE ENVIRONMENT

Undoubtedly, the Supreme Creator conquered nature for man so that he can use it and profit from it. Verses from the Holy Quran clearly emphasize this point: Certainly god has honoured the Children of Adam, and carried them over land and sea, and provided them with all the good things, and preferred them with a complete preference over many of those god has created(Javadi Amoli, 2020). Since the earth and what is in it constitutes the human environment and is a part of the public blessings and damage to it endangers the entire society, the ruler must protect the environment by establishing the necessary laws and guaranteeing its implementation. Take the initiative as this method exists in the government history of the Prophet of Islam and Imam Ali and it seems that this covenant was between God and the rulers to prevent corruption on earth (Makarem Shirazi, 2020).

The main goal of the Islamic government in the environment sector is to ensure stability and fair and efficient use of natural and human resources. For this purpose, the Islamic government undertakes different measures in the four main areas of planning, legislation, supervision and public ownership of resources. Based on these four areas, the most important duties of the Islamic government regarding the environment are implementation of environmental laws based on divine principles and regulations; implementation of laws to fight against land corrupters and nature destroyers to provide public benefits; attention to environmental issues and environmental ethics in Islamic education and environmental culture based on heavenly teachings; utilization of competent people in the environment administration pyramid who have the full power to deal with corruptors; taking advantage of new environmental practices compatible with the Islamic system; taking advantage of macroeconomic and cultural policies for sustainable development combined with the protection of natural resources and the environment; drawing and presenting strategies for transitioning from the existing state of the environment to its desired state; efforts to empower researchers and produce research information and develop network management in order to improve the quality and quantity of human, industrial and medical sciences and develop the knowledge of new technologies centered on the environment(Maraghi, 2020).

ENVIRONMENT-BASED JURISPRUDENTIAL PRINCIPLES

The Quran, narraions, logic and consensus are the four main sources of jurisprudence and they have very high capacities for establishing environmental protection regulations. In general, in this direction, you can seek help from jurisprudence in two parts, general and special. General provisions are as follows: Many verses and narrations in general, positively and negatively, express the rules of the environment in the form of obligation, respect, love and dislike, in the form of obligatory and situational:

Rule of Sanctity of Corruption in the Land

The Holy Qur'an has repeatedly prohibited corruption in the land, including in Al-Aaraf/85 and Al-Bagharah/205. These verses are used to mean that corruption is not developmental corruption; Because no one is involved in some of the corruptions. The world of corruption and the ecstasy of conflict in survival is not meant, but this corruption means legislative corruptions; Corruption that is created by human hands (Mashhadi, 2020). In the same way (and destroy crops and cattle) is also in the position of expressing one of the examples of corruption in the earth, which is the destruction of the type of plants and the generation of beings in the world, and since it is absolute, it can be said to include all generations of beings, animals and plants. and forests and fields and does not refer to human beings. Therefore, it can be said that destroying any kind of plowing and generation, whether animal, plant, or vegetable, is one of the examples of this verse, and it is considered corruption, and it can be issued a ruling of sanctity in jurisprudence. By referring to the word "Allah does not love corruption" such corruption is abhorrent to God, it makes it easy to express such a ruling and it can be concluded that corruption is absolutely forbidden on earth (Mesbah Yazdi, 2020).

Rule of Need to Maintain Balance in Nature

This fact is mentioned in some Quranic verses of Al-Talaq/3 and Al-Hajar/19-21. According to these verses, God's will be to maintain balance and harmony in the system of creation, and for this reason, God has asked man to maintain balance and not disturb it with his actions (Mianji, 2020). Since the will of god is the harmony with the nature, it can be inferred that any behavior that disturbs harmony of the nature leads to a religiously forbidden act.

Rule of Harmlessness

Although a lot has been said about the rule (harmlessness) and various possibilities have been proposed, each of them expands the main meaning of the rule, which indicates the illegitimacy of harm in Islam. This rule, which is used in the issue of protecting the environment and preventing its destruction, is considered a solid jurisprudential basis for legislation and law enforcement. And therefore, any occupation and change in human society and nature that causes the violation of the rights of others is prohibited and rejected in Islam. Many environmental issues can be examined in the form of the harmlessness rule (Mostafapour, 2019).

Rule of Justice

Justice is the great honor of God and it was the circuit of the creation of the system of nature and that of the legal system. It is stated in the Quranic verses of Al-Rahman/7-9 and a narration by prophet Muhammad (peace be upon him) which states that the stability of the heavens and the earth is due to justice.

Justice in the system of existence means that everything is placed in its proper place, and justice is that every creature's right is paid as it deserves; It is considered as the main pillar of existence. Therefore, in order to preserve this pillar, God Almighty orders man to establish justice and has sent his prophets to men for this purpose. Justice requires that human occupations in the surrounding environment are not cruel occupations and take into account the rights of future nations. So as to consider the rights of other people around the world (Naraqi, 2021).

The first point: according to the verses and traditions, it is obligatory for every person to observe the principle of justice and fairness in his behavior and character, and to reach the heights of justice, fairness takes the side of fairness in benefiting from the God-given nature, and the same with the environment around him. And God's blessings will meet him as he likes to be treated like that (Noori et al., 2017).

Second point: Just as the rules of Islam are universal and eternal, they also include human beings who have not yet entered the field of existence. Possession of natural gifts is not exclusive to one generation, that is, in addition to intragenerational justice, intergenerational justice also has its place. Based on this, any use of natural resources that leads to their destruction or pollutes the environment and makes life difficult for other people is far from justice and against fairness and is prohibited and guaranteed in terms of Islamic logic (Sadr, 2021).

Rule of More Wealth, More Greed

According to the narrations related to the mortgage, the rule of more wealth, more greed is used. Also, the famous prophetic narration abscess by guarantee indicates it. According to this rule, whoever benefits from something must compensate for its losses (Saghafi Tehrani, 2019). According to this principle, every person who benefits from an activity must also accept the damages caused by it. Therefore, since the users of the environment, including the builders of facilities and factories, who in some way produce products by benefiting from natural resources, are subject to this ruling of Shariah guarantee, they are obliged to compensate for the defects and to compensate, it is necessary for the Islamic government to establish laws for these guarantees and to deal with government instruments. The jurists should also state the guarantee of such people by identifying the subject of the verdict (Shoushtari, 2021).

Rule of Addition of Rights and Duties

The meaning of the right here is the right against the obligation; Something that is for the benefit of the individual and for the benefit of others, and duty is something that is for the benefit of the individual and for the benefit of others. In other words, the right belongs to the person who gives the right, the right to claim, and the obligation belongs to the person who is obligated (Sobhani, 2022). About the issue of rights and obligations, Islamic scholars have discussed many topics and dedicated independent and valuable works to it. According to this principle, rights and obligations are two sides of the same coin, in such a way that wherever the jurisprudence defines the law for someone, it also explains the duties and obligations. In other words, a right cannot be assumed without an obligation, and the existence of an obligation cannot be assumed without the existence of a right.

According to the principle that God created nature and the environment for humans and gave humans the right to possess and use it, this right exists for all humans and at all times. Almighty God has clarified this Muslim principle in the Holy Quran verses of Al-Bagharah/29, Al-Rahman/10, Al-Hood/61 and Al-Aaraf/85. Therefore, even though humans have the right to use the environment, they should also respect this right for others, lest by abusing their right, they violate the right of others to benefit from a healthy environment (Tamimi Amadi, 2021).

Rule of Causing

Causing rule is one of the effective jurisprudential rules regarding guarantee. According to this rule, if a person causes damage to another through an intermediary, he will be the guarantor if the damage is attributed to him (Tayyib, 2018). The documentation of this rule is narration and consensus that the relevant hadiths play the biggest role in proving the rule. According to this theory, anyone who causes damage or loss of another's property must be responsible for the damage, so in this theory, proving the relationship of causation is of particular importance, which from the perspective of religious teachings, any cause that is due to the imbalance of the environment and if its destruction occurs, according to this rule, the causer is responsible. But according to some researchers, causation is not a suitable rule for full, quick and appropriate compensation of environmental damages, because in most cases related to environmental damages, it is very difficult to document the relationship of causation. Islamic jurists have also criticized this rule.

In spite of the fact that the theory of wastage is more compatible with the objectives of environmental protection and can provide the maximum of this situation, but it is necessary to mention a rival theory in Islamic jurisprudence, which refers to the famous rule of "Causing". Here, similar to the theory of risk or based on fault, the evidence of the act is important to the cause of damage, which is very critical from the point of view of the teachings of environmental law (Tayyib, 2018)

This theory, which is also accepted in the current laws of many countries, is not a suitable theory for full, quick and proper compensation of environmental damages. Because in most of the cases related to environmental damages, the documentation of causality relationship is facing many difficulties. Islamic jurists have also criticized this theory. The owner of the jewel does not consider such a rule to be derived from scattered narrations and believes that the cause of civil liability is waste, not attribution, and in the absence of waste, the criteria and rules are numerous traditions that civil liability can only be eliminated by abolishing the character. Similar cases spread. Some jurists, while acknowledging the origin of the loss for the guarantee, have only made the "customary realization of the loss" the basis of responsibility. Of course, some jurists have also considered the Causing rule as an independent rule (Mesbah Yazdi, 2020).

Rule of System Disorder

Based on this rule, no Shari'a rule or principal rule that will disrupt the livelihood system and human life has been forged in Islam. What they mean by the life system is the human livelihood system, the livelihood system that organizes human life, in turn, has smaller systems that are inside it, such as the political system, the economic system, the social and cultural system, and cases of such as this. Undoubtedly, man can continue his life if the system of nature and the natural cycle work properly, and whenever a part of it is disturbed and the natural cycle and the biological system governing nature suffer problems or its movement slows down, the human will not be able to maintain his life system and without a doubt, a huge disruption and challenge will dominate his life system (Makarem Shirazi, 2020). If human actions, whether individual or collective, damage and destroy the environment, then this excellent and unique system has been disrupted. The name of this rule can be called the rule of maintaining balance in nature, and the reasons for it can be considered to be reason and tradition, including the Quran and the Sunnah, and it can be considered a part of the system disorder rule. In both cases, since it is a new issue, the jurists have not addressed it. The difference between the two is that in the rule of necessity to maintain the balance in nature, it is forbidden to destroy nature and change it, even if it does not reach the stage of massive disruption and destruction (Makarem Shirazi, 2020).

Rule of Loss

The rule of wasting other people's property is one of the most important rules used by Islamic jurists regarding guarantee (Mashhadi, 2020). From all relevant sources, jurists have taken a rule and interpreted it as whoever wastes other people's property will be a guarantor and must pay for the damage caused to them (Mesbah Yazdi, 2020).

According to the rule of waste, anyone who wastes another's property is a guarantor. The rule of waste can play an appropriate and prominent role in the cases of applying the right to the environment and compensating for the damages caused to it. Based on this, whoever harms the environment is a guarantor to its restoration. Some jurists have considered adherence to the rule of waste as a rational and self-evident rule. In examining the rule of waste, the issue of ownership is also raised. In this case, it can be said that due to the recognition and legality of three types of ownership (personal ownership, ownership of the Islamic government and national ownership) in jurisprudence, the ownership of the environment in some cases such as Anfal (seas, beaches, public waters, forests, pastures, air and space, etc.) are under the state ownership of the Islamic government, and some others, such as lands, are defined under national ownership, which belong to all people and cannot be transferred to private sector.

Therefore, it can be said that any kind of pollution and harm to the environment is a cause to invoke this rule, because the environment belongs to the people or is under the management of the Islamic government(Javadi Amoli, 2020). Thus, the well-known principle "the polluter must pay the damage" or the principle "the polluter pays the damage" can be justified based on this rule.

Rule of People's Right

In traditional Islamic culture, two types of rights are known: the rights of Allah and the rights of people. What is meant by the rights of Allah is to do the duties and leave the sins that are common in one main axis, which is the relationship with God. As an example, prayer and fasting are considered great duties and rights of God, and if a person does not perform these duties in a proper and appropriate manner, he deserves divine punishment, because he has committed a sin. In contrast to this right is the right of people. The scope of these rights includes economic, social, political and cultural issues(Javadi Amoli, 2020). Knowing the rights of others is a necessary thing in the system of religious education. Undoubtedly, one of the indeniable rights of human beings is to enjoy a healthy environment, the violation of which causes the creation of human rights and causes the violation of the rights of other species. With this basis, regulating the relationship between man and the environment becomes a double necessity (Ansari, 2021).

Special provisions are as follow: There are many capacities in Islamic verses and traditions about special regulations, in such a way that they can be used in various fields of the environment, including water, air, soil, the growth and development of animals and plants, and the stability of generations. Some of them are mentioned below:

Rule of the Environment Pollution

Islam holds humans responsible for their environment. It is mentioned in several narraions by Imams and prophet. Therefore, it is forbidden to destroy and harm the environment of people and different countries, including land, buildings, and cattle, and such prohibitions are meant in sanctity, unless there is a reason for their disgust (Ansari, 2021). In Islam, polluting the environment is forbidden. Defilements are among the polluting factors that believers should avoid. Narrations forbid defilement of the living environment; Places such as around mosques, by rivers, orchards and under fruit trees, in front of and around people's houses, crossings and running into standing water (Amid Zanjani, 2019).

Rule of Extravagance and Wastage

In a general sense, it is said to be any kind of excess (extravagance), but this word is often used in relation to consumption and expenses. Tabir also means to scatter. In the Quranic verse of Al-Ana'am/14 Almighty God has repeatedly warned against extravagance. Also a narration of Imams proves this rule(Javadi Amoli, 2020).

Rule of Self Harm

The Holy Quran forbids weakening and harming the body in verse Al-Bagharah/195. One of the most important orders of Islam is to preserve and protect the health of the body, for this purpose, in various fields, from eating, sleeping, and dressing to work and activities, it has brought customs and programs for human beings, whose observance will lead to growth, health, physical height and length. Human life and benefiting more from God's blessings have a vital effect and all of this depends on a healthy environment free from any pollution (Amid Zanjani, 2019).

Rule of Protecting Animals

The awakening of animals on the Day of Resurrection is one of the issues mentioned in the Holy Quran verse Al-Ana'am/38 and also narrations of prophet. According to Islam, animals have rights, especially animals that humans have direct contact with. The mentioned cases are only about the carrier animals. Regarding other wild and domestic animals, many cases of animal rights have been mentioned in the hadiths of the Prophet (Dashti, 2021).

Harassment of divine creatures is strictly prohibited in Islam, to the extent that afterlife punishments are considered for harming animals, and humans are encouraged to be tolerant of animals' disobedience and not to harm them. Also, they are patient against the harassment that comes from animals, and they are promised a reward. Some narrations from the Imams regarding the rewards that God gives to his servants for keeping and protecting animals and forgiving sins from Him. Islam does not allow the hunting of animals except in special cases. Hunting animals during Hajj and in the area of Mecca's shrine is considered forbidden, and expiation has also been determined for it(Amid Zanjani, 2019).

Rule of Protecting Plants

According to Islamic traditions, watering trees is equal to watering a thirsty person. Another important issue that endangers the environment the most today is cutting down trees and destroying pastures and forests. According to Islam, this act is very ugly and causes divine punishment. Interestingly, it can be seen in the life and speech of religious elders that in addition to fighting against cutting down trees, if necessary, they considered themselves to be committed to planting another tree in its place(Javadi Amoli, 2020).

CONCLUSION

Considering the current conditions of the environment and its increasing destruction, it can be said that the environmental crisis, like other crises arising from materialistic culture, has philosophical and moral grounds and supports. Currently, due to the continuation of environmental crises, the world's view of religion has changed and it has been considered as one of the effective factors in reducing environmental problems. In such a situation, examining religion's view of the environment becomes doubly important. In order to understand the place of the environment in the system of religious teachings, it is possible to consider a fourfold division of all human relationships, the relationship between man and God, the relationship between man and himself, the relationship between man and other humans, and the relationship between man and nature. How humans interact with nature and natural resources, water, soil, air, forest, animals, etc., is defined in the fourth field of human relations, and the rulings of these relations are explained and explained in the jurisprudential view of this field.

According to the main question of this article, "Can jurisprudence rules define the requirements for environmental protection?", it can be concluded from the presented material that in the jurisprudential examination of environmental protection, various evidences from Quran, Sunnah and Reason is important, but in the meantime, due to the special characteristics of the jurisprudence rule in being independent and referring to the person obliged to it, and that it can determine the verdict of minor issues, and on the other hand, it is inclusive and comprehensive, the examination of the jurisprudence rules in the field of environment has this result and practical result for the oblige, the result of which is the proof of guarantee (according to some jurisprudence rules) as a result of any act of destruction in the environment.

The rule of harm, the rule of attrition, the rule of Causing, the rule of greed, the rule of right and duty, the rule of justice, the rule of disruption of the system, and the rule of the right of the people, are eight rules that have been examined in this article and each of them from the angle they require responsibilities in human interaction with nature.

Harmless is a rule that can be relied upon to compensate damage caused by encroachment on environmental assets by accepting its affirmative role in the Shari'a rulings that some jurists have acknowledged. The key point in examining the harmless rule is its rule over all sentences and verbs. For this reason, even in one's personal use and within the limits of one's private property, one is not absolute and is required to respect one's environmental rights. In the field of environmental law, the application of the harmless rule leads to the prohibition of any act or omission, as well as any law that has harmful effects on the environment. According to the rule of waste, whoever destroys the environment is a guarantor against it.

According to the Causing rule, anyone who causes damage or loss of another's property must be responsible for the damage, therefore, in this rule, proving the relationship of causation is of particular importance, which from the perspective of religious teachings, any cause that leads to the imbalance of the environment and if its destruction occurs, the cause is responsible. According to the rule of Greed, those who exploit the environment in various ways are subject to this ruling of Sharia guarantee and are obliged to compensate for the damages.

Taking advantage of nature also includes a wide range of actions. Things like construction of manufacturing industries, factories, dams, construction in forests and pastures, exploitation of river boundaries, cutting trees, etc. The rule of strengthening the right and duty, the rule of justice, the rule of disruption of the system and the rule of human rights, each of them outline legal and moral duties for humans and define the frameworks that can be used to preserve the environment under strict implementation.

At the end, it is necessary to mention that one of the important considerations in jurisprudence studies is the resolution of the conflict between the narrative evidences, especially in the hadiths, and the examination and summation between them. Allocation and specialization etc. have formed. But in this article, considering that the basis of the research was in the field of jurisprudence, these rules support each other in the requirements of environmental protection, and according to what was concluded in the research process, jurisprudence rules are not only in conflict with each other in explaining the requirements of environmental protection. They did not have, but they supported and strengthened each other. Regarding the overlapping of these rules with each other, it is necessary to mention that increasing the number of evidences that support a jurisprudential conclusion does not mean that those evidences overlap with each other, but rather helps to strengthen the result.

REFERENCES

Abdillah, A., Mastuti, A. G., Rijal, M., & Rahman, M. A. (2020). Students' intuitive and analytical thinking in the mathematics study through the integration of STAD and environmental islamic jurisprudence (fiqh). Al-Jabar. *Jurnal Pendidikan Matematika*, *11*(1), 49–60.

Amid Zanjani, A. A. (2019). *Political Jurisprudence*. Amir Kabir Publication.

Ansari, M. (2021). *Faraed Al-Usul* (1st ed.). Majmaolfekr Alislamii Publication.

Begum, A., Jingwei, L., Haider, M., Ajmal, M. M., Khan, S., & Han, H. (2021). Impact of environmental moral education on Pro-environmental behaviour: Do psychological empowerment and Islamic religiosity matter? *International Journal of Environmental Research and Public Health*, *18*(4), 1604. doi:10.3390/ijerph18041604 PMID:33567647

Dashti, M. (2021). *Nahj Al-Balaghah*. Nahj Al-Balaghah Institute.

Ghaemi Khargh, M. (2022). Lack of Governance in Political Jurisprudence; Challenges & Requirements. *Public Law Knowledge Quarterly*, *11*(35), 1–24.

Hancock, R. (2019). Ecology in Islam. In Oxford Research Encyclopedia of Religion. Oxford. doi:10.1093/acrefore/9780199340378.013.510

Haq, Z. A., Imran, M., Ahmad, S., & Farooq, U. (2020). Environment, Islam, and women: A study of eco-feminist environmental activism in Pakistan. *Journal of Outdoor and Environmental Education*, *23*(3), 275–291. doi:10.100742322-020-00065-4

Jamil, S. (2021). Halal wastewater recycling: environmental solution or religious complication? In *Religious Environmental Activism* (pp. 93–111). Routledge.

Javadi Amoli, A. (2020). *Islam and the Environment*. Esra Publishing Center.

Maghrebi, A. H. N. M. T. (2020). *Daim Al-Islam* (2nd ed.). Al-Bayt Institute.

Majlisi Esfahani, M. B. (2020). *Merat Aloqul fi Sharah Akhbar Aal Alrasu* (2nd ed.). Dar Al-Kitab Al-Islamiya.

Makarem Shirazi, N. (2020). The Rules of Jurisprudence. Third Edition. Qom: Published by the School of Imam Ali ibn Abi Talib.

Maraghi, A. (2020). *Jurisprudence Titles* (1st ed.). Islamic Publications Office.

Marandi, S. M. R. (2020). Being a Jurist as a Condition for the Islamic Ruler in the Sunni Political Jurisprudence. *Journal of Contemporary Research on Islamic Revolution*, 21-38.

Mashhadi, A. (2020). *The Right to a Healthy Environment (Iranian-French Model)* (1st ed.). Mizan Publication.

Mesbah Yazdi, M. T. (2020). *Teaching Philosophy*. Benalmelal Publication.

Meyer, C. A. (2021). Taking Lessons from Silent Spring: Using Environmental Literature for Climate Change. *Literature*, *1*(1), 2–13. doi:10.3390/literature1010002

Mianji, A. A. (2020). *Captive in Islam* (1st ed.). Islamic Publications Office.

Miller, A. C., Khan, A. M., Hebishi, K., Bigalli, A. A. C., & Vahedian-Azimi, A. (2020). Ethical issues confronting Muslim patients in perioperative and critical care environments: A survey of Islamic jurisprudence. *Anesthesiology Clinics*, *38*(2), 379–401. doi:10.1016/j.anclin.2020.01.002 PMID:32336391

Mostafapour, M. R. (2019). The Intrinsic Value of Nature from the Viewpoint of Sadr al-Mutaallehin Shirazi. *Esra Wisdom*, *12*, 113–130.

Naraqi, A. M. (2021). Avead Alayam fi Bayan Qavaeid Alahkam. First Edition. Qom: Publications of the Office of Islamic teachings.

Noori, V., Zamani, S. Q., Raee, M., & Abdollahi, M. (2017). The Principle of Proportion in Armed Conflicts in the Light of Documents and Procedures of International Criminal Courts. *Journal of Criminal Law Teachings, 12,* 91-.611

Sadr, M. B. (2021). *Our Economy* (1st ed.). Center for Research and Specialized Studies of the Martyr Imam Al-Sadr.

Saghafi Tehrani, M. (2019). *Javid Fluent Commentary* (3rd ed.). Borhan Publication.

Shoushtari, M. H. M. (2021). *New Perspectives on Law*. Mizan Publication Center.

Siyavooshi, M., Foroozanfar, A., & Sharifi, Y. (2019). Effect of Islamic values on green purchasing behavior. *Journal of Islamic Marketing*, *10*(1), 125–137. doi:10.1108/JIMA-05-2017-0063

Smith, A. E., & Veldman, R. G. (2020). Evangelical Environmentalists? Evidence from Brazil. *Journal for the Scientific Study of Religion*, *59*(2), 341–359. doi:10.1111/jssr.12656

Sobhani, J. (2022). *Al-Wasit* (1st ed.). Imam Sadegh Publication.

Soleymani, I. (2019). Right to Environment in the Islamic sources. *Anthropogenic Pollution*, *3*(1), 33–38.

Tamimi Amadi, A. W. (2021). *The Ghurar Al-Hikam wa Durar Al-Kalim* (1st ed.). Islamic Propaganda Office.

Tayyib, S. A. H. (2018). *Atiab Al-Bayan fi Tafsir Al-Quran* (2nd ed.). Islamic Publication.

Tortell, P. D. (2020). Earth 2020: Science, society, and sustainability in the Anthropocene. *Proceedings of the National Academy of Sciences of the United States of America*, *117*(16), 8683–8691. doi:10.1073/pnas.2001919117 PMID:32312801

Trivedi, D. (2019). Glimpses of Green Consumerism and Steps Towards Sustainability. [JOM]. *Journal of Management*, *6*(3), 35–41. doi:10.34218/JOM.6.3.2019.005

Yusuf, M. (2020). Developing environmental awareness and actualizing complete piety based on quran. *International Journal of Advanced Science and Technology*, *29*(5), 2039–2050.

ADDITIONAL READING

Abdillah, A., Mastuti, A. G., Rijal, M., & Rahman, M. A. (2020). Students' intuitive and analytical thinking in the mathematics study through the integration of STAD and environmental islamic jurisprudence (fiqh). Al-Jabar. *Jurnal Pendidikan Matematika*, *11*(1), 49–60.

Amid Zanjani, A. A. (2019). *Political Jurisprudence*. Amir Kabir Publication.

Ansari, M. (2021). *Faraed Al-Usul* (1st ed.). Majmaolfekr Alislamii Publication.

Begum, A., Jingwei, L., Haider, M., Ajmal, M. M., Khan, S., & Han, H. (2021). Impact of environmental moral education on Pro-environmental behaviour: Do psychological empowerment and Islamic religiosity matter? *International Journal of Environmental Research and Public Health*, *18*(4), 1604. doi:10.3390/ijerph18041604 PMID:33567647

Dashti, M. (2021). *Nahj Al-Balaghah*. Nahj Al-Balaghah Institute.

Ghaemi Khargh, M. (2022). Lack of Governance in Political Jurisprudence; Challenges & Requirements. *Public Law Knowledge Quarterly*, *11*(35), 1–24.

Hancock, R. (2019). Ecology in Islam. In Oxford Research Encyclopedia of Religion. Oxford. doi:10.1093/acrefore/9780199340378.013.510

Haq, Z. A., Imran, M., Ahmad, S., & Farooq, U. (2020). Environment, Islam, and women: A study of eco-feminist environmental activism in Pakistan. *Journal of Outdoor and Environmental Education*, *23*(3), 275–291. doi:10.100742322-020-00065-4

Jamil, S. (2021). Halal wastewater recycling: environmental solution or religious complication? In *Religious Environmental Activism* (pp. 93–111). Routledge.

KEY TERMS AND DEFINITIONS

Equity: Defined by UNEP to include intergenerational equity - "the right of future generations to enjoy a fair level of the common patrimony" - and intragenerational equity - "the right of all people within the current generation to fair access to the current generation's entitlement to the Earth's natural resources" - environmental equity considers the present generation under an obligation to account for long-term impacts of activities and to act to sustain the global environment and resource base for future generations. Pollution control and resource management laws may be assessed against this principle.

Polluter pays principle: The polluter pays principle stands for the idea that "the environmental costs of economic activities, including the cost of preventing potential harm, should be internalized rather than imposed upon society at large." All issues related to responsibility for environmental remediation costs and compliance with pollution control regulations involve this principle.

Precautionary principle: One of the most commonly encountered and controversial principles of environmental law, the Rio Declaration formulated the precautionary principle: To protect the environment, the precautionary approach shall be widely applied by States according to their capabilities. Where there are threats of serious or irreversible damage, lack of complete scientific certainty shall not be used as a reason for postponing cost-effective measures to prevent environmental degradation. The principle may play a role in any debate over the need for environmental regulation.

Prevention: The concept of prevention can perhaps better be considered an overarching aim that gives rise to a multitude of legal mechanisms, including prior assessment of environmental harm, licensing or authorization that set out the conditions for operation and the consequences for violation of the conditions, as well as the adoption of strategies and policies. Emission limits and other product or process standards, the use of best available techniques, and similar techniques can all be seen as applications of the concept of prevention.

Public participation and transparency: identified as necessary conditions for "accountable governments,... industrial concerns," and organizations generally, public participation and transparency are presented by UNEP as requiring "effective protection of the human right to hold and express opinions and to seek, receive and impart ideas,... a right of access to appropriate, comprehensible and timely information held by governments and industrial concerns on economic and social policies regarding the sustainable use of natural resources and the protection of the environment, without imposing undue financial burdens upon the applicants and with adequate protection of privacy and business confidentiality," and "effective judicial and administrative proceedings." These principles are present in environmental impact assessment, laws requiring publication and access to relevant environmental data, and administrative procedures.

Transboundary responsibility: Defined in the international law context as an obligation to protect one's environment and prevent damage to neighboring environments, UNEP considers transboundary responsibility at the international level as a potential limitation on the sovereign state's rights. Laws that limit externalities imposed upon human health and the environment may be assessed against this principle.

ENDNOTES

[1]   Therefore, ethics is the dynamic, evolving activity of applying, balancing, and modifying principles in light of new facts, new technology, new social attitudes and changing economic and political conditions.

[2]   Imams are possessed of supernatural knowledge, authority, and infallibility ('Iṣmah) as well as being part of the Ahl al-Bayt(we hail them), the family of Muḥammad (peace be upon him).

Conclusion

The development and codification of international law are one of the main tasks of the United Nations Commission on International Law, which is specified in paragraph 1 of Article 13 of the Charter of the United Nations with the same title. In addition, the United Nations Environment Program has played an important role in developing and drafting international environmental law. Since its inception, it has proposed several binding and non-binding instruments to governments for ratification. In addition, governments and other international organizations, both governmental and non-governmental, have played important roles in developing and developing international environmental law in recent years. Environmental protection, at least in its modern sense, which began in the 1970s and continues to this day, encompassed various areas such as air, soil, forest, and water and ratified various conventions. The context reflects the importance of contemporary international law in this regard. Undoubtedly, environmental protection is one of the concerns and concerns of today's human societies like explosive population growth, unreasonable exploitation of natural resources, degradation and degradation of biodiversity, increasing pollution that has affected the air, soil, and water of the world in various ways, and finally degrading the natural quality of human life. As a result, the disturbance of the balance and appropriateness of the environment has led governments, organizations, and international forums to formulate and implement laws and regulations to prevent pollution and environmental degradation. The development of binding environmental principles and rules has gradually led to the development of environmental law, both nationally and internationally. In the form of non-governmental organizations, global public opinion calls on their governments to take the necessary measures to protect the region and the environment. It has been accepted as a valid criterion by the competent authority in society. The obligatory nature of the law and the guarantee of its implementation provide the enforcement of legal rules in a way that prevents harmful behavior to the environment. Thus, efforts to regulate environmental protection were made in the 1970s, especially after environmentalists warned of their widespread destruction, especially by industrialized nations.

Because international environmental law for the global protection of the environment has been developed by the international community through the enactment and implementation of binding and non-binding legal rules and has developed in terms of content, form, and structure in recent decades, however, despite international efforts to protect the environment, environmental challenges persist and have increased significantly in many areas. Today's environmental concerns and threats go beyond the predictions of experts and scientists present at the first Stockholm International Human and Environment Conference in 1972. Ozone depletion, climate change, air pollution, soil and water, biodiversity loss, deforestation, and desertification are the most important problems facing human beings today. Thus, despite global efforts to enact and enforce international environmental law, the lack of reduction in environmental degradation and the increase in environmental pollution indicate the inefficiency and inefficiency of this legal field. This article uses a critical approach to address the obstacles and shortcomings of international environmental law while showing the existing development capacities and proposing legal and executive gaps to address them. There are three important challenges to the substantive development of international environmental law:

- Reluctance of governments to delegate national sovereignty

Regardless of the concept of sovereignty in public international law and international environmental law, one of the main obstacles to developing international environmental law is the unwillingness of governments to delegate or limit sovereignty in favor of international environmental organizations. The political structure of government always tends to focus and has no interest in delegating it to other centers of power and decision-making. On the other hand, the conflict of interests between the main actors of international law (states) regarding environmental protection, which arises from the national sovereignty of states, challenges the development and expansion of international law. This conflict of interest can include political, economic, commercial, etc. In addition, the conflict of interest of developing countries and developed countries in applying and enforcing the rules and regulations of international environmental law is also one of the limitations of this field of law. Despite the principle of shared but distinct responsibility, in many sources of international environmental law, conflict is seen between these two groups of developed and developing states. The International Court of Justice's advisory opinion on the legitimacy of the threat and use of nuclear weapons in 1996 is a clear example of the pressure exerted on this international body by countries with nuclear weapons. However, one of the most important principles enshrined in the Stockholm Declaration of 1972 so far in international environmental declarations is the principle of sovereignty, which redefines the sovereignty of states and a new definition of the concept of sovereignty in international environmental law based on the concept of "use Fair and reasonable "of the land, but the shadow of absolute sovereignty still weighs heavily on international environmental law. But the main problem in this regard is how governments participate in exercising this type of governance, different positions of countries towards the environmental prospects of the world, including the northern countries, which can influence and direct international budgets and accelerate the process. Have been a source of concern, for example, the refusal of the United States to ratify the major agreement in recent governments and as a result of tensions with Europe and Japan and the protests of southern countries can have a significant impact on the process of international environmental governance. Therefore, the refusal of a country such as the United States from its international environmental obligations will have catastrophic consequences in the validity and application of environmental governance policies and policies invented by other northern countries. The legitimacy of these countries in Among the countries of the South as recipients of environmental assistance due to lack of cooperation and coordination among donor countries and the risk of non-cooperation between institutions and the suspension of aid, this will be the case with the Convention on Diversity.

Biology and Kyoto Protocol as well The United Nations is quite obvious. The crises and challenges of international environmental governance Governance in pluralistic management should consider specific social and environmental policies and activities by integrating experience and knowledge in various institutions and effective social factors. And apply. Increased environmental problems in terms of climate change, biodiversity loss and natural ecosystem degradation, increased pollution, precautionary principle, and transgenic organisms, the risk of nuclear radiation, freshwater scarcity on a very large scale as threats should be factors in exacerbating the crisis and Economic development prospects in different countries and regions are limited, controlled and managed. The measures taken and ongoing to protect the environment are insufficient against the warnings given by the scientific community, and the necessary reforms to improve this protection process require time, energy, financial resources, and most importantly, diplomatic negotiations and consultations. Currently, the serious crisis of environmental degradation and the inability to deal with it unanimously has become a topic for all countries. But persistent differences and slow progress in organizing this crisis are serious governance challenges.

- Multiplicity and diversity of sources of international environmental law

Another substantive challenge to developing international environmental law is the multiplicity and diversity of binding and non-binding resources. Soft or non-binding rights are resources that are not binding and do not guarantee specific implementation, which include: statements, resolutions, agendas, action plans, and so on. Their main purpose is to express the principles and rules that guide governments. Although these resources are not binding in themselves, they have a very important impact on the development of international environmental law. The most important examples of these sources are the 1972 Stockholm Declaration on Man and the Environment, the Universal Charter of Nature adopted by the 1982 UN General Assembly, the Rio Declarations of 1992, Johannesburg 2002, and finally, the Rio + 20 Declaration of 2012. They had a law. In addition to the above, Agenda 21 adopted in 1992 on the sidelines of the Rio Summit is a non-binding legal document that contains important guidelines for the protection, protection, and management of the environment. UN General Assembly resolutions and statements by the UNEP Board of Governors can also be considered non-binding instruments in international environmental law. Many international conventions were formed during this period, including the Convention on International Trade in Endangered Species of Wild Fauna and Flora, the 1973 Washington Convention on the Law of the Sea, the 1982 Monte Gobi Convention, and many other conventions. They usually had an organization and structure for that convention. Today, most of the sources of international law are contained in the first paragraph of Article 38 of the Statute of the International Court of Justice, and according to paragraph 1 of Article 38 of the Statute, the Court International treaties, general (public) or special (private) treaties that establish rules (legal) and are explicitly recognized by the parties to the dispute; International custom as the reason for the general procedure that is accepted as a legal rule; General principles of law recognized by civilized nations; According to the provisions of Article 59 of the Statute of Judicial Decisions and the opinions of the most competent public law experts of different nations, it is considered as a subsidiary (auxiliary) knowledge of legal rules; Article 38 also stipulates that the provisions of this article do not prejudice the discretion of the court in issuing a judgment in accordance with the (rule of fairness and discretion) clause, provided that the litigants agree with it. However, today Article 38 of the Statute of the Court does not provide a complete picture of the sources of international law, but rather the international rules, unilateral legal actions of countries, and binding resolutions of international organizations, each of which is effective in regulating international legal relations and can serve as a source. International law to play a role. Despite their thematic diversity and geographical scope, international environmental treaties have common features, use similar legal techniques, and are often interrelated. Features such as the absence of retaliation in commitments, related materials or references from one document to another, structural protocol agreements, the creation of new institutions or the facilitation of existing or former institutions to promote continuous cooperation, the existence of rituals and innovative practices related to adherence and Non-compliance of the contracting members is the existence of a mechanism to correct and review their common funds. However, more than 300 multilateral treaties and 700 bilateral environmental agreements have led to the sequence and multiplicity of this source of law, which is considered one of the essential challenges in the development of international environmental law. Custom is also an important source of international environmental law, and many issues, such as the international responsibility of governments, are rooted in custom, especially in transboundary pollution. The number of other international treaties and instruments that enshrine the same legal rules related to the environment in international treaties is increasing. The work of the UN Commission on International Law shows that the inclusion of customary rules in many international instruments can be considered as substantial development of international environmental law. Although not sufficient to protect the environment, due to unsystematic development, treaty law remains in force. Principles of unity, resources, and responsibilities The common heritage of humanity is also one of the basic rules in this branch of law. On the other hand, the issuance of judicial rulings plays an important role in developing this field of law. International environmental law, in its short life, has been able to study and study solutions in terms of the potential importance of the environment; Issues such as the international responsibility of polluters, the obligations of governments towards their ratified treaties, and finally, the criminal liability for environmental degradation are among the important issues of international environmental law. But in international environmental law, despite the large volume of binding and non-binding documents and adherence to international custom, fulfilling the international responsibility of governments is not difficult. As many legal opinions and procedures have shown, adherence is not customary in limited cases.

Thus, the custom seems to have lost its traditional effectiveness in international environmental law and to make custom effective in international law. It is necessary to consider the behaviors and practices of governments in the field of environmental protection. Therefore, due to the novelty of this legal field, sources in international environmental law In addition to these sources, other sources such as international rules of procedure, international obligations of states and international organizations, and UN Security Council resolutions are available and can be cited. This proliferation of resources, especially in the large number of international treaties that sometimes have similar or sometimes conflicting obligations, has challenged the implementation of international environmental law.

- Insufficient guarantee for the implementation of international environmental law

Another essential challenge of international environmental law is the inadequacy of enforcement guarantees. After going through the rules to implement a legal system, it must be thought that its implementation guarantees an important role. This issue in the world community; has different forms because governments are never willing to jeopardize their national interests against each other. Even in some cases, there are shortcomings in the domestic laws to support this issue. Due to the sovereignty of governments and their role, there is an obstacle to the effective allocation and enforcement of these documents. On the other hand, the existing enforcement mechanisms will sometimes be forced to retreat in response to pressure from governments. Unlike the domestic law system of countries, international law lacks coherent enforcement mechanisms such as the executive, the administrative police, or the judiciary. Existing classic safeguards, such as criminal or civil warranties, are not as effective as many cross-border and non-refundable environmental issues.

- Structural Challenges of International Environmental Law

With the growing importance of global environmental protection since the 1970s, this mission has been directly or indirectly reflected in many international organizations' goals. In the meantime, international governmental organizations, both global and regional, and even non-governmental organizations have tried to play their role in environmental protection by committing their members to international environmental regulations to support Take a step away from the environment. Undoubtedly, as active subjects of international law, international organizations have a serious task to achieve in this regard. The number of these organizations that have some environmental competence today is significant. Almost every international governmental organization has somehow gotten into the subject of the environment. The world community is aware today that the position of environmental protection is very important for the well-being of the world community and the health of ecosystems, as well as sustainable development. For this reason, the issue of environmental protection and sustainable economic development is at the center of the plan of international and regional international organizations. In addition to the United Nations, other international organizations also offer programs in this direction. In this regard, one of the important reflections of the UN Conference on Human and Environment Stockholm 1972 was establishing the United Nations Environment Program as the global and executive arm of the organization in the field of environment and its preservation under this The organization was established. Accordingly, it was tasked with assessing the state of the global environment, formulating global environmental programs, and providing the necessary funding for these matters. It is one of the United Nations-affiliated bodies. In other words, the only UN mechanism that has the authority to policy, coordinate, and encourages various environmental issues between governments and specialized agencies of the United Nations and other organizations has established international governmental and non-governmental organizations. The United Nations Environment Program was established to monitor the state of the global environment and the impact of national and international environmental policies and measures. The General Assembly in 1997, while emphasizing UNEP as the United Nations Chief Agent in the field of environment, called on it to be the main custodian of the global environment and set the world environmental plan and promote and coherently implement the environmental dimensions of sustainable development. The program is headquartered in Nairobi, the capital of Kenya. It is responsible for directing programs and managing the work of other offices that were later established in other countries. The establishment of the United Nations Environment Program's headquarters in Africa will help it understand the pattern of environmental issues facing developing countries. The main goal of the United Nations Environment Program is to establish and encourage international coordination and cooperation for the protection of the environment. UNEP is also a center for informing the governments and peoples of the world and a means of improving their standard of living without harming the rights of future generations. In other words, UNEP is an institution that advocates the prudent use of the environment and sustainable development. UNEP works with the United Nations, international organizations, governments, non-governmental organizations, the private sector, and civil society to achieve its goals. Accordingly, the United Nations Environment Program, as one of the most fundamental structures of global environmental protection, has a distinct position compared to other international organizations. However, according to their field of work in the field of environmental programs, other organizations act according to authority, which all these structures have been created with one purpose, and that: to The current challenges facing the environmental system at the international level reveal the need to review the institutions' structure. Although all these efforts are to address environmental sustainability concerns, they are still witnessing the process of environmental degradation and the non-implementation of decisions and ideals of related institutions. The lack of coordination in the environmental system is due to the structural weakness of international environmental law. However, the structural development of international environmental law has been challenged by international competition and conflict between governments, especially developed and developing countries. Although the United Nations Environment Program is a United Nations body that oversees the environmental activities of its members, it continues to operate as a program to become the "World Environment Organization," which has Universal and more complete powers that face limitations. The United Nations Environment Program, as the United Nations Main Program for Environmental Protection, despite its many successes in protecting the global environment, does not have the authority to enforce the rules and regulations of international environmental law on a global scale. Hence, the need to revise this program to improve its current position or change the type of dependence of this institution on the United Nations is inevitable.

In recent decades, the gradual development of international environmental law has been based more on human-environmental needs and necessities than anything else. Economic growth and increasing technological advances in the contemporary period have caused major environmental damage. In response to these needs and requirements, international law has sought to enforce international laws and regulations to oblige governments to environmental protection. Although the international community has made great efforts to enforce binding and non-binding legal instruments for global environmental protection, these efforts have failed to stop the widespread environmental degradation in the world. In terms of the content of the law, creating a comprehensive system of international environmental law can reduce the legal gaps created by the diversity of environmental legal documents. In addition, one of the ways to reduce the existing challenges is to adopt an approach of "internationalization" of environmental protection, which is based on two legal bases. On the one hand, the basic rules and regulations of international environmental law are binding on members of the international community. The legal basis of these rules and regulations is mainly "customary" and is based on customary law, some of which are now known as "international rules." Some of the basic principles of international environmental law, such as the principle of "prohibition of harm to other lands," the principle of "fair and rational use of land," the principle of " cooperation," and the principle of "prevention" should be considered in the protection of the global environment. To be placed. Therefore, international law's conceptual and substantive development can be considered one of the effective solutions to solve the limitations and obstacles to international environmental law. On the other hand, "internationalization" of environmental protection on the principle of "institutionalization" of international environmental law. Contemporary based. Institutionalization of international law means creating and expanding international organizations and institutions for the effective and efficient protection of the environment. Also, participation in the domestic dimension includes public participation in environmental decision-making and implementation of environmental decisions and participation in the international dimension overseeing the participation of all governments, both rich and poor, developed and developing, industrial and non-industrial, in decision-making. Internationalization can be one of the important strategies for the development of international environmental law.

Also, the criminalization of large-scale environmental degradation and the passage of "crime against humanity" through the recognition of "crime against future generations" can be a sign of progress in the sense of double responsibility for the future, future conditions, and protection of other living species and Be considered the environment. International criminal law does not yet have specific legal solutions to criminalize environmental degradation as an international crime. However, recognizing "crime against future generations" as an international crime could contribute to the development of international environmental law.

Structurally, transforming the United Nations Environment Program, which currently has limited capacity, into a "global environmental protection organization" with the necessary authority and power, could be an important step in the structural development of international law. Consider the environment. The issues related to establishing this global organization will not be outside the conflict of developing and developed countries.

Hussein Movahedian
Islamic Azad University, UAE & Department of Private Law, Islamic Studies and Law Faculty, Imam Sadiq University, Tehran, Iran

Nima Norouzi
Islamic Azad University, UAE & Law and Political Science Department, University of Tehran, Tehran, Iran

Compilation of References

Aall, C. (2014). Sustainable tourism in practice: Promoting or perverting the quest for a sustainable development? *Sustainability*, *6*(5), 2562–2583. doi:10.3390u6052562

Abdillah, A., Mastuti, A. G., Rijal, M., & Rahman, M. A. (2020). Students' intuitive and analytical thinking in the mathematics study through the integration of STAD and environmental islamic jurisprudence (fiqh). Al-Jabar. *Jurnal Pendidikan Matematika*, *11*(1), 49–60.

Abu Dhabi National Oil Company (ADNOC). (2020). *General Terms and Conditions, for the Sale of Crude Oil / Condensate and Liquefied Petroleum Gas,* P.31. ADNOC. https://www.adnoc.ae/-/media/adnoc-v2/files/adnoc_crude-and-lpg_gtcs_january-2020-edition-final_v1.ashx?la=en&hash=C9551678CC5CBBBAB30DFE83A495800E8AD540A1

Akinyemi, O. E., Osabuohien, E. S., Alege, P. O., & Ogundipe, A. A. (2017). Energy security, trade and transition to green economy in Africa. *International Journal of Energy Economics and Policy*, *7*(3), 127–136.

Alexander, D. (1993). Some Themes in Intellectual Property the Environment and the Environment. *Rev. Eur. Comp. & Int'l Envtl. L.*, *2*(2), 113–120. doi:10.1111/j.1467-9388.1993.tb00100.x

Alexandroff, A. S., & Cooper, A. F. (Eds.). (2010). *Rising states, rising institutions: Challenges for global governance*. Brookings Institution Press.

Alvarez, J. E. (2002). The WTO as linkage machine. *The American Journal of International Law*, *96*(1), 146–158. doi:10.2307/2686131

Amani, M. (2010). *International oil contract law*. Imam Sadiq University Press.

Amid Zanjani, A. A. (2019). *Political Jurisprudence*. Amir Kabir Publication.

Anderson, J. E. (2009). *Law, knowledge, culture: The production of indigenous knowledge in intellectual property law*. Edward Elgar Publishing. doi:10.4337/9781848447196

Ansari, M. (2021). *Faraed Al-Usul* (1st ed.). Majmaolfekr Alislamii Publication.

Arney, C. (2010). Hot, Flat, And Crowded: Why We Need A Green Revolution-And How It Can Renew America. *Mathematics and Computer Education*, *44*(2), 180.

Arrow, K., Jodha, N., Jentoft, S., McCay, B., McKean, M., Sanderson, S., & Young, O. (1996). *Rights to nature: ecological, economic, cultural, and political principles of institutions for the environment*. Island Press.

Bande, L. C. (2017). *Criminal law in Malawi*. Juta.

Barbier, E. B. (2012). The green economy post Rio+ 20. *Science*, *338*(6109), 887–888. doi:10.1126cience.1227360 PMID:23161980

Barry, J. (2007). *Environment and social theory*. Routledge. doi:10.4324/9780203946923

Begum, A., Jingwei, L., Haider, M., Ajmal, M. M., Khan, S., & Han, H. (2021). Impact of environmental moral education on Pro-environmental behaviour: Do psychological empowerment and Islamic religiosity matter? *International Journal of Environmental Research and Public Health*, *18*(4), 1604. doi:10.3390/ijerph18041604 PMID:33567647

Bennett, P. (2000). Anti-Trust? European Competition Law and Mutual Environmental Insurance. *Economic Geography*, *76*(1), 50–67. doi:10.2307/144540

Berkhout, F., & Smith, A. (2003). Carbon flows between the EU and Eastern Europe: Baselines, scenarios and policy options. *International Environmental Agreement: Politics, Law and Economics*, *3*(3), 199–219. doi:10.1023/B:INEA.0000005624.46391.96

Birnie, P. (2002). Salman MA Salman and Kishor Uprety, Conflict and Cooperation on South Asia's International Rivers. The Hague & Law International.

Boer, B. (1998). The Rise of Environmental Law in the Asian Region. *U. Rich. L. Rev.*, *32*, 1503.

Bolla, A. J., & McDorman, T. L. (1999). *Comparative Asian environmental law anthology*. Carolina Academic Press.

Bosselman, F. (2011). A Brief history of energy law in United States law schools: An introduction to the symposium. Chi.-. *Kent L. Rev.*, *86*, 3.

Bourdeau, P. (2004). The man– nature relationship and environmental ethics. *Journal of Environmental Radioactivity*, *72*(1-2), 9–15. doi:10.1016/S0265-931X(03)00180-2 PMID:15162850

Bovenberg, A. L., & De Mooij, R. A. (1994). Environmental levies and distortionary taxation. *The American Economic Review*, *84*(4), 1085–1089.

Bovenberg, A. L., & van der Ploeg, F. (1994). Environmental policy, public finance and the labour market in a second-best world. *Journal of Public Economics*, *55*(3), 349–390. doi:10.1016/0047-2727(93)01398-T

Bown, C. P. (2017). Mega-regional Trade Agreements and the Future of the WTO. *Global Policy*, *8*(1), 107–112. doi:10.1111/1758-5899.12391

Boyle, A., & Birnie, P. (1995). *Basic documents on international law and the environment*. Clarendon Press.

Bradbrook, A. J. (2011). Creating law for next generation energy technologies. *Geo. Wash. J. Energy & Envtl. L.*, *2*, 17.

Brickey, K. F. (1996). Environmental crime at the crossroads: The intersection of environmental and criminal law theory. *Tul. L. Rev.*, *71*, 487.

Brunch, C., Coker, W., & VanArsdale, C. (2001). Constitutional environmental law: Giving force to fundamental principles in Africa. *Colum. J. Envtl. L.*, *26*, 131.

Buchanan, A., & Keohane, R. O. (2006). The legitimacy of global governance institutions. *Ethics & International Affairs*, *20*(4), 405–437. doi:10.1111/j.1747-7093.2006.00043.x

Bucher, H., Drake-Brockman, J., Kasterine, A., & Sugathan, M. (2014). *Trade in environmental goods and services: Opportunities and challenges*. ITC.

Campbell-Lendrum, D., & Corvalán, C. (2007). Climate change and developing-country cities: Implications for environmental health and equity. *Journal of Urban Health*, *84*(1), 109–117. doi:10.100711524-007-9170-x PMID:17393341

Carson, R. (1962). Silent spring III. *New Yorker (New York, N.Y.)*, 23.

Charnovitz, S. (2002). A World Environmental Organization. *Colum. J. Envtl. L.*, *27*, 323.

Cho, B. S. (2000). Emergence of an international environmental criminal law. *UCLA J. Envtl. L. & Pol'y*, *19*(1), 11. doi:10.5070/L5191019216

Cioffo, G. D., Ansoms, A., & Murison, J. (2016). Modernising agriculture through a 'new'Green Revolution: The limits of the Crop Intensification Programme in Rwanda. *Review of African Political Economy*, *43*(148), 277–293. doi:10.10 80/03056244.2016.1181053

Clarke, P. (2008). Supporting conservation and sustainable livelihoods in the Pacific: The IUCN Regional Environmental Law Programme for Oceania 2007-2008. *NATIONAL ENVIRONMENTAL LAW REVIEW*, (2), 49–55.

Cooper, D. E., & James, S. P. (2017). *Buddhism, virtue and environment*. Routledge. doi:10.4324/9781315261195

Cropper, M. L., & Oates, W. E. (1992). Environmental economics: A survey. *Journal of Economic Literature*, *30*(2), 675–740.

Cullet, P. (2001). *Intellectual property and environment: impacts of the TRIPS agreement on environmental law making in India*. In Global environmental change and the nation state. Proceedings of the 2001 Berlin Human Dimensions of Global Environmental Change Conference, Postdam Institute for Climate Impacts Research, Postdam.

Cullet, P. (2005). *Intellectual property protection and sustainable development. LexisNexis*. Butterworths.

Cullet, P., & Raja, J. (2004). Intellectual property rights and biodiversity management: The case of India. *Global Environmental Politics*, *4*(1), 97–114. doi:10.1162/152638004773730239

Dabiri, F., & Poorhashemi, S. A. (2009). A Study of the Principles and Concepts of International Environmental Law with a Look at Sustainable Development. *Journal of Environmental Science and Technology*, *11*(3), 220.

Dallal, A. (2010). *Islam, science, and the challenge of history*. Yale University Press.

Dashti, M. (2021). *Nahj Al-Balaghah*. Nahj Al-Balaghah Institute.

Daugbjerg, C., & Svendsen, G. T. (2001). Designing Green Taxation. In Green Taxation in Question (pp. 117-135). Palgrave Macmillan, London. doi:10.1057/9780230595538_5

de Aguiar Patriota, A. (2008). Introduction to Brazilian Environmental Law. *Geo. Wash. Int'l L. Rev.*, *40*, 611.

De StefanoG. (2020). Measurable Environmental Protection As A Necessity For Competition Law. Available at SSRN 3533499. doi:10.2139/ssrn.3533499

Delgado, R. (1985). Rotten social background: Should the criminal law recognize a defense of severe environmental deprivation. *Law & Inequality*, *3*, 9.

Dessus, B., Thomas, J. P., & Tillerson, K. (1999). *MDP and priorities in African development; MDP et priorites du developpement en Afrique*. Liaison Energie-Francophonie.

DiMento, J. F. (2021). *Book Review: Philosophies of Polar Law. International Environmental Agreements: Politics.* Law and Economics.

Dincer, I. (1999). Environmental impacts of energy. *Energy Policy*, *27*(14), 845–854. doi:10.1016/S0301-4215(99)00068-3

Dreyfuss, R. C., & Strandburg, K. J. (Eds.). (2011). *The law and theory of trade secrecy: a handbook of contemporary research*. Edward Elgar Publishing. doi:10.4337/9780857933072

Du Rées, H. (2001). Can criminal law protect the environment? *Journal of Scandinavian Studies in Criminology and Crime Prevention, 2*(2), 109–126. doi:10.1080/140438501753737606

Dung, M. T., Khoa, N. M., & Thi Thu Huong, P. (2021). Integrating Environmental Requirements into Vietnamese Sectoral Laws: Some Legal Issues. *Environmental Policy and Law, 51*(5), 343–350. doi:10.3233/EPL-201010

Dunlap, R. E., & Brulle, R. J. (Eds.). (2015). *Climate change and society: Sociological perspectives.* Oxford University Press. doi:10.1093/acprof:oso/9780199356102.001.0001

Dutfield, G. (2000). *Intellectual property rights trade and biodiversity.* Routledge. doi:10.4324/9781849776233

Epstein, I. (1945). *Judaism.* Epworth Press.

Erickson, P., van Asselt, H., Koplow, D., Lazarus, M., Newell, P., Oreskes, N., & Supran, G. (2020). Why fossil fuel producer subsidies matter. *Nature, 578*(7793), E1–E4. doi:10.103841586-019-1920-x PMID:32025022

Eskeland, G. S. (1993). *A presumptive pigovian tax on gasoline: analysis of an air pollution control program for Mexico City (No. 1076).* The World Bank.

Farhad, D., & Khalatbari, Y. (2018). Achieving sustainable development from the perspective of international environmental law. *Journal of Human and Environment, 16*(44).

Faruque, A. (2006). Validity and Efficacy of Stabilisation Clauses: Legal Protection vs. Functional Value. *J. Int'l Arb., 23*, 317.

Faure, M. (2004). European environmental criminal law: do we really need it?. European Energy and Environmental Law Review, 13(1).

Faure, M. G. (2010). *Vague notions in environmental criminal law.*

Faure, M. (2017). The development of environmental criminal law in the EU and its member states. *Review of European, Comparative & International Environmental Law, 26*(2), 139–146. doi:10.1111/reel.12204

Faure, M. G. (2016). A paradigm shift in environmental criminal law. In *Fighting Environmental Crime in Europe and Beyond* (pp. 17–43). Palgrave Macmillan. doi:10.1057/978-1-349-95085-0_2

Faure, M. G. (2016). The revolution in environmental criminal law in Europe. *Va. Envtl. LJ, 35*, 321.

Faure, M. G., & Zhang, H. (2011). Environmental criminal law in China: A critical analysis. *Envtl. L. Rep. News & Analysis, 41*, 10024.

Fisher, D. (2014). *Australian environmental law: norms, principles and rules.* Thomson Reuters Professional (Austalia) Limited.

Floyd, R. (2008). The environmental security debate and its significance for climate change. *The International Spectator, 43*(3), 51–65. doi:10.1080/03932720802280602

Fortney, D. C. (2002). Thinking Outside the Black Box: Tailored Enforcement in Environmental Criminal Law. *Texas Law Review, 81*, 1609.

Fowler, R. (2020). Environmental Law in Singapore, written by Joseph Chun and Lye Lin Heng. *Chinese Journal of Environmental Law, 4*(1), 111–114. doi:10.1163/24686042-12340052

Framarin, C. (2014). *Hinduism and environmental ethics: law, literature, and philosophy.* Routledge. doi:10.4324/9781315852522

Frihy, O. E. (2001). The necessity of environmental impact assessment (EIA) in implementing coastal projects: Lessons learned from the Egyptian Mediterranean Coast. *Ocean and Coastal Management, 44*(7-8), 489–516. doi:10.1016/S0964-5691(01)00062-X

Fuentes, X. (2002). International law-making in the field of sustainable development: The unequal competition between development and the environment. *International Environmental Agreement: Politics, Law and Economics, 2*(2), 109–133. doi:10.1023/A:1020990026398

Gada, M. Y. (2014). Environmental ethics in Islam: Principles and perspectives. *World Journal of Islamic History and Civilization, 4*(4), 130–138.

Gao, Z., & Kao, C. K. (1998). *Environmental regulation of oil and gas* (Vol. 11). Kluwer Law International BV.

Gehring, M. W. (2006). Competition for sustainability: Sustainable development concerns in national and EC competition law. *Review of European Community & International Environmental Law, 15*(2), 172–184. doi:10.1111/j.1467-9388.2006.00519.x

Gervais, D. J. (2005). Intellectual Property, Trade & (and) Development: The State of Play. *Fordham Law Review, 74*, 505.

Ghaemi Khargh, M. (2022). Lack of Governance in Political Jurisprudence; Challenges & Requirements. *Public Law Knowledge Quarterly, 11*(35), 1–24.

Gienapp, P., Teplitsky, C., Alho, J. S., Mills, J. A., & Merilä, J. (2008). Climate change and evolution: Disentangling environmental and genetic responses. *Molecular Ecology, 17*(1), 167–178. doi:10.1111/j.1365-294X.2007.03413.x PMID:18173499

Gillies, D. (2013). A guide to EC []. Routledge.]. *Environmental Law (Northwestern School of Law), 9*.

Gomez, C. M. G. (2001). On optimal environmental taxation and enforcement: Information, monitoring and efficiency. *Natural Resource Modeling, 14*(1), 5–30. doi:10.1111/j.1939-7445.2001.tb00048.x

Gottlieb, R. S. (Ed.). (2006). *The Oxford handbook of religion and ecology.* OUP USA. doi:10.1093/oxfordhb/9780195178722.001.0001

Gray, J. S., Bakke, T., Beck, H. J., & Nilssen, I. (1999). Managing the environmental effects of the Norwegian oil and gas industry: From conflict to consensus. *Marine Pollution Bulletin, 38*(7), 525–530. doi:10.1016/S0025-326X(99)00004-1

Griffin, K. (2003). Economic globalization and institutions of global governance. *Development and Change, 34*(5), 789–808. doi:10.1111/j.1467-7660.2003.00329.x

Habibi, M. H. (2005). *Environmental Law.* Tehran University Press.

Hancock, R. (2019). Ecology in Islam. In Oxford Research Encyclopedia of Religion. Oxford. doi:10.1093/acrefore/9780199340378.013.510

Haq, Z. A., Imran, M., Ahmad, S., & Farooq, U. (2020). Environment, Islam, and women: A study of eco-feminist environmental activism in Pakistan. *Journal of Outdoor and Environmental Education, 23*(3), 275–291. doi:10.100742322-020-00065-4

Hatami, A., & Karimiyan, E. (2014). *Foreign Investment Law in Light of Investment Act and Contracts.* Teesa Publication.

Hay, J. (2010). How efficient can international compensation regimes be in pollution prevention? A discussion of the case of marine oil spills. *International Environmental Agreement: Politics, Law and Economics, 10*(1), 29–44. doi:10.100710784-009-9096-8

Helfer, L. R. (2004). Mediating Interactions in an Expanding International Intellectual Property Regime. *Case W. Res. J. Int'l L.*, *36*, 123.

Henderson, P. G. (2001). Some thoughts on distinctive principles of South African environmental law. *South African Journal of Environmental Law and Policy*, *8*(2), 139–184.

Hey, E. (2001). The Climate Change Regime: An Enviro-Economic Problem and International Administrative Law in the Making. *International Environmental Agreement: Politics, Law and Economics*, *1*(1), 75–100. doi:10.1023/A:1010117910664

Hilf, M. (2001). Power, rules and principles-which orientation for WTO/GATT law? *Journal of International Economic Law*, *4*(1), 111–130. doi:10.1093/jiel/4.1.111

Hinkle, J., & Rosencranz, A. (2008). *Jon Birger Skjærseth and Tora Skodvin, Climate Change and the Oil Industry: Common Problem.* Varying Strategies.

Ismail, O. S., & Umukoro, G. E. (2012). Global impact of gas flaring. *Energy and Power Engineering*, *4*(4), 290–302. doi:10.4236/epe.2012.44039

Ismail, O. S., & Umukoro, G. E. (2016). Modelling combustion reactions for gas flaring and its resulting emissions. *Journal of King Saud University-Engineering Sciences*, *28*(2), 130–140. doi:10.1016/j.jksues.2014.02.003

Ivry, A. (1997). Jewish philosophy. *Bulletin de Philosophie Medievale*, *39*, 45–48. doi:10.1484/J.BPM.3.534

Jacobs, R. (1993). EEC Competition Law and the Protection of the Environment. Legal Issues of Eur. *Integration (Tokyo, Japan)*, *20*, 37.

Jaffe, A. B., & Stavins, R. N. (1995). Dynamic incentives of environmental regulations: The effects of alternative policy instruments on technology diffusion. *Journal of Environmental Economics and Management*, *29*(3), S43–S63. doi:10.1006/jeem.1995.1060

Jam, F., & Blake, J. (2017). Global environmental governance system: Challenges and solutions. *Environmental Sciences*, *15*(1), 141–156.

Jamil, S. (2021). Halal wastewater recycling: environmental solution or religious complication? In *Religious Environmental Activism* (pp. 93–111). Routledge.

Javadi Amoli, A. (2020). *Islam and the Environment.* Esra Publishing Center.

Johnston, A. (2012). The Interface between EU Energy, Environmental and Competition Law in the UK. *Oil, Gas & Energy Law*, *10*(4).

Jura, C. (2012). Considerations on a New Branch of Public International Law–Energy International Law. *Procedia: Social and Behavioral Sciences*, *62*, 801–805. doi:10.1016/j.sbspro.2012.09.135

Karns, M. P., & Mingst, K. A. (2010). International Organizations: The Politics and Process.

Kasa, S. (2000). Policy networks as barriers to green tax reform: The case of CO2-taxes in Norway. *Environmental Politics*, *9*(4), 104–122. doi:10.1080/09644010008414553

Kennedy, K. (2012). The environmental Law Framework of the Democratic Republic of the Congo and the Balancing of Interests. *The balancing of interests in environmental law in Africa.*

Keohane, R. O. (2011). Global governance and legitimacy. *Review of International Political Economy*, *18*(1), 99–109. doi:10.1080/09692290.2011.545222

Khajehpour, H., Ahmady, M. A., Hosseini, S. A., Mashayekhi, A. N., & Maleki, A. (2017). Environmental policy-making for Persian Gulf oil pollution: A future study based on system dynamics modeling. *Energy Sources. Part B, Economics, Planning, and Policy*, *12*(1), 17–23. doi:10.1080/15567249.2015.1004382

Khor, M. (2002). Rethinking intellectual property rights and TRIPS. In *Global Intellectual Property Rights* (pp. 201–213). Palgrave Macmillan. doi:10.1057/9780230522923_12

Kimerling, J. (1990). Disregarding environmental law: Petroleum development in protected natural areas and indigenous homelands in the Ecuadorian Amazon. *Hastings Int'l & Comp. L. Rev.*, *14*, 849.

King, C. S., Feltey, K. M., & Susel, B. O. N. (1998). The question of participation: Toward authentic public participation in public administration. *Public Administration Review*, *58*(4), 317–326. doi:10.2307/977561

Kiss, A. C., Shelton, D., & Shelton, D. (1991). *International environmental law* (Vol. 3). Transnational Publishers.

Kiss, A., & Shelton, D. (1997). *Manual of European environmental law*. Cambridge University Press.

Kistaubayev, S. U. (2020). Zoroastrianism And The Expression Of The Relationship Between Nature And Man In Islam. *Theoretical & Applied Science*, *84*(4), 624–626. doi:10.15863/TAS.2020.04.84.104

Kloosterhuis, E., & Mulder, M. (2015). Competition law and environmental protection: The Dutch agreement on coal-fired power plants. *Journal of Competition Law & Economics*, *11*(4), 855–880. doi:10.1093/joclec/nhv017

Koh, K. L. (2008). Regional and state level environmental governance. ASEAN's environment governance: An evaluation. In *UNITAR/Yale Conference on Environmental Governance and Democracy*, New Haven, CT.

Kopczuk, W., Marion, J., Muehlegger, E., & Slemrod, J. (2013). *Do the laws of tax incidence hold? point of collection and the pass-through of state diesel taxes (No. w19410)*. National Bureau of Economic Research. doi:10.3386/w19410

Kotzé, L. J. (2014). Rethinking global environmental law and governance in the Anthropocene. *Journal of Energy & Natural Resources Law*, *32*(2), 121–156. doi:10.1080/02646811.2014.11435355

Krass, D., Nedorezov, T., & Ovchinnikov, A. (2013). Environmental taxes and the choice of green technology. *Production and Operations Management*, *22*(5), 1035–1055. doi:10.1111/poms.12023

Kuik, O. (2003). Climate change policies, energy security and carbon dependency trade-offs for the European Union in the longer term. *International Environmental Agreement: Politics, Law and Economics*, *3*(3), 221–242. doi:10.1023/B:INEA.0000005625.44125.54

Kumamoto, N. (1989). Japanese environmental law and ocean resources. *Ecology Law Quarterly*, *16*, 267.

Lazarus, R. J. (2001). The Greening of America and the Graying of United States Environmental Law: Reflections on Environmental Law's First Three Decades in the United States. *Virginia Environmental Law Journal*, 75-106.

Lazarus, R. J. (1994). Meeting the demands of integration in the evolution of environmental law: Reforming environmental criminal law. *Geological Journal*, *83*, 2407.

Lazarus, R. J. (1995). Mens rea in environmental criminal law: Reading supreme court tea leaves. *Fordham Envtl. LJ*, *7*, 861.

Lazarus, R. J. (2008). *The making of environmental law*. University of Chicago Press.

Lilliestam, J., Patt, A., & Bersalli, G. (2021). The effect of carbon pricing on technological change for full energy decarbonization: A review of empirical ex-post evidence. *Wiley Interdisciplinary Reviews: Climate Change*, *12*(1), e681. doi:10.1002/wcc.681

Lindhout, P. E., & Van den Broek, B. (2014). The polluter pays principle: Guidelines for cost recovery and burden sharing in the case law of the European court of justice. *Utrecht Law Review*, *10*(2), 46. doi:10.18352/ulr.268

List, J. A., & Qui, L. D. (2004). Intellectual property rights, environmental regulations, and foreign direct investment. *Land Economics*, *80*(2), 153–173. doi:10.2307/3654736

Luppi, B., Parisi, F., & Rajagopalan, S. (2012). The rise and fall of the polluter-pays principle in developing countries. *International Review of Law and Economics*, *32*(1), 135–144. doi:10.1016/j.irle.2011.10.002

Maghrebi, A. H. N. M. T. (2020). *Daim Al-Islam* (2nd ed.). Al-Bayt Institute.

Maisin, J. B., & Meagher, M. (2020). Sustainable development and competition law: Towards a Green Growth regulatory osmosis. In Sustainable development and competition law.

Majlisi Esfahani, M. B. (2020). *Merat Aloqul fi Sharah Akhbar Aal Alrasu* (2nd ed.). Dar Al-Kitab Al-Islamiya.

Makarem Shirazi, N. (2020). The Rules of Jurisprudence. Third Edition. Qom: Published by the School of Imam Ali ibn Abi Talib.

Maraghi, A. (2020). *Jurisprudence Titles* (1st ed.). Islamic Publications Office.

Marandi, S. M. R. (2020). Being a Jurist as a Condition for the Islamic Ruler in the Sunni Political Jurisprudence. *Journal of Contemporary Research on Islamic Revolution*, 21-38.

Mashhadi, A. (2020). *The Right to a Healthy Environment (Iranian-French Model)* (1st ed.). Mizan Publication.

Matthews, D. (2003). *Globalising intellectual property rights: the TRIPS Agreement*. Routledge. doi:10.4324/9780203165683

McHugh, S., Maruca, S. D., Lilien, J., & Manning, A. (2006). Environmental, social, and health impact assessment (ESHIA) process. In SPE International Health, Safety & Environment Conference. OnePetro.

Meganck, R. A., & Saunier, R. E. (2012). *Dictionary and introduction to global environmental governance*. Routledge. doi:10.4324/9781849771009

Mesbah Yazdi, M. T. (2020). *Teaching Philosophy*. Benalmelal Publication.

Meyer, C. A. (2021). Taking Lessons from Silent Spring: Using Environmental Literature for Climate Change. *Literature*, *1*(1), 2–13. doi:10.3390/literature1010002

Mianji, A. A. (2020). *Captive in Islam* (1st ed.). Islamic Publications Office.

Miller, A. C., Khan, A. M., Hebishi, K., Bigalli, A. A. C., & Vahedian-Azimi, A. (2020). Ethical issues confronting Muslim patients in perioperative and critical care environments: A survey of Islamic jurisprudence. *Anesthesiology Clinics*, *38*(2), 379–401. doi:10.1016/j.anclin.2020.01.002 PMID:32336391

Miller, C. A. (2007). Democratization, international knowledge institutions, and global governance. *Governance: An International Journal of Policy, Administration and Institutions*, *20*(2), 325–357. doi:10.1111/j.1468-0491.2007.00359.x

Mistura, A. (2018). Is There Space for Environmental Crimes under International Criminal Law: The Impact of the Office of the Prosecutor Policy Paper on Case Selection and Prioritization on the Current Legal Framework. *Colum. J. Envtl. L.*, *43*, 181.

Mitsilegas, V., Fitzmaurice, M., & Fasoli, E. (2016). The relationship between EU criminal law and environmental law. In *Research handbook on EU criminal law*. Edward Elgar Publishing. doi:10.4337/9781783473311.00024

Mohamed, L., & Al-Thukair, A. A. (2009). Environmental Assessments in the Oil and Gas Industry. *Water Air and Soil Pollution Focus*, *9*(1-2), 99–105. doi:10.100711267-008-9190-x

Mol, A. P. (2006). Environment and modernity in transitional China: Frontiers of ecological modernization. *Development and Change*, *37*(1), 29–56. doi:10.1111/j.0012-155X.2006.00468.x

Mol, A. P. (2008). *Environmental Reform in the Information Age. The Contours of Informational Governance*. Cambridge University Press. doi:10.1017/CBO9780511491030

Momtaz, D. (1996). The United Nations and the protection of the environment: From Stockholm to Rio de Janeiro. *Political Geography*, *15*(3-4), 261–271. doi:10.1016/0962-6298(95)00109-3

Monti, G. (2020). Four options for a greener competition law. *Journal of European Competition Law & Practice*, *11*(3-4), 124–132. doi:10.1093/jeclap/lpaa007

Monti, G., & Mulder, J. (2017). Escaping the clutches of EU competition law. *European Law Review*, *42*(5), 635–656.

Mostafapour, M. R. (2019). The Intrinsic Value of Nature from the Viewpoint of Sadr al-Mutaallehin Shirazi. *Esra Wisdom*, *12*, 113–130.

Motloch, J. L. (2000). *Introduction to landscape design*. John Wiley & Sons.

Movahedian, H., Norouzi, N., & Ataei, E. (2021). *Energy law and environmental law in oil and gas contracts with an emphasis on the MENA region*.

Nagel, T. (2017). The problem of global justice. In *Global Justice* (pp. 173–207). Routledge. doi:10.4324/9781315254210-9

Najam, A. (2003). The case against a new international environmental organization. *Global Governance*, *9*(3), 367–384. doi:10.1163/19426720-00903008

Nanda, V. (1995). *International environmental law & policy*. Brill Nijhoff.

Nanda, V., & Pring, G. R. (2012). *International environmental law and policy for the 21st century*. Martinus Nijhoff Publishers.

Nannerup, N. (2001). Equilibrium pollution taxes in a two industry open economy. *European Economic Review*, *45*(3), 519–532. doi:10.1016/S0014-2921(99)00028-8

Naraqi, A. M. (2021). Avead Alayam fi Bayan Qavaeid Alahkam. First Edition. Qom: Publications of the Office of Islamic teachings.

Nijar, G. S. (2013). The Nagoya–Kuala Lumpur Supplementary Protocol on Liability and Redress to the Cartagena Protocol on Biosafety: An analysis and implementation challenges. *International Environmental Agreement: Politics, Law and Economics*, *13*(3), 271–290. doi:10.100710784-012-9187-9

Nomani, M. Z. M., & Hussain, Z. (2020). Innovation technology in health care management in the context of Indian environmental planning and sustainable development. *International journal on emerging technologies, 11*(2), 560-564.

Noori, V., Zamani, S. Q., Raee, M., & Abdollahi, M. (2017). The Principle of Proportion in Armed Conflicts in the Light of Documents and Procedures of International Criminal Courts. *Journal of Criminal Law Teachings, 12,* 91-.611

Norouzi, N. (2022). A Practical and Analytic View on Legal Framework of Circular Economics as One of the Recent Economic Law Insights: A Comparative Legal Study. *Circular Economy and Sustainability,* 1-26.

Norouzi, N. (2022). Regulating Sustainable Economics: A Legal and Policy Analysis in the Light of the United Nations Sustainable Development Goals. In Handbook of Research on Changing Dynamics in Responsible and Sustainable Business in the Post-COVID-19 Era (pp. 266-287). IGI Global. doi:10.4018/978-1-6684-2523-7.ch013

Norouzi, N. (2022b). A Practical and Analytic View on Legal Framework of Circular Economics as One of the Recent Economic Law Insights: A Comparative Legal Study. *Circular Economy and Sustainability*, 1-26.

Norouzi, N., & Fani, M. (2021). Monopoly and competition in the energy market: A legal analysis. *Global Journal of Business Management, 15*(2), 001-007.

Norouzi, N. (2021a). Post-COVID-19 and globalization of oil and natural gas trade: Challenges, opportunities, lessons, regulations, and strategies. *International Journal of Energy Research*, *45*(10), 14338–14356. doi:10.1002/er.6762 PMID:34219899

Norouzi, N., & Ataei, E. (2021). Covid-19 Crisis and Environmental law: Opportunities and challenges. *Hasanuddin Law Review*, *7*(1), 46–60. doi:10.20956/halrev.v7i1.2772

Norouzi, N., & Ataei, E. (2021). Environmental Protection Regulations in the Light of Public Law and Social Obligations. *Research Journal of Ecology and Environmental Sciences*, *1*(1), 1–16.

Norouzi, N., & Fani, M. (2022). Globalization and the oil market: An overview on considering petroleum as a trade commodity. *Journal of Energy Management and Technology*, *6*(1), 54–62.

Norouzi, N., Fani, M., & Talebi, S. (2022). Green tax as a path to greener economy: A game theory approach on energy and final goods in Iran. *Renewable & Sustainable Energy Reviews*, *156*, 111968. doi:10.1016/j.rser.2021.111968

Norouzi, N., & Sheikhi, M. (2021). Achieving Sustainable Development from the Perspective of International Environmental Law. *Eurasian Journal of Environmental Research*, *5*(1), 1–13.

Norouzi, N., Sheikhi, M., Jafari, M., Kalantari, S., Narani, S. V., & Shaebani, A. (2021a). Green Victimology View in Iranian Criminology System. *Research Journal of Ecology and Environmental Sciences*, *1*(2), 82–95.

Norouzi, N., Sheikhi, M., Jafari, M., Kalantari, S., Narani, S. V., & Shaebani, A. (2021b). Criminal Legislative Policy in the Protection of Water Resources with Regard to International Treaties: A case for Iranian Legal System. *Universal Journal of Social Sciences and Humanities*, *1*(1), 67–79.

Nouzha, C. (2000). Réflexions sur la contribution de la Cour Internationale de Justice à la protection des ressources naturelles. *Revue juridique de l'Environnement, 25*(3), 391-420.

Nowag, J. (2016). *Environmental integration in competition and free-movement laws*. Oxford University Press. doi:10.1093/acprof:oso/9780198753803.001.0001

O'brien, R., Goetz, A. M., Scholte, J. A., & Williams, M. (2000). *Contesting global governance: Multilateral economic institutions and global social movements* (Vol. 71). Cambridge University Press. doi:10.1017/CBO9780511491603

O'Connor, R. E., Bard, R. J., & Fisher, A. (1999). Risk perceptions, general environmental beliefs, and willingness to address climate change. *Risk Analysis*, *19*(3), 461–471. doi:10.1111/j.1539-6924.1999.tb00421.x

Oates, W. E. (1995). Green taxes: Can we protect the environment and improve the tax system at the same time? *Southern Economic Journal*, *61*(4), 915–922. doi:10.2307/1060731

Oates, W., & Baumol, W. (1975). The instruments for environmental policy. In *Economic analysis of environmental problems* (pp. 95–132). NBER.

Odudu, O. (2010). The wider concerns of competition law. *Oxford Journal of Legal Studies*, *30*(3), 599–613. doi:10.1093/ojls/gqq020

Orts, E. W. (1994). Reflexive environmental law. *Nw. UL Rev.*, *89*, 1227.

Outka, U. (2012). Environmental law and fossil fuels: Barriers to renewable energy. *Vand. L. Rev.*, *65*, 1679.

Palmer, G. (1992). New ways to make international environmental law. *The American Journal of International Law*, *86*(2), 259–283. doi:10.2307/2203234

Pedersen, K. P. (2018). Environmental ethics in interreligious perspective. In *Explorations in global ethics* (pp. 253–290). Routledge. doi:10.4324/9780429500626-12

Penn, D. J. (2003). The evolutionary roots of our environmental problems: Toward a Darwinian ecology. *The Quarterly Review of Biology*, *78*(3), 275–301. doi:10.1086/377051 PMID:14528621

Podder, S. (2014). Mainstreaming the non-state in bottom-up state-building: Linkages between rebel governance and post-conflict legitimacy. *Conflict Security and Development*, *14*(2), 213–243. doi:10.1080/14678802.2014.889878

Poorhashemi, S. A. (2013). *International Environmental Law*. Dadgostar Press.

Posner, E. A., & Sykes, A. O. (2013). *Economic foundations of international law*. Harvard University Press. doi:10.2307/j.ctt2jbtsp

Pūraitė, A. (2012). *Origin of Environmental Regulation*. Mykolas Romeries University Press, Faculty of Public Security. https://intranet.mruni.eu/upload/iblock/b7c/014_puraite.pdf

Qian, W., & Zhu, Y. (2001). Climate change in China from 1880 to 1998 and its impact on the environmental condition. *Climatic Change*, *50*(4), 419–444. doi:10.1023/A:1010673212131

Raleigh, C., & Urdal, H. (2007). Climate change, environmental degradation and armed conflict. *Political Geography*, *26*(6), 674–694. doi:10.1016/j.polgeo.2007.06.005

Ramazani, G. M. H. (2013). A Comparative Study of "Precautionary Principle" in Opinions and Decisions of Internationals Tribunals. *Public Law Journal*, *15*(40), 143.

Reid, C. T. (2016). Brexit and the future of UK environmental law. *Journal of Energy & Natural Resources Law*, *34*(4), 407–415. doi:10.1080/02646811.2016.1218133

Riazati, S. (2006). A closer look: Professor seeks stronger UN. *The Daily Bruin, 21.* dailybruin.com/2006/10/17/a-closer-look-professor-seeks.

Risse, T. (2004). Global governance and communicative action. *Government and Opposition*, *39*(2), 288–313. doi:10.1111/j.1477-7053.2004.00124.x

Rizzuto, F. (2010). The Private Enforcement of European Competition Law: What Next? *Global Competition Litigation Review*, *3*(2), 57–68.

Robertson, V. H. (2022). Sustainability: A World-First Green Exemption in Austrian Competition Law. *Journal of European Competition Law & Practice*.

Rosenbaum, K. L. (1993). Sustainable Environmental Law: Integrating Natural Resource and Pollution Abatement Law from Resources to Recovery. Environmental Law Institute, 575-674.

Sabour, M. R. (2015). *Alternative Energy*. Khaje Nasir Toosi University Press.

Sadr, M. B. (2021). *Our Economy* (1st ed.). Center for Research and Specialized Studies of the Martyr Imam Al-Sadr.

Saghafi Tehrani, M. (2019). *Javid Fluent Commentary* (3rd ed.). Borhan Publication.

Saleh, S. M. H., & Hasani, S. A. (2017). Critical study of the models of environmental Theology in the new Christianity. *Religions and Mysticism*, *50*(1), 103–127.

Salmon, P., & Grinlinton, D. P. (Eds.). (2015). *Environmental Law in New Zealand*. Thomson Reuters.

Salzman, J., & Ruhl, J. B. (2000). Currencies and the commodification of environmental law. *Stanford Law Review*, *53*(3), 607. doi:10.2307/1229470

Salzman, J., & Thompson, B. H. Jr. (2010). *Environmental Law and Policy: Concepts and Insights*. Foundation Press.

Samimi, A. J., Ahmadpour, M., & Ghaderi, S. (2012). Governance and environmental degradation in MENA region. *Procedia: Social and Behavioral Sciences*, *62*, 503–507. doi:10.1016/j.sbspro.2012.09.082

Sands, P., & Peel, J. (2012). *Principles of international environmental law*. Cambridge University Press. doi:10.1017/CBO9781139019842

Saniotis, A. (2012). Muslims and ecology: Fostering Islamic environmental ethics. *Contemporary Islam*, *6*(2), 155–171. doi:10.100711562-011-0173-8

Schlosberg, D. (2009). *Defining environmental justice: Theories, movements, and nature*. Oxford University Press.

Schlosberg, D., & Collins, L. B. (2014). From environmental to climate justice: Climate change and the discourse of environmental justice. *Wiley Interdisciplinary Reviews: Climate Change*, *5*(3), 359–374. doi:10.1002/wcc.275

Schroeder, C. H. (1993). Cool Analysis Versus Moral Outrage in the Development of Federal Environmental Criminal Law. *Wm. & Mary L. Rev.*, *35*, 251.

Setness, K. H. (1996). Statutory Interpretation of Clean Water Act Section 1319 (C)(2)(A)'s Knowledge Requirement: Reconciling the Needs of Environmental and Criminal Law. *Ecology Law Quarterly*, *23*(2), 447–494.

Shaffer, G. C. (2001). The world trade organization under challenge: Democracy and the law and politics of the WTO's treatment of trade and environment matters. *Harv. Envt'l L. Rev.*, *25*, 1.

Shepardson, D. P., Niyogi, D., Roychoudhury, A., & Hirsch, A. (2012). Conceptualizing climate change in the context of a climate system: Implications for climate and environmental education. *Environmental Education Research*, *18*(3), 323–352. doi:10.1080/13504622.2011.622839

Shiravi, A.-H. (2016). *Oil and Gas Law*. Mizan Press.

Shohani, A., Ataei, E., & Norouzi, N. (2021). Prevention and Suppression of Environmental Crimes in the Light of the Actions of Non-Governmental Organizations in the Iranian Legal System. *Research Journal of Ecology and Environmental Sciences*, *1*(1), 57–70.

Shohani, A., Ataei, E., & Norouzi, N. (2021a). Environmental Constitutionalism in Latin America. *Universal Journal of Social Sciences and Humanities*, *1*(1), 54–66.

Shohani, A., Ataei, E., & Norouzi, N. (2021b). The Extent of the Researcher's Liability for Environmental Damage Caused by Academic Research. *Research Journal of Ecology and Environmental Sciences*, *1*(2), 71–81.

Shoushtari, M. H. M. (2021). *New Perspectives on Law*. Mizan Publication Center.

Siyavooshi, M., Foroozanfar, A., & Sharifi, Y. (2019). Effect of Islamic values on green purchasing behavior. *Journal of Islamic Marketing*, *10*(1), 125–137. doi:10.1108/JIMA-05-2017-0063

Skjærseth, J. B. (2013). Governance by EU emissions trading: Resistance or innovation in the oil industry? *International Environmental Agreement: Politics, Law and Economics*, *13*(1), 31–48. doi:10.100710784-012-9201-2

Smith, T. (2012). Creating a framework for the prosecution of environmental crimes in international criminal law. Ashgate Publishers.

Smith, A. E., & Veldman, R. G. (2020). Evangelical Environmentalists? Evidence from Brazil. *Journal for the Scientific Study of Religion*, *59*(2), 341–359. doi:10.1111/jssr.12656

Smits, C. C., van Tatenhove, J. P., & van Leeuwen, J. (2014). Authority in Arctic governance: Changing spheres of authority in Greenlandic offshore oil and gas developments. *International Environmental Agreement: Politics, Law and Economics*, *14*(4), 329–348. doi:10.100710784-014-9247-4

Sobhani, J. (2022). *Al-Wasit* (1st ed.). Imam Sadegh Publication.

Sofronova, E., Holley, C., & Nagarajan, V. (2014). Environmental non-governmental organizations and Russian environmental governance: Accountability, participation and collaboration. *Transnational Environmental Law*, *3*(2), 341–371. doi:10.1017/S2047102514000090

Soleymani, I. (2019). Right to Environment in the Islamic sources. *Anthropogenic Pollution*, *3*(1), 33–38.

Sornarajah, M. (2006). A law for need or a law for greed?: Restoring the lost law in the international law of foreign investment. *International Environmental Agreement: Politics, Law and Economics*, *6*(4), 329–357. doi:10.100710784-006-9016-0

Staiger, R. W. (2004). *Report on the international trade regime for the International Task Force on Global Public Goods*.

Steward, F. (2012). Transformative innovation policy to meet the challenge of climate change: Sociotechnical networks aligned with consumption and end-use as new transition arenas for a low-carbon society or green economy. *Technology Analysis and Strategic Management*, *24*(4), 331–343. doi:10.1080/09537325.2012.663959

Stone, D. (2008). Global public policy, transnational policy communities, and their networks. *Policy Studies Journal: the Journal of the Policy Studies Organization*, *36*(1), 19–38. doi:10.1111/j.1541-0072.2007.00251.x

Stubbs, R. (2008). The ASEAN alternative? Ideas, institutions and the challenge to 'global' governance. *The Pacific Review*, *21*(4), 451–468. doi:10.1080/09512740802294713

Swartz, D. (1996). Jews, Jewish texts, and nature: A brief history. *This sacred earth: Religion, nature, environment*, 87-103.

Tamimi Amadi, A. W. (2021). *The Ghurar Al-Hikam wa Durar Al-Kalim* (1st ed.). Islamic Propaganda Office.

Tarlock, A. D. (1993). The nonequilibrium paradigm in ecology and the partial unraveling of environmental law. *Loy. LAL Rev.*, *27*, 1121.

Tayyib, S. A. H. (2018). *Atiab Al-Bayan fi Tafsir Al-Quran* (2nd ed.). Islamic Publication.

Tertrais, B. (2020). French Nuclear Deterrence Policy, Forces, And Future: A Handbook. Fondation pour la recherche stratégique. *Recherches & Documents*, *4*, 12.

Thomas, C. (2000). *Global governance, development and human security: the challenge of poverty and inequality*. Pluto.

Thomas, C. (2001). Global governance, development and human security: Exploring the links. *Third World Quarterly*, *22*(2), 159–175. doi:10.1080/01436590120037018

Tienhaara, K. (2011). Foreign investment contracts in the oil & gas sector: A survey of environmentally relevant clauses. *Sustainable Development Law & Policy*, *11*(3), 6.

Tortell, P. D. (2020). Earth 2020: Science, society, and sustainability in the Anthropocene. *Proceedings of the National Academy of Sciences of the United States of America*, *117*(16), 8683–8691. doi:10.1073/pnas.2001919117 PMID:32312801

Tóth, T. (2018). Life after Menarini: The conformity of the Hungarian Competition Law enforcement system with human rights principles. [YARS]. *Yearbook of Antitrust and Regulatory Studies*, *11*(18), 35–60. doi:10.7172/1689-9024. YARS.2018.11.18.2

Tranter, B. (2011). Political divisions over climate change and environmental issues in Australia. *Environmental Politics*, *20*(1), 78–96. doi:10.1080/09644016.2011.538167

Trivedi, D. (2019). Glimpses of Green Consumerism and Steps Towards Sustainability. [JOM]. *Journal of Management*, *6*(3), 35–41. doi:10.34218/JOM.6.3.2019.005

Trombetta, M. J. (2008). Environmental security and climate change: Analysing the discourse. *Cambridge Review of International Affairs*, *21*(4), 585–602. doi:10.1080/09557570802452920

Turner, R. K. (1992). *Environmental policy: An economic approach to the polluter pays principle*. CSERGE.

Uhlmann, D. M. (2011). After the spill is gone: The Gulf of Mexico, environmental crime, and the criminal law. *Michigan Law Review*, *109*(8), 1413–1461.

Van Calster, G., & Reins, L. (2017). *EU environmental law*. Edward Elgar Publishing. doi:10.4337/9781782549185

Van Rooij, B. (2006). Implementation of Chinese environmental law: Regular enforcement and political campaigns. *Development and Change*, *37*(1), 57–74. doi:10.1111/j.0012-155X.2006.00469.x

Vedder, H. (2003). *Competition law and environmental protection in Europe: towards sustainability?* (Vol. 3). Europa Law Publishing.

VedderH. H. (2000). Voluntary Agreements and Competition Law: What are, and What Should Be the Boundaries to Va's Imposed by Competition Law? SSRN 253333. doi:10.2139/ssrn.253333

Von Stein, J. (2008). The international law and politics of climate change: Ratification of the United Nations Framework Convention and the Kyoto Protocol. *The Journal of Conflict Resolution*, *52*(2), 243–268. doi:10.1177/0022002707313692

Warner, K., Hamza, M., Oliver-Smith, A., Renaud, F., & Julca, A. (2010). Climate change, environmental degradation and migration. *Natural Hazards*, *55*(3), 689–715. doi:10.100711069-009-9419-7

Watling, T. (2009). *Ecological imaginations in the world religions: An ethnographic analysis*. A&C Black.

Wawryk, A. S. (2002). Adoption of international environmental standards by transnational oil companies: Reducing the impact of oil operations in emerging economies. *Journal of Energy & Natural Resources Law*, *20*(4), 402–434. doi:10.1080/02646811.2002.11433308

Weiss, T. G., & Thakur, R. (2010). *Global governance and the UN: An unfinished journey*. Indiana University Press.

White, L. (2004). The historical roots of our ecological crisis. *This sacred earth: religion, nature, environment*, (179), 192.

White, L. Jr. (1967). The historical roots of our ecologic crisis. *Science*, *155*(3767), 1203–1207. doi:10.1126cience.155.3767.1203 PMID:17847526

Wiener, J. B. (2000). Something borrowed for something blue: Legal transplants and the evolution of global environmental law. *Ecology Law Quarterly*, *27*, 1295.

Wildermuth, A. J. (2011). The Next Step: The Integration of Energy Law and Environmental Law. *Utah Envtl. L. Rev.,* *31,* 369.

Wood, S., Tanner, G., & Richardson, B. J. (2010). What ever happened to Canadian environmental law. *Ecology Law Quarterly, 37,* 981.

Yoreh, T. S. (2014). *The Jewish Prohibition Against Wastefulness: The Evolution of an Environmental Ethic.*

Young, M. A. (2017). Energy transitions and trade law: Lessons from the reform of fisheries subsidies. *International Environmental Agreement: Politics, Law and Economics, 17*(3), 371–390. doi:10.100710784-017-9360-2

Yusuf, M. (2020). Developing environmental awareness and actualizing complete piety based on quran. *International Journal of Advanced Science and Technology, 29*(5), 2039–2050.

Zhang, K. M., & Wen, Z. G. (2008). Review and challenges of policies of environmental protection and sustainable development in China. *Journal of Environmental Management, 88*(4), 1249–1261. doi:10.1016/j.jenvman.2007.06.019 PMID:17767999

About the Authors

Nima Norouzi was born in Iran, and graduated his primary, high school, and college education from the NODET (National organization for Intellectual people) primary Education system. he took the national university entrance exam and, with rank 365 (among 164000), entered the Amirkabir University of Technology and started his education in system management and engineering in the B.Sc degree which he continued this filed in his M.Sc degree in the same university. After his graduation with his B.Sc and M.Sc degree, and because of his various educational and research field successes, he has been appreciated many times by the university. While studying his M.Sc Degree, Nima started a L.L.B cource in Islamic Azad university. Then Nima got his S.J.D in Private and Islamic law from university of Tehran. Nima currently does reading and research in energy and environmental law. He is currently a researcher in the environmental field at Islamic Azad University and University of Tehran.

Hussein Movahedian graduated with a Juris Doctor and Associate Master in Private Law in Imam Sadiq University. He is currently affiliated with Islamic Azad University's senior researcher and is well-established in environmental law. Several well-respected papers of his are published in this field. His main research field is the environmental responsibility of governments and corporations in contracts.

Index

A

air pollution 4, 10, 19, 33, 51, 65-66, 72, 88-89, 107, 110, 113, 151-152, 197, 201, 223
animal species 37, 46-47, 107
anticompetitive effects 137, 139, 141, 143

B

biodiversity 4-5, 19, 21-22, 25, 28, 40-41, 49, 65-67, 70-72, 88, 105, 121, 133-134, 145, 198, 200, 217, 223, 228
border crossings 19, 27

C

carbon monoxide 110, 112, 223
circular economy 79, 100, 114-115, 137-138, 145
civil society 2, 8, 13, 18, 20-22, 25-28, 57, 137, 186-187, 192
climate change 4-5, 11, 19, 22-23, 25, 28, 39-41, 46, 61-62, 65-66, 71, 77-82, 110, 113, 117-118, 123, 125, 129-133, 136-137, 146, 152, 155, 157, 163-164, 170-171, 180, 183, 185, 193-194, 197-198, 200, 211, 216, 223-224, 228, 240
Competition law 136-140, 142-149
Concession Contracts 151-153, 155
Criminal Court 48, 92, 98
criminal matters 57, 173-175

D

decision-making processes 108, 112
decision-making systems 22, 26
developed countries 5, 14, 38, 84, 89-91, 97-98, 110, 112, 120, 122, 125, 130-132, 151, 183-185, 187
developing countries 5, 8-9, 12, 23-25, 41, 85-91, 97-98, 111, 117-133, 151-152, 154, 156, 158, 162, 170, 184-189, 191, 193-195, 216, 219

development plan 3, 104-105, 110, 160-161
Dichlorodiphenyltrichloroethane (DDT) 228
domestic laws 7, 152-153, 159, 162

E

Ecological Law 167, 197
emissions 33, 41, 51, 65, 72, 78, 88-89, 103, 110, 112, 117, 131, 137, 139, 144, 146, 151, 158, 164-165, 170, 174, 182-184, 189-193, 216-217, 223
Endangered Species 6, 11, 13, 39, 59, 65, 71-72, 228
Environmental degradation 4, 6-7, 9, 11, 13-14, 16, 19, 25, 27, 31, 37-38, 40, 43, 45-46, 48, 61-63, 68, 73, 80-82, 85, 90-92, 101, 107-108, 112, 115, 135, 150, 153, 157-159, 162, 166, 169-170, 172, 178, 195, 214, 219, 223, 227, 243
Environmental Governance 5, 18-29, 67, 80, 85
environmental issues 2, 8, 13, 22, 25, 33-34, 46, 48, 51, 61-62, 65-67, 73-74, 78, 80-81, 93, 95, 97, 109, 111, 138, 145, 152, 162, 172, 175, 180-182, 188, 190, 220, 222, 228, 232-234
Environmental Law 4-9, 13-16, 18, 23, 29-34, 37, 39-43, 45, 47-49, 52, 56-57, 60, 63-75, 77-82, 84, 90-93, 98-101, 103-104, 106, 114-115, 117, 133-136, 139, 143, 148, 150-151, 153, 155-156, 160-161, 163-164, 166-169, 172, 176-181, 183, 188, 190, 193-195, 197, 214-216, 218-226, 228, 235, 239, 242
environmental matters 48-54, 57-60, 170, 172, 174-175
Environmental Policy 1, 18, 21, 32, 45, 52, 64, 70, 72, 78, 84, 103, 105, 113-114, 117, 136, 151, 154, 167, 172, 179, 194, 197, 215, 228
environmental pollution 3-4, 12-13, 33, 38, 70, 91, 103-105, 107-109, 112, 159, 162, 201, 232
environmental problems 3, 5, 20, 22, 26, 28, 39, 45, 54, 56-57, 70, 90, 95-96, 104, 114, 127, 167-170, 174-175, 181, 187, 189, 192, 203, 212-213, 216, 222, 238
Environmental Protection 1, 3-5, 7-9, 11, 13-14, 17-18,

23, 27, 31-32, 35, 38, 41-42, 44-47, 49, 51-53, 57, 63-64, 66-70, 72-74, 83-84, 86, 89, 91-92, 99, 102-106, 109-110, 112, 114, 116-117, 123, 136, 138-139, 143, 146, 148-149, 151-156, 159-161, 167, 169-173, 179, 181, 186-187, 193, 197, 201, 210, 215, 218-223, 228-231, 233, 235, 239

environmentally friendly 73, 84-85, 88-89, 91, 96-97, 106, 117-120, 122-130, 132, 137, 142, 199, 217, 219

environmentally responsible 118-123, 125-126, 128-129, 132

Equity 16, 30, 43, 61-62, 78, 81-82, 101, 115, 131, 134, 150, 166, 178, 195, 213, 226, 242

European Green Deal 136-138, 140, 147

external costs 103, 109, 112

F

financial burden 108, 110

fundamental role 39, 42

G

general policy 180, 189

global community 85, 98

governance system 18-21, 23-28, 85

Green Deal 136-138, 140-143, 147

green economy 84-90, 99, 137-138, 142, 147, 171, 179-194

green investment 192

Green Law 18, 32, 45, 64, 84, 103, 117, 136, 151, 167, 179, 197, 215, 228

Green Policy 18, 32, 45, 64, 84, 103, 117, 136-137, 167, 179, 197, 215, 228

Green taxes 103-104, 109-110, 112, 114

greenhouse gas emissions 41, 78, 88, 131, 151, 182-184, 189, 192-193, 223

H

hazardous waste 36, 60, 91, 98

Historical Roots 197, 211-213

Holy Quran 199, 204-205, 230, 232, 235, 237-238

Horizontal Cooperation 137, 139-140

human relationships 229, 238

I

inappropriate language 19, 21

inflation 18, 26-27

Intellectual Property Rights 87, 93, 96, 99, 117-119,

122, 129-134

interconnected nature 84-85

international community 4, 14, 19-20, 36, 39, 49, 76, 85, 99, 168-169, 179-181, 183-184, 186, 188-189, 193, 219-220

international law 4-8, 14-17, 20, 28, 31, 33-34, 37, 41-44, 48-50, 55, 57, 63, 65, 74-77, 83, 85, 99-102, 115-116, 135, 150, 158, 164-166, 178, 187-190, 194-196, 214, 220-221, 225-227, 243

International trade 6, 24-26, 38-39, 65, 71, 84-88, 90-92, 94-95, 98-100, 154, 181, 186-187

J

jurisprudence rules 229-230, 239

K

Kyoto Protocol 5, 41, 65, 170-171, 183, 187, 192-194

M

managerial issues 108, 110

Metropolitan Commission 32, 64, 72

N

natural resources 4, 11-13, 16-17, 19, 25, 28, 30-32, 34, 40, 43-46, 49, 62, 64, 67, 70, 72-73, 80, 82, 88, 92, 94-96, 98, 101, 105, 108, 115-116, 128, 134-135, 138, 145, 150, 153, 156-158, 160, 165-166, 169, 173-174, 178-180, 182, 187, 189-190, 195-196, 210, 213-219, 225-227, 229, 233-235, 238, 242-243

negative constraints 218

non-economic reasons 108, 110

non-governmental actors 18, 25-27, 179, 192, 220, 223

non-renewable resources 103, 179, 188, 193

P

Plant Health 86, 95, 99

Polluter pays principle 16, 30, 43, 63, 82, 101, 115, 135, 150, 153-154, 161, 164, 166, 178, 194-195, 214, 226, 242

pollution damage 52, 103, 109, 112

post-industrial era 167, 175

Precautionary principle 5, 16, 31, 43, 57, 63, 82, 101, 115, 135, 150, 153, 164-166, 178, 195, 214, 226, 242

Prevention 14, 16, 31, 33, 37-38, 42, 44, 50, 54, 57, 60,

63, 67-69, 73, 82, 90, 92, 101, 105, 114-115, 135, 150, 153-154, 156-157, 163, 166, 176, 178, 182, 191, 196, 200-201, 214, 220-221, 224, 227, 243
Public participation and transparency 17, 31, 44, 82, 101, 116, 135, 150, 166, 178, 196, 214, 227, 243

R

resource productivity 84-85

S

sanitary goods 91, 98
soil erosion 37, 46-47, 107, 111, 198
Staff Working Document 137, 140
Stockholm Conference 40, 169-172
Stockholm Declaration 5-6, 34, 41, 65, 75, 85, 154, 158, 169, 220, 222
summer heat 32, 64, 72
Sustainable development 2, 8-10, 12-13, 17-23, 25-27, 31, 40-42, 44, 47, 63, 65, 67-68, 73-74, 79, 83-86, 92, 97-103, 107-109, 112-117, 121, 123, 128, 133-134, 136, 138, 142-143, 148, 153-155, 163-165, 170-171, 174, 179-182, 184-194, 215, 220, 222-223, 225, 228-229, 233

T

Technical Barriers 86, 95, 99
Trade-Related Aspects 87, 96, 99
Transboundary responsibility 17, 31, 44, 63, 83, 102, 116, 135, 150, 166, 178, 196, 214, 227, 243

U

underdeveloped countries 91, 132
United Nations 4-6, 8-9, 14, 17, 21-23, 25-27, 31, 34-35, 39-42, 44-45, 63, 65, 74, 76, 83-84, 97, 99-100, 102, 113-114, 116-118, 129, 131-134, 136, 154-155, 158, 169-172, 175, 180, 183, 185, 188, 190-191, 193-194, 220, 222, 228
United Nations Conference 21, 65, 76, 97, 154, 158, 169-171
United Nations Environment Program 4, 8-9, 14, 22-23, 25, 84, 97, 171, 188, 190-191, 193

W

world governance 20

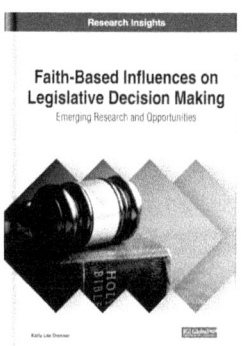

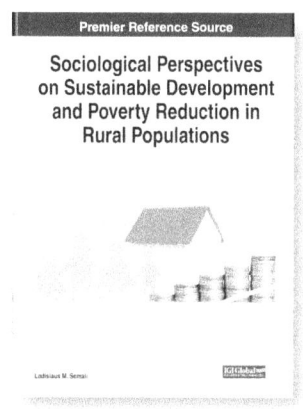

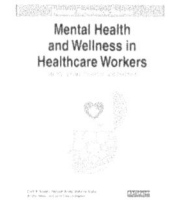

Ingram Content Group UK Ltd.
Milton Keynes UK
UKHW032259080523
421436UK00008B/310